T0332113

MANY-BODY THEORY OF CONDENSED MATTER SYSTEMS

In this primer to the many-body theory of condensed matter systems, the authors introduce the subject to the nonspecialist in a broad, concise, and up-to-date manner. A wide range of topics are covered including the second quantization of operators, coherent states, quantum-mechanical Green's functions, linear response theory, and Feynman diagrammatic perturbation theory. Material is also incorporated from quantum optics, low-dimensional systems such as graphene, and localized excitations in systems with boundaries as in nanoscale materials. More than 100 problems are included at the end of chapters, which are used both to consolidate concepts and to introduce new material. This book is suitable as a teaching tool for graduate courses and is ideal for nonspecialist students and researchers working in physics, materials science, chemistry, or applied mathematics who want to use the tools of many-body theory.

MICHAEL G. COTTAM is a professor of physics in the Department of Physics and Astronomy at the University of Western Ontario. He has previously been the chair of the Department of Physics and Astronomy, the director of Western University's Institute for Nanomaterials Science, and the associate dean in the Faculty of Science.

ZAHRA HAGHSHENASFARD holds PhDs from both the University of Isfahan in quantum optics and nonlinear processes, and from the University of Western Ontario in nonlinear processes for the magnetization dynamics in nanowires.

MANY-BODY THEORY OF CONDENSED MATTER SYSTEMS

An Introductory Course

MICHAEL G. COTTAM

University of Western Ontario

ZAHRA HAGHSHENASFARD

University of Western Ontario

CAMBRIDGE
UNIVERSITY PRESS

CAMBRIDGE
UNIVERSITY PRESS

University Printing House, Cambridge CB2 8BS, United Kingdom

One Liberty Plaza, 20th Floor, New York, NY 10006, USA

477 Williamstown Road, Port Melbourne, VIC 3207, Australia

314–321, 3rd Floor, Plot 3, Splendor Forum, Jasola District Centre, New Delhi – 110025, India

79 Anson Road, #06–04/06, Singapore 079906

Cambridge University Press is part of the University of Cambridge.

It furthers the University's mission by disseminating knowledge in the pursuit of education, learning, and research at the highest international levels of excellence.

www.cambridge.org
Information on this title: www.cambridge.org/9781108488242
DOI: 10.1017/9781108762366

First published 2020

Printed in the United Kingdom by TJ International Ltd, Padstow Cornwall

A catalogue record for this publication is available from the British Library.

Library of Congress Cataloging-in-Publication Data
Names: Cottam, Michael G., author. | Haghshenasfard, Zahra, 1978– author.
Title: Many-body theory of condensed matter systems : an introductory course /
Michael G. Cottam, Zahra Haghshenasfard.
Description: Cambridge ; New York, NY : Cambridge University Press, 2020. |
Includes bibliographical references and index.
Identifiers: LCCN 2019055996 (print) | LCCN 2019055997 (ebook) |
ISBN 9781108488242 (hardback) | ISBN 9781108762366 (epub)
Subjects: LCSH: Many-body problem. | Condensed matter–Mathematical models. |
Geometric quantization. | Green's functions.
Classification: LCC QC174.17.P7 C684 2020 (print) | LCC QC174.17.P7
(ebook) | DDC 530.4/1–dc23
LC record available at https://lccn.loc.gov/2019055996
LC ebook record available at https://lccn.loc.gov/2019055997

ISBN 978-1-108-48824-2 Hardback

Additional resources for this publication at www.cambridge.org/9781108488242

We dedicate this book to our parents, in loving memory:

William Cottam and *Elizabeth Caldow Good*

Ali Haghshenasfard and *Sakineh Haghshenas*

Contents

Preface

In the last few years there have been rapid advances in the research into low-dimensional materials (such as graphene) and artificially engineered materials (such as photonic crystals or band-gap materials) that involve sample growth and fabrication on the nanometer or micrometer length scales. The role of surfaces and interfaces becomes emphasized in these systems. Such developments have led to a renewed need for a wider use and application of advanced theoretical techniques, such as those for the many-body theoretical methods for condensed matter physics. However, the increased interest in fundamentals of the interactions between light and matter has placed a focus on many concepts and models developed in the field of quantum optics (e.g., those involving coherent states).

Our intention in writing this book is to provide a text that is genuinely introductory in scope and is at a suitable level for the nonspecialists (not just for those in theoretical physics) wanting to use the tools of many-body theory. Hence, although the book has been kept relatively concise, we have also incorporated "new" topics such as those mentioned previously that are not typically found in textbooks on many-body theory. Our book allows its readers to learn the basics of a range of techniques and to take in a number of applications and examples before eventually moving on to one of the excellent texts that cover material at an advanced or specialized level. In keeping with this pedagogical approach, we have provided problems (more than 100 in total) at the end of each chapter to introduce additional applications. As an extra teaching tool, a solutions manual is made available to course instructors.

The material in this book is organized as follows. The operator methods of second quantization, including those applied to coherent states and their time evolution, are the main topics of Chapters 1 and 2. Examples of some many-body systems and the techniques of solving for the wavelike excitations are given. Green's functions are formally introduced in Chapter 3 in both the real-time and imaginary-time techniques, and important results are established connecting them (and their frequency Fourier transforms) to the correlation functions and excitation spectra relevant to experiments. Then Chapters 4 and 5 cover the evaluation of Green's function by the equation-of-motion method, either exactly (within a theoretical model) or more typically using so-called decoupling approximations. Next, Chapter 6 involves linear response theory and the connection between Green's functions and response functions (or generalized susceptibilities). Chapter 7 deals with the application of the preceding techniques to systems exhibiting localization of the excitations (e.g., at the surfaces or interfaces in a finite sample or due to the structural symmetrybreaking by the presence of impurities). In Chapters 8 and 9 the development and application of the imaginary-time Green's functions to diagrammatic perturbation theory (Feynman diagrams) are presented. The treatment is mostly carried out for boson and fermion systems, but examples of unconventional diagram methods, such as those required for spin systems at general temperatures, are also included.

The target readership of this book includes physicists, chemists, materials scientists, applied mathematicians, and engineers. They could be based in universities, industries and/or research laboratories. The book's level of presentation is appropriate for graduate students and researchers.

We are indebted to our many colleagues, collaborators, and mentors who directly or indirectly have influenced this book and provided ideas. Among others, it is a pleasure to give special mention to A. Akbari-Sharbaf, E. L. Albuquerque, M. Babiker, R. E. Camley, N. N. Chen, R. Loudon, A. A. Maradudin, E. Meloche, M. H. Naderi, H. T. Nguyen, T. M. Nguyen, M. Soltanolkotabi, and D. R. Tilley.

Abbreviations

1D	one dimension or one dimensional
2D	two dimensions or two dimensional
3D	three dimensions or three dimensional
AC	armchair
AFMR	antiferromagnetic resonance
b.c.c.	body-centered cubic
BCS	Bardeen–Cooper–Schrieffer (theory)
BE	Bose–Einstein (distribution function)
BEC	Bose–Einstein condensation
BLS	Brillouin light scattering
DE	Damon–Eshbach (mode)
DF	drone-fermion
EM	electromagnetic or electromagnetism
f.c.c.	face-centered cubic
FD	Fermi–Dirac (distribution function)
FHO	forced harmonic oscillator
FMR	ferromagnetic resonance
GF	Green's function
GNR	graphene nanoribbon
H.c.	Hermitian conjugate
HF	Hartree–Fock (theory)
HP	Holstein–Primakoff (transformation)
HTSC	high-temperature superconductor
INS	inelastic neutron scattering
JC	Jaynes–Cummings (model)
PBG	photonic band gap
QM	quantum mechanics or quantum mechanical (system)

RLS	Raman light scattering
RPA	random phase approximation
RWA	rotating wave approximation
s.c.	simple cubic
SHO	simple harmonic oscillator
SW	spin wave
TDM	tridiagonal matrix
Tr	trace
ZZ	zigzag

1

Introduction to Second Quantization

In many-body theory we are concerned with studying systems that have a large number of particles interacting with one another. Some examples of many-body systems in condensed matter physics, many of which we will explore later in this book, are

- The interacting fermion gas (e.g., electrons in a metal or semiconductor);
- The interacting boson gas (e.g., phonons in a crystalline solid; or optical excitations, such as polaritons, in solids);
- Magnetic systems (e.g., spin waves, or magnons, due to interacting spins in a ferromagnet);
- Excitations in quantum fluids, such as liquid helium (^{3}He and ^{4}He), including properties related to superfluidity and superconductivity;
- Electronic excitations and topological excitations in novel low-dimensional systems, such as graphene;
- Plasma excitations in solids, where the charged particles may undergo a collective behavior (called plasmons) distinct from the individual particle behavior; and
- The interaction of light with atoms (as in quantum optics) and the scattering of light by solids.

Additionally, with the technological advances in materials science in recent years for producing finite-sized elements (often in arrays) with one or more of the dimensions on the nanometer scale, there is now an immense interest in systems with surfaces and interfaces that play an important role in the properties of these artificial materials, including photonic crystals, superlattices, and so forth.

The usual first-order or second-order perturbation theory of elementary quantum mechanics in terms of the wave functions (see, e.g., [1, 2]) is generally of limited use for such systems because we typically do not know the wave functions and also, even if we did know them, the interactions may be too strong to proceed.

Alternatively, the number of particles in the system may be so large as to render the method impractical for summations over the states of the particles (e.g., in a macroscopic-sized metallic system the number of "free" electrons may be of the order of Avogadro's number, or $\sim 10^{23}$, depending on the size of the system). Therefore, we need new mathematical tools and a different formulation of perturbation theory: This brings us, in particular, to the methods of second quantization of the operators and to quantum-mechanical Green's functions.

We will make use of the standard results of statistical mechanics, where applicable. Usually (but not always) this will be in the context of equilibrium statistical mechanics, where there are the standard methods for calculating equilibrium averages for grand-canonical and canonical ensembles (see, e.g., [3, 4]). Suppose we have a system of N particles in a volume V. Often the system will be sufficiently large that we can regard it as being infinite, that is, we can take $N \to \infty$ and $V \to \infty$, such that the average particle density N/V is finite (the usual "thermodynamic limit"). In other contexts we may need to study localized excitations near the boundaries of finite systems.

We want the theory to provide predictions about properties such as the ground-state energy and also the excitations of the system from its ground state. This might bring up issues of finding the best set of coordinate variables to describe the excitations. We want to carry out these calculations at any temperature, in general, and so considerations of the phase transitions in the system will be relevant. Some examples of the latter are the Curie temperature T_C in a ferromagnet, the transition temperature to superconductivity in some materials, and the so-called λ-transition in liquid ^{4}He between helium I and the lower-temperature helium II phase that exhibits superfluidity. Other applications that may involve the use of many-body theory include the scattering of light or particles (such as in Raman light scattering [RLS], inelastic neutron scattering [INS], etc.), transport-related properties (including electrical conductivity, thermal conductivity, quantum Hall effect, etc.), and other collision problems (e.g., electrons at surfaces or in random or impure media).

The main topics of this introductory chapter are *second quantization* and examples of its basic applications to systems. Regarding terminology, first quantization usually refers to the elementary quantum mechanics as taught in an undergraduate physics course: The classical position and momentum are replaced by operators, which may be noncommuting in general. The analysis proceeds through the use of a "wave" equation (Schrödinger's equation) resulting in quantization of the energy levels, for example. In second quantization we go one step further because the elementary approach becomes impractical for many-body systems, as already mentioned. The main outcome is that the field variables for the system are raised to having the status of operators, for example, as demonstrated by P. A. M. Dirac in

1927 for the electromagnetic field. A description emerges in terms of particle-like entities (the photons in the electromagnetic field case) that can be acted upon by creation and annihilation operators.

The creation and annihilation operators are introduced in the next section, first through a mathematical example and then for the electromagnetic field in free space. Generalizations to bosons and fermions in second quantization are presented in Section 1.2. Then in Section 1.3 we describe the creation and annihilation operators required for coherent states, as frequently encountered in quantum optics. The remaining sections of this chapter are taken up with examples of Hamiltonians in second quantization that are of interest in many-body theory, and we show how to obtain solutions in some simple cases by using operator techniques.

1.1 Creation and Annihilation Operators

1.1.1 Quantum-Mechanical Simple Harmonic Oscillator

We start by considering, just as a mathematical example, the quantum-mechanical simple harmonic oscillator (SHO) in one dimension (1D). It is well known, of course, that the energy eigenvalues and wave functions can be found by solving the time-independent Schrödinger equation [1]. As an alternative approach, however, we review here the use of creation and annihilation operators for this model system. The Hamiltonian can be written as

$$\mathcal{H} = \frac{p^2}{2m} + \frac{1}{2}m\omega^2 x^2, \tag{1.1}$$

where $p = -i\partial/\partial x$ is the momentum operator, x is the position along the 1D axis, and we employ units such that $\hbar = 1$. The constants in the potential energy term (proportional to x^2) are m as the mass of the particle and ω as the classical angular frequency of the oscillations.

We now define new operators $a^\dagger$ and a by

$$a^\dagger = (2m\omega)^{-1/2}(m\omega x + ip),$$
$$a = (2m\omega)^{-1/2}(m\omega x - ip). \tag{1.2}$$

By using the commutation property $[x, p] = i$ from elementary quantum mechanics (QM), it is easy to prove that

$$[a, a^\dagger] = 1 \tag{1.3}$$

for the new operators, while the Hamiltonian can be expressed as

$$\mathcal{H} = \omega\left(a^\dagger a + \frac{1}{2}\right). \tag{1.4}$$

Here we are employing a standard notation that $[A, B] = AB - BA$ for the commutator between any two operators A and B.

Several other convenient properties of the $a^\dagger$ and a operators can readily be established, as follows.

- If ψ_α denotes an eigenfunction of the Hamiltonian $\mathcal{H}$ with an eigenvalue written as $\alpha\omega$, then $a^\dagger\psi_\alpha$ is also an eigenfunction of $\mathcal{H}$, and it has the eigenvalue $(\alpha + 1)\omega$.

Proof We are given the property that $\mathcal{H}\psi_\alpha = \alpha\omega\,\psi_\alpha$ and are asked to prove that $\mathcal{H}a^\dagger\psi_\alpha = (\alpha + 1)\,\omega a^\dagger\psi_\alpha$. The starting point is to note from Equation (1.4) that $\mathcal{H}a^\dagger = \omega\left(a^\dagger a a^\dagger + \frac{1}{2}a^\dagger\right)$ and $a^\dagger\mathcal{H} = \omega\left(a^\dagger a^\dagger a + \frac{1}{2}a^\dagger\right)$. Therefore, by subtraction we have

$$\mathcal{H}a^\dagger - a^\dagger\mathcal{H} = \omega a^\dagger\left(aa^\dagger - a^\dagger a\right) = \omega a^\dagger.$$

This leads to $\mathcal{H}a^\dagger\psi_\alpha = a^\dagger\mathcal{H}\psi_\alpha + \omega a^\dagger\psi_\alpha = \alpha\omega a^\dagger\psi_\alpha + \omega a^\dagger\psi_\alpha = (\alpha + 1)\,\omega a^\dagger\psi_\alpha$, which proves the result.

- The quantity $a\psi_\alpha$ is another eigenfunction (provided it is nonzero) of the Hamiltonian $\mathcal{H}$, but it has the eigenvalue $(\alpha - 1)\omega$.

Proof This is similar to the previous property, and the result follows after we have shown that $\mathcal{H}a - a\mathcal{H} = -\omega a$.

- The eigenvalues of the Hamiltonian correspond to $\alpha = \left(n + \frac{1}{2}\right)$, where integer $n = 0, 1, 2, \ldots$.

Proof We suppose ψ_0 denotes the ground-state eigenfunction and the smallest eigenvalue is $\alpha_0\omega$. By definition, we must have $a\psi_0 = 0$ because there is no lower state. Its eigenvalue is easily found by considering

$$\mathcal{H}\psi_0 = \omega\left(a^\dagger a + \frac{1}{2}\right)\psi_0 = \omega a^\dagger a\psi_0 + \frac{1}{2}\omega\psi_0 = \frac{1}{2}\omega\psi_0.$$

Therefore, the ground state ψ_0, henceforth denoted using Dirac's "ket" notation as $|0\rangle$ (see [5]), has the eigenvalue $\frac{1}{2}\omega$. It then follows that the first excited state $|1\rangle$ is proportional to $a^\dagger|0\rangle$ and has the eigenvalue $(1 + \frac{1}{2})\omega$, while the second excited state $|2\rangle$, which is proportional to $(a^\dagger)^2|0\rangle$, has the eigenvalue $(2 + \frac{1}{2})\omega$, and so on. The nth excited state is

$$|n\rangle \propto (a^\dagger)^n|0\rangle$$

and has the eigenvalue $(n + \frac{1}{2})\omega$. Hence, the eigenvalue equation becomes

$$\mathcal{H}|n\rangle = \left(n + \frac{1}{2}\right)\omega|n\rangle. \tag{1.5}$$

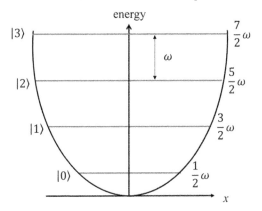

Figure 1.1 Representation of energy levels for the first four eigenstates of a quantum-mechanical simple harmonic oscillator in 1D. The horizontal axis is the displacement x and the parabolic curve represents the potential energy term $\frac{1}{2}m\omega^2 x^2$.

The energy levels for a 1D harmonic oscillator for the first few eigenstates are represented schematically in Figure 1.1. We note that the energy levels are evenly spaced.

- The normalized eigenfunctions have the properties that

$$a^\dagger|n\rangle = \sqrt{n+1}|n+1\rangle \quad \text{and} \quad a|n\rangle = \sqrt{n}|n-1\rangle. \tag{1.6}$$

Proof From the previous properties we know that

$$a^\dagger|n\rangle = c_n|n+1\rangle.$$

This follows because they have the same eigenvalue. To find the constant c_n, we premultiply each side by its Hermitian conjugate, noting that the Hermitian conjugate of the ket $|n\rangle$ is the bra $\langle n|$. This yields $\langle n|aa^\dagger|n\rangle = (c_n)^2\langle n+1|n+1\rangle$, and so

$$(c_n)^2 = \langle n|aa^\dagger|n\rangle = \langle n|(1+a^\dagger a)|n\rangle = \langle n|\left(\omega^{-1}\mathcal{H} + \frac{1}{2}\right)|n\rangle$$

$$= \langle n|\left(n + \frac{1}{2} + \frac{1}{2}\right)|n\rangle = n+1,$$

which proves that $c_n = \sqrt{n+1}$ (ignoring any complex phase factor). The second result can be proved in a similar way.

One can show (e.g., by iteration) that the eigenstates after normalization are

$$|n\rangle = \frac{(a^\dagger)^n}{\sqrt{n!}}|0\rangle. \tag{1.7}$$

- The combination $a^\dagger a$ has the eigenvalue n corresponding to the eigenfunction $|n\rangle$ of the nth state.

Proof We use $a^\dagger a|n\rangle = (\omega^{-1}\mathcal{H} - \frac{1}{2})|n\rangle = (n + \frac{1}{2} - \frac{1}{2})|n\rangle = n|n\rangle$. This means that $a^\dagger a$ behaves as a *number operator*. In other words, when it operates on the nth excited state it gives us the number n as the eigenvalue.

To summarize, the main conclusions are as follows. Rather than working in terms of the original QM operators p and x, we have found new operators $a^\dagger$ and a such that $a^\dagger$ takes us from one state to the next highest. In other words, it acts as a creation operator for an excitation. Conversely the operator a takes us from one state to the next lowest. So, it acts as an annihilation operator for an excitation. Also we have shown that a and $a^\dagger$ satisfy a simple commutation relation and that the operator combination $a^\dagger a$ acts as a number operator.

1.1.2 Electromagnetic Field in Free Space

As a second example we now show that creation and annihilation operators are also useful concepts in electromagnetism. In this case, the second-quantization techniques were introduced by Dirac [6] in 1927 for the electromagnetic fields. Quantization of the electromagnetic (EM) field, along with the introduction of the concept of photons, are crucial ideas in quantum optics. We now outline the arguments to show that, after quantization, the EM field can be seen as being equivalent to a collection of harmonic oscillators.

We start with Maxwell's equations in vacuum (i.e., in the absence of charges and currents), which can be stated as (see, e.g., [7, 8])

$$\nabla \cdot \mathbf{B} = 0, \qquad \nabla \times \mathbf{H} = \frac{\partial \mathbf{D}}{\partial t},$$

$$\nabla \cdot \mathbf{D} = 0, \qquad \nabla \times \mathbf{E} = -\frac{\partial \mathbf{B}}{\partial t}. \tag{1.8}$$

The magnetic flux density $\mathbf{B}$ and the magnetic intensity $\mathbf{H}$ are related by $\mathbf{B} = \mu_0 \mathbf{H}$ in SI units and the relation between the electric displacement field $\mathbf{D}$ and electric field $\mathbf{E}$ is $\mathbf{D} = \varepsilon_0 \mathbf{E}$. Here μ_0 and ε_0 are the magnetic permeability and electric permittivity of free space, respectively, obeying $\mu_0\varepsilon_0 = c^{-2}$ where c is the speed of light in vacuum. The EM fields in vacuum can be conveniently determined by using the magnetic vector potential $\mathbf{A}(\mathbf{r},t)$ in a specific gauge. Normally, the Coulomb gauge, for which $\nabla \cdot \mathbf{A} = 0$, is chosen. In this case, both $\mathbf{B}$ and $\mathbf{E}$ can be obtained from $\mathbf{A}(\mathbf{r},t)$ by taking

$$\mathbf{B} = \nabla \times \mathbf{A}, \qquad \mathbf{E} = -\frac{\partial \mathbf{A}}{\partial t}. \tag{1.9}$$

Substituting Equation (1.9) into (1.8) gives the wave equation for $\mathbf{A}(\mathbf{r}, t)$ in the form

$$\nabla^2 \mathbf{A}(\mathbf{r}, t) = \frac{1}{c^2} \frac{\partial^2 \mathbf{A}(\mathbf{r}, t)}{\partial t^2}. \tag{1.10}$$

We may now separate $\mathbf{A}(\mathbf{r}, t)$ into two terms that involve the normalized eigenmodes $\mathbf{u}_j(\mathbf{r})$ by writing

$$\mathbf{A}(\mathbf{r}, t) = \sum_j \sqrt{\frac{1}{2\omega_j \varepsilon_0}} \left[a_j(t) \mathbf{u}_j(\mathbf{r}) + a_j^\dagger(t) \mathbf{u}_j^*(\mathbf{r}) \right]. \tag{1.11}$$

Here j is an index labeling the modes and $a_j(t)$ are complex coefficients. Substituting Equation (1.11) into (1.10) gives

$$\left(\nabla^2 + \frac{\omega_j^2}{c^2} \right) \mathbf{u}_j(\mathbf{r}) = 0 \tag{1.12}$$

and

$$\frac{\partial^2 a_j(t)}{\partial t^2} + \omega_j^2 a_j(t) = 0. \tag{1.13}$$

The second expression leads to a characteristic time dependence like $a_j(t) = a_j \exp(-i\omega_j t)$. Also the eigenmodes $\mathbf{u}_j(\mathbf{r})$ satisfy the transversality condition

$$\nabla \cdot \mathbf{u}_j(\mathbf{r}) = 0, \tag{1.14}$$

and they form a complete orthonormal set:

$$\int \mathbf{u}_j^*(\mathbf{r}) \mathbf{u}_{j'}(\mathbf{r}) dV = \delta_{j, j'},$$

where $\delta_{j, j'}$ is the *Kronecker delta*, defined as being equal to unity if $j = j'$ and zero otherwise. If we assume periodic boundary conditions for $\mathbf{A}(\mathbf{r}, t)$, which is restricted to the interior of a cubical box with sides of length L (and hence volume $V = L^3$), it follows that the solutions for $\mathbf{u}_j(\mathbf{r})$ are expressible in a plane-wave form as

$$\mathbf{u}_j(\mathbf{r}) = \frac{1}{\sqrt{V}} \hat{\mathbf{e}}_{j, \lambda} e^{i \mathbf{k}_j \cdot \mathbf{r}}.$$

Here $|\mathbf{k}_j|^2 = \omega_j^2/c^2$ and $\hat{\mathbf{e}}_{j, \lambda}$ (with $\lambda = 1, 2$ as labels for the two transverse directions) is the unit vector that specifies the polarization of the field. The components of the wave vector $\mathbf{k}_j$ must satisfy

$$k_{jx} = \frac{2\pi}{L} n_{jx} \quad \text{with} \quad n_{jx} = 0, \pm 1, \pm 2, \dots,$$

with similar results for the y and z components.

Equation (1.14) implies that $\hat{\mathbf{e}}_{j,\lambda} \cdot \mathbf{k}_j = 0$, which means that the vectors $\{\hat{\mathbf{e}}_1, \hat{\mathbf{e}}_2, \mathbf{k}\}$ are an orthogonal basic set. Hence, the form of $\mathbf{A}(\mathbf{r}, t)$ is

$$\mathbf{A}(\mathbf{r}, t) = \sum_j \sum_{\lambda=1,2} \sqrt{\frac{1}{2\omega_j \varepsilon_0 V}} \hat{\mathbf{e}}_{j,\lambda} \left[a_j e^{i(\mathbf{k}_j \cdot \mathbf{r} - \omega_j t)} + a_j^\dagger e^{-i(\mathbf{k}_j \cdot \mathbf{r} - \omega_j t)} \right]. \quad (1.15)$$

From Equation (1.9) we see that the magnetic and electric fields can be written as

$$\mathbf{H}(\mathbf{r}, t) = \frac{-i}{c\mu_0} \sum_j \sum_{\lambda=1,2} \sqrt{\frac{\omega_j}{2\varepsilon_0 V}} (\hat{\mathbf{e}}_{j,\lambda} \times \mathbf{k}_j) \left[a_j e^{i(\mathbf{k}_j \cdot \mathbf{r} - \omega_j t)} - a_j^\dagger e^{-i(\mathbf{k}_j \cdot \mathbf{r} - \omega_j t)} \right],$$

$$\mathbf{E}(\mathbf{r}, t) = i \sum_j \sum_{\lambda=1,2} \sqrt{\frac{\omega_j}{2\varepsilon_0 V}} \hat{\mathbf{e}}_{j,\lambda} \left[a_j e^{i(\mathbf{k}_j \cdot \mathbf{r} - \omega_j t)} - a_j^\dagger e^{-i(\mathbf{k}_j \cdot \mathbf{r} - \omega_j t)} \right]. \quad (1.16)$$

Finally, the Hamiltonian of the multimode EM field is found from

$$\mathcal{H} = \frac{1}{2} \int \left(\varepsilon_0 |\mathbf{E}|^2 + \mu_0 |\mathbf{H}|^2 \right) dV,$$

which leads to

$$\mathcal{H} = \frac{1}{2} \int \left(\varepsilon_0 \left| \frac{\partial \mathbf{A}}{\partial t} \right|^2 + \frac{1}{\mu_0} |\nabla \times \mathbf{A}|^2 \right) dV$$

$$= \frac{1}{2} \sum_j \omega_j \left(a_j a_j^\dagger + a_j^\dagger a_j \right). \quad (1.17)$$

In the preceding expression the a_j and $a_j^\dagger$ are scalar complex numbers, but we have been careful to preserve their order in products. Then, to accomplish the EM field quantization, we choose a_j and $a_j^\dagger$ to be mutually adjoint operators. The photons are bosons, and so the quantization rules (as in the previous example for the SHO) are

$$\left[a_j, a_{j'}^\dagger \right] = \delta_{j,j'}, \quad [a_j, a_{j'}] = \left[a_j^\dagger, a_{j'}^\dagger \right] = 0. \quad (1.18)$$

Therefore, the Hamiltonian may be rewritten as

$$\mathcal{H} = \sum_j \omega_j \left(a_j^\dagger a_j + \frac{1}{2} \right), \quad (1.19)$$

which has the same form as Equation (1.4) for a single quantum-mechanical SHO in 1D, except that there is a summation over the mode index j. Thus, we interpret a_j and $a_j^\dagger$ as annihilation and creation operators for mode j of the EM field in this oscillator description.

1.2 Second Quantization for Bosons and Fermions

We now demonstrate how the essential features of the operator formalism developed in the previous section can be generalized through the process of second quantization to apply to systems of N identical boson or fermion particles occupying a volume V.

1.2.1 Boson System

Here we suppose that the complete set of single-particle wave functions are specified as $\{\phi_1, \phi_2, \phi_3, \ldots\}$. Then, for the total system we assume there are n_1 particles in state 1, n_2 particles in state 2, and so forth, such that $N = n_1 + n_2 + \cdots$ gives the total number of particles.

If we were to follow a standard description in terms of wave functions, the argument would go broadly as follows. We could denote the total wave function for the system of N particles as $\Phi_{n_1, n_2, \ldots}(x_1, x_2, \ldots, x_N)$, where x_i denotes the set of coordinates describing the ith particle. This function must be symmetric with respect to the interchange of any two particles. Next, we could attempt to construct an expression for Φ in terms of the single-particle wave functions as

$$\Phi_{n_1, n_2, \ldots}(x_1, x_2, \ldots, x_N) = \mathcal{A} \sum_P \phi_{i_1}(x_1) \phi_{i_2}(x_2) \cdots \phi_{i_N}(x_N),$$

where the set $\{i_1, i_2, \ldots, i_N\}$ of integers has label 1 occurring n_1 times, label 2 occurring n_2 times, and so forth. The summation is over all the permutations P of the particle coordinates $x_1, x_2, \ldots, x_N$ for the N identical particles, and $\mathcal{A} = (n_1! n_2! \cdots / N!)^{1/2}$ is a normalization constant. It is easy to realize that this is not helpful because often we do not know the one-particle wave functions $\{\phi_i\}$, and in any case it would not be practical to construct Φ as mentioned in the preceding text because of the very large numbers typically involved.

Therefore, we look for a different approach, based on second quantization and guided by the previous calculations for the simple harmonic oscillator and EM field. We start by recognizing that one, potentially useful, way to specify Φ is by giving the number of particles in each available single-particle state (as n_1 particles in state 1, and so on). Thus, we may choose to denote

$$\Phi = |n_1, n_2, \ldots, n_i, \ldots\rangle, \tag{1.20}$$

where n_i is the occupation number of the ith state. This is often called the *occupation number representation*. We will require the wave function to be orthogonalized and normalized, so that

$$\langle n_1', n_2', \ldots | n_1, n_2, \ldots \rangle = \delta_{n_1, n_1'} \delta_{n_2, n_2'} \cdots \tag{1.21}$$

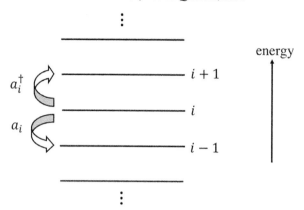

Figure 1.2 Schematic representation for the role of the $a_i^\dagger$ and a_i operators on the ladder of energy levels for the ith state of a boson system.

for any two states. We next define creation and annihilation operators for the ith state (by analogy with the relationships for the SHO) as

$$a_i^\dagger |n_1, n_2, \ldots, n_i, \ldots\rangle = \sqrt{n_i + 1} |n_1, n_2, \ldots, n_i + 1, \ldots\rangle \qquad \text{"creation,"}$$
$$a_i |n_1, n_2, \ldots, n_i, \ldots\rangle = \sqrt{n_i} |n_1, n_2, \ldots, n_i - 1, \ldots\rangle \qquad \text{"annihilation."}$$
$$(1.22)$$

The effects of the $a_i^\dagger$ and a_i operators may be envisioned schematically in terms of a ladder of energy levels as represented in Figure 1.2. Of course, the levels may not be equally spaced in general. Notice that the occupation numbers for all other states are unaffected in Equation (1.22).

From the preceding definitions we can deduce some of the properties of the operators. First we see that

$$a_i^\dagger a_i |n_1, n_2, \ldots, n_i, \ldots\rangle = \sqrt{n_i} a_i^\dagger |n_1, n_2, \ldots, n_i - 1, \ldots\rangle$$
$$= n_i |n_1, n_2, \ldots, n_i, \ldots\rangle,$$

so the number operator for the ith state is $a_i^\dagger a_i$. Also it follows that

$$a_i a_i^\dagger |n_1, n_2, \ldots, n_i, \ldots\rangle = \sqrt{n_i + 1} a_i |n_1, n_2, \ldots, n_i + 1, \ldots\rangle$$
$$= (n_i + 1) |n_1, n_2, \ldots, n_i, \ldots\rangle,$$

and so by subtraction $(a_i a_i^\dagger - a_i^\dagger a_i) |n_1, n_2, \ldots, n_i, \ldots\rangle = |n_1, n_2, \ldots, n_i, \ldots\rangle$. Therefore, we have the simple result that $[a_i, a_i^\dagger] = 1$ for the commutator. The arguments may be extended to show that any two operators referring to different states i and j commute, and so the same commutation relations as in Equation (1.18) are applicable.

It should be remarked that the preceding results are general in the sense that we have not specified the labels i and j in physical terms (they are just labels for the states). In any specific application we would *choose* the most convenient labels (e.g., position, wave vector, spin projection).

1.2.2 Fermion System

The essential difference now relates to the intrinsic symmetry of the particles. For fermions the total wave function Φ has to be antisymmetric with respect to interchanging the coordinates of any two identical particles in the system. As in Equation (1.20), we use the occupation number representation, which is now assumed to be correctly antisymmetrized, as well as being orthogonalized and normalized. According to the Pauli exclusion principle, no two identical particles can occupy the same quantum state, so the only possibilities for the occupation numbers are $n_i = 0$ or 1 for any single-particle state i.

One part of the definition of the creation operator for state i must be that

$$a_i^\dagger |n_1, n_2, \ldots, n_i = 1, \ldots\rangle = 0. \qquad (1.23)$$

for an occupied state because it is impossible to add a particle to an already occupied state. For the creation operator acting on an unoccupied state (with $n_i = 0$) the corresponding definition must be of the form

$$a_i^\dagger |n_1, n_2, \ldots, n_i = 0, \ldots\rangle = c_i |n_1, n_2, \ldots, n_i = 1, \ldots\rangle, \qquad (1.24)$$

where c_i is a constant. The analogy with the boson case would suggest taking $c_i = \sqrt{n_i + 1} = 1$ for $n_i = 0$, but this choice turns out to have inconvenient consequences later. Instead we choose (as described, e.g., in [9]) to define $c_i = (-1)^{l_i}$ where l_i is an integer defined by

$$l_i = \sum_{j=1}^{i-1} n_j = \text{number of particles with } j < i. \qquad (1.25)$$

The corresponding definition of the annihilation operator for the ith state becomes

$$a_i |n_1, n_2, \ldots, n_i, \ldots\rangle = \begin{cases} 0 & \text{if } n_i = 0, \\ (-1)^{l_i} |n_1, n_2, \ldots, n_i - 1, \ldots\rangle & \text{if } n_i = 1. \end{cases} \qquad (1.26)$$

With these chosen definitions we next examine the properties of the operators in the fermion case. Initially it seems inconvenient (and arbitrary) to have the $(-1)^{l_i}$ factors, but we shall eventually see that it does not matter.

First, it is easy to verify the property that

$$a_i^\dagger a_i |n_1, n_2, \ldots, n_i, \ldots\rangle = n_i |n_1, n_2, \ldots, n_i, \ldots\rangle, \quad (n_i = 0, 1), \qquad (1.27)$$

so the number operator for the ith state is again given by $a_i^\dagger a_i$, just as in the boson case. Also it can be checked that

$$a_i a_i^\dagger |n_1, n_2, \ldots, n_i, \ldots\rangle = (1 - n_i)|n_1, n_2, \ldots, n_i, \ldots\rangle, \quad (n_i = 0, 1).$$

Then, by adding these last two expressions we conclude that

$$\left(a_i a_i^\dagger + a_i^\dagger a_i\right) |n_1, n_2, \ldots, n_i, \ldots\rangle = |n_1, n_2, \ldots, n_i, \ldots\rangle, \qquad (1.28)$$

which leads to the property that $a_i a_i^\dagger + a_i^\dagger a_i = 1$. In particular, we note the appearance of a plus sign on the left-hand side. Accordingly, we introduce the shorthand *anticommutator* notation that $\{A, B\} = AB + BA$ for any two operators, and so our preceding result is $\{a_i, a_i^\dagger\} = 1$. This may be extended to show that any two operators referring to different states anticommute, and so

$$\{a_i, a_j^\dagger\} = \delta_{i,j}, \quad \{a_i, a_j\} = \{a_i^\dagger, a_j^\dagger\} = 0. \qquad (1.29)$$

In summary, an important difference between the two types of statistics is that boson operators satisfy commutation relations, while the fermion operators satisfy anticommutation relations. Otherwise, the operators have the analogous creation and annihilation properties, and the same $a^\dagger a$ combination acts as a number operator.

1.3 Coherent States

It is helpful to pursue further the role of creation and annihilation operators, along with the analogies to SHOs, by introducing the concept of *coherent* states. The pioneering work in the field of quantum optics was carried out by R. J. Glauber [10–12], who received the 2005 Nobel Prize in physics for his contributions to the quantum theory of optical coherence. As we shall see, there are some interesting analogies between coherent states and wave packets in QM, including the property that specific minimum-uncertainty conditions are obtained for coherent states.

Subsequently, coherent states have found applications in other areas of physics where there is wavelike behavior and superposition. Some examples of wavelike excitations, which we will discuss later, include spin waves (or magnons), phonons, and polaritons in condensed matter systems. Some textbooks giving detailed accounts of coherent states, mainly with a focus on quantum optics, are [13–17].

1.3.1 Basic Definitions for Coherent States

Coherent states are specifically QM states of quantum harmonic oscillators that have dynamics similar to the wavelike behavior of classical harmonic oscillators.

They may be introduced in various alternative ways, but here we start by defining a coherent state $|\alpha\rangle$ as the unique (apart from the overall phase) eigenstate of the annihilation operator a. Thus, we write

$$a|\alpha\rangle = \alpha|\alpha\rangle, \tag{1.30}$$

with normalization such that $\langle\alpha|\alpha\rangle = 1$. Because the operator a turns out not to be Hermitian, any eigenvalue α may in general be a complex number that we write as $\alpha = |\alpha|e^{i\theta}$. This quantity will behave analogously to the complex wave amplitude in classical optics. We note that the previous zero-occupation state (or vacuum state) denoted by $|0\rangle$ is, in fact, a coherent state with the special property $\alpha = 0$.

More generally, coherent states (including those with $\alpha \neq 0$) can be easily generated using the *coherent-state displacement operator* $D(\alpha)$ as introduced by Glauber [10]. This operator is defined by

$$D(\alpha) = \exp(\alpha a^\dagger - \alpha^* a), \tag{1.31}$$

where α is an arbitrary complex number, while $a^\dagger$ and a are the creation and annihilation operators. Now, using the convenient operator identity that

$$e^{(A+B)} = e^A e^B e^{-1/2[A, B]}, \tag{1.32}$$

which is valid provided the commutator $[A, B]$ commutes with both A and B individually, we can rewrite $D(\alpha)$ in an alternative form as a product of exponential terms. The result is easily found to be

$$D(\alpha) = e^{-|\alpha|^2/2} e^{\alpha a^\dagger} e^{-\alpha^* a}. \tag{1.33}$$

As an aside, we comment that Equation (1.32) is a particular case of the Zassenhaus formula [18], for which the reader is taken through a proof in Problem 1.4.

Starting from Equation (1.33), the displacement operator is easily seen to be unitary because $D^\dagger(\alpha) = D^{-1}(\alpha) = D(-\alpha)$. We also note three important properties of the displacement operator as follows:

$$D^\dagger(\alpha)aD(\alpha) = a + \alpha, \quad D^\dagger(\alpha)a^\dagger D(\alpha) = a^\dagger + \alpha^*, \text{ and}$$
$$D(\alpha + \beta) = D(\alpha)D(\beta)\exp[-i\,\mathrm{Im}(\alpha\beta^*)]. \tag{1.34}$$

A mathematical procedure to generate a coherent state $|\alpha\rangle$ from the vacuum state can now be established by using the displacement operator. We consider the product $aD(-\alpha)|\alpha\rangle$, for which it can be shown that

$$aD(-\alpha)|\alpha\rangle = D(-\alpha)D^\dagger(-\alpha)aD(-\alpha)|\alpha\rangle$$
$$= D(-\alpha)(a - \alpha)|\alpha\rangle = 0.$$

Here we have used the unitary property of $D(\alpha)$ as well as Equations (1.30) and (1.34). It follows, therefore, that $D(-\alpha)|\alpha\rangle = 0$ can be rearranged as

$$D(\alpha)|0\rangle = |\alpha\rangle. \tag{1.35}$$

This gives us the useful result that a coherent state is equivalent to the vacuum state displaced by the $D(\alpha)$ operator.

1.3.2 Simple Properties of Coherent States

The wave aspect of the coherent state $|\alpha\rangle$ is evident from the previous subsection. Now, by contrast, we want to illustrate the particle aspect of the coherent state by expanding the coherent state in the so-called Fock basis of number states $|n\rangle$ of the SHO. We write the expansion as

$$|\alpha\rangle = \sum_{n=0}^{\infty} c_n |n\rangle,$$

where the c_n are scalar coefficients to be determined. It is straightforward to see that

$$\langle n|\alpha\rangle = \sum_{n'=0}^{\infty} c_{n'} \langle n|n'\rangle = c_n,$$

which follows as a consequence of the completeness property for the states $|n\rangle$. Thus, we can write

$$|\alpha\rangle = \sum_{n-0}^{\infty} |n\rangle \langle n|\alpha\rangle, \tag{1.36}$$

so we next need to evaluate the quantities $\langle n|\alpha\rangle$.

By using Equation (1.7) we obtain

$$\langle n|\alpha\rangle = \frac{\alpha^n}{\sqrt{n!}} \langle 0|\alpha\rangle, \tag{1.37}$$

and therefore

$$|\alpha\rangle = \langle 0|\alpha\rangle \sum_{n=0}^{\infty} \frac{\alpha^n}{\sqrt{n!}} |n\rangle.$$

From the condition for normalization, the overall scalar coefficient $\langle 0|\alpha\rangle$ in the preceding expression can be determined to be $e^{-|\alpha|^2/2}$. Therefore, the expansion of a coherent state in terms of the states $|n\rangle$ is finally obtained as

$$|\alpha\rangle = e^{-|\alpha|^2/2} \sum_{n=0}^{\infty} \frac{\alpha^n}{\sqrt{n!}} |n\rangle. \tag{1.38}$$

The substitution of Equation (1.7) into the preceding equation establishes the connection with the vacuum state $|0\rangle$ as being

$$|\alpha\rangle = e^{-|\alpha|^2/2} \sum_{n=0}^{\infty} \frac{(\alpha a^\dagger)^n}{n!} |0\rangle$$
$$= e^{\alpha a^\dagger - |\alpha|^2/2} |0\rangle. \tag{1.39}$$

Some other useful properties of the coherent states of the EM field are summarized in the following text for later reference.

- The probability $P(n)$ of having n photons in a coherent state $|\alpha\rangle$ is given by a Poisson distribution:

$$P(n) = |\langle n|\alpha\rangle|^2 = \frac{|\alpha|^{2n} e^{-|\alpha|^2}}{n!}. \tag{1.40}$$

This follows from the analysis given in the preceding text after ensuring the normalization for a probability function.

- The coherent states satisfy a minimum-uncertainty relation in terms of position and momentum. The proof follows from Equation (1.2), which can be rewritten as

$$x = (2m\omega)^{-1/2}(a + a^\dagger),$$
$$p = -i(m\omega/2)^{1/2}(a - a^\dagger). \tag{1.41}$$

By taking the expectation values (averages) over a coherent state, we find

$$\langle x\rangle_\alpha = (2m\omega)^{-1/2}(\alpha + \alpha^*),$$
$$\langle x^2\rangle_\alpha = (2m\omega)^{-1}\left[1 + (\alpha + \alpha^*)^2\right],$$

and so for the mean square deviation in the position we have

$$(\Delta x)^2_\alpha = \langle x^2\rangle_\alpha - \langle x\rangle^2_\alpha = \frac{1}{2m\omega}.$$

In a similar manner, it may be shown for the mean square deviation in the momentum that $(\Delta p)^2_\alpha = (m\omega/2)$, and so

$$(\Delta x)^2_\alpha (\Delta p)^2_\alpha = 1/4. \tag{1.42}$$

The preceding result establishes the special property of the coherent states that they correspond to the minimum-uncertainty states.

- The coherent states are not orthogonal because they are easily shown to have the property that

$$|\langle \beta|\alpha\rangle|^2 = e^{-|\beta - \alpha|^2} \neq 0. \tag{1.43}$$

- The coherent states form an overcomplete set of states (see, e.g., [17]), satisfying in general

$$\int |\alpha\rangle\langle\alpha| \, d^2\alpha = \pi. \tag{1.44}$$

The factor of π is a consequence of the normalization. It follows that any coherent state can be expanded in terms of other coherent states by using

$$|\beta\rangle = \frac{1}{\pi} \int |\alpha\rangle\langle\alpha|\beta\rangle \, d^2\alpha. \tag{1.45}$$

1.4 Model Hamiltonians for Interacting Boson or Fermion Particles

So far, we have introduced the creation and annihilation operators in some relatively straightforward cases like the quantum-mechanical SHO and the vacuum EM field, but we have not discussed the form of the Hamiltonian in second quantization for other, often more complicated, systems. That is the purpose of the present section, where we begin in Subsection 1.4.1 with a description of single- and two-particle operators as a preliminary before giving a specific derivation of a model Hamiltonian of wider applicability in Subsection 1.4.2. Other examples will follow later.

1.4.1 Single- and Two-Particle Operators

Here we will explore some general properties of single- and two-particle operator contributions to the Hamiltonian in terms of the creation and annihilation operators $a_l^\dagger$ and a_l for a particle labeled l $(= 1, 2, \ldots, N)$ in a many-body system of N identical particles. We start by considering an operator $\mathcal{H}^{(1)}$ for the system defined as the sum of *single-particle* operators $h(l)$, meaning operators related to that one particle only, as in

$$\mathcal{H}^{(1)} = \sum_{l=1}^{N} h(l). \tag{1.46}$$

For instance, if the Hamiltonian $\mathcal{H}^{(1)}$ consists only of the kinetic energy contributions, then $h(l) = -(1/2m)\nabla_l^2$ is the corresponding single-particle operator with m denoting the particle mass. More generally, $h(l)$ might also contain potential energy contributions due to external influences, such as for electrons moving in a periodic potential for a lattice of ions or for atoms in a lattice vibrating about their fixed equilibrium positions.

Thus, if h is the single-particle operator for any one particle and its matrix elements (in the basis set $|i\rangle$ of one-particle states) are written as $h_{ij} = \langle i|h|j\rangle$, then its contribution to $\mathcal{H}^{(1)}$ is just

$$\sum_{i,j} h_{ij} |i\rangle\langle j|.$$

Therefore, on extending the argument to a system consisting of N particles we have simply

$$\mathcal{H}^{(1)} = \sum_{i,j} h_{ij} \sum_{l=1}^{N} |i\rangle_l \langle j|_l,$$

or, expressed in terms of the creation and annihilation operators introduced by analogy with the previous sections, we obtain

$$\mathcal{H}^{(1)} = \sum_{i,j} h_{ij} a_i^\dagger a_j. \tag{1.47}$$

As mentioned, such a term may have various physical interpretations. For example, if h_{ij} has only diagonal terms (i.e., when $h_{ij} = \epsilon_i \delta_{i,j}$) we have

$$\mathcal{H}^{(1)} = \sum_i \epsilon_i a_i^\dagger a_i, \tag{1.48}$$

which is just like the quadratic term in the previous SHO examples. However, if some off-diagonal terms of h_{ij} are nonzero, the contribution to $\mathcal{H}^{(1)}$ describes the creation of a particle in state i and the annihilation of a particle in a different state j, which is a process sometimes called *hopping* in electronic systems with lattice potentials. There is a further discussion of this in Subsection 1.4.3 (and then later in Chapters 2 and 5).

Similarly, we may next introduce two-particle operators, which involve a sum over terms depending on distinct pairs of operators, as in

$$\mathcal{H}^{(2)} = \frac{1}{2} \sum_{l,l'(\neq l)} f(l,l').$$

Here $f(l,l')$ might represent, for example, the Coulomb potential energy between two charged particles. Following the procedure used in the single-particle case, this may be reexpressed in terms of the creation and annihilation operators as

$$\mathcal{H}^{(2)} = \frac{1}{2} \sum_{i,j,i',j'} \langle ij|f|i'j'\rangle a_i^\dagger a_j^\dagger a_{i'} a_{j'}. \tag{1.49}$$

The detailed steps to arrive at such a result are shown for the following example in Subsection 1.4.2, but we note that there are now four creation and annihilation operators (two of each). This is a consequence of there being a double summation over particles in the original expression for $\mathcal{H}^{(2)}$, rather than the single summation as for $\mathcal{H}^{(1)}$.

In summary, the anticipated form of a total Hamiltonian that comprises single- and two-particle operators will be the sum of expressions with the same form as in Equations (1.47) and (1.49).

1.4.2 Gas of Bosons or Fermions with Pairwise Interactions

To take a specific case we now consider a gas of N identical boson or fermion particles in a volume V, where the particles can move freely except for their interactions with one another (i.e., there is no external potential energy for the present). In this example the Hamiltonian is given by the sum of two parts as

$$\mathcal{H} = \hat{T} + \hat{W}. \tag{1.50}$$

The first term represents the total kinetic energy, which can be expressed as

$$\hat{T} = -\frac{1}{2m} \sum_{l=1}^{N} \nabla_l^2 \tag{1.51}$$

for particles of mass m. The second term represents two-particle pairwise potential interactions, with each energy term taken to depend only (for simplicity) on the distance apart of the pair of particles, as in

$$\hat{W} = \frac{1}{2} \sum_{l,s(\neq l)}^{N} w\big(|\mathbf{r}_l - \mathbf{r}_s|\big). \tag{1.52}$$

The factor of $\frac{1}{2}$ appearing here avoids double counting.

We now transform each of the contributions in Equation (1.50) into the second-quantization notation, bearing in mind the preliminary results of the previous subsection. Here we will focus on macroscopically large systems, for which it is typically convenient to employ a wave-vector representation. As an often-made approximation, we assume the wave functions to have the form of plane waves appropriate to particles in a box of volume V. Thus, we write

$$\Phi_{\mathbf{k}}(\mathbf{r}) = \frac{1}{\sqrt{V}} e^{i\mathbf{k}\cdot\mathbf{r}}, \tag{1.53}$$

and these wave functions satisfy the orthonormality property that

$$\frac{1}{V} \int d^3r \, \Phi_{\mathbf{k}}^*(\mathbf{r}) \Phi_{\mathbf{k}'}(\mathbf{r}) = \delta_{\mathbf{k},\mathbf{k}'}. \tag{1.54}$$

The allowed values for the components of the wave vector $\mathbf{k}$ can be written down by assuming periodic boundary conditions as was done in Subsection 1.1.2. Proceeding now by analogy with Equation (1.47) in the discussion of single-particle operators, we have for the kinetic energy term in the Hamiltonian

$$\hat{T} = \sum_{\mathbf{k},\mathbf{k}'} T_{\mathbf{k},\mathbf{k}'}\, a_{\mathbf{k}'}^{\dagger} a_{\mathbf{k}}, \tag{1.55}$$

where

$$T_{\mathbf{k},\mathbf{k}'} = \frac{1}{V} \int d^3 r\, e^{i(\mathbf{k}-\mathbf{k}')\cdot\mathbf{r}}\, \frac{k^2}{2m}.$$

However, making use of the following mathematical identity:

$$\frac{1}{V} \int d^3 r\, e^{i(\mathbf{k}-\mathbf{k}')\cdot\mathbf{r}} = \delta_{\mathbf{k},\mathbf{k}'}, \tag{1.56}$$

which can be shown to hold when discrete wave vectors are involved (see Problem 1.6), we conclude that the Hamiltonian contribution in terms of creation and annihilation operators is simply

$$\hat{T} = \sum_{\mathbf{k}} \frac{k^2}{2m} a_{\mathbf{k}}^{\dagger} a_{\mathbf{k}}. \tag{1.57}$$

This is just the result that might be expected because $a_{\mathbf{k}}^{\dagger} a_{\mathbf{k}}$ is the number operator for particles that each have a kinetic energy equal to $k^2/2m$.

For the pairwise potential energy part, we need to proceed cautiously because of the commuting (or anticommuting) properties of the boson (or fermion) operators. We first rewrite Equation (1.52) by separating out the excluded double summation term. Then we change from summations over individual particles to integrations over the (macroscopically large) volume of the system, giving

$$\hat{W} = \frac{1}{2} \sum_{l=1}^{N} \sum_{s=1}^{N} w(|\mathbf{r}_l - \mathbf{r}_s|) - \frac{1}{2} \sum_{s=1}^{N} w(0)$$

$$= \frac{1}{2} \int \int \rho(\mathbf{r}) w(|\mathbf{r} - \mathbf{r}'|)\rho(\mathbf{r}')d^3 r\, d^3 r' - \frac{1}{2} \int \rho(\mathbf{r}') w(0)d^3 r',$$

where $\rho(\mathbf{r})$ is the density of particles at the position $\mathbf{r}$. Next, we may introduce $a^{\dagger}(\mathbf{r})$ and $a(\mathbf{r})$ as the operators creating and annihilating a particle at the position $\mathbf{r}$, so that

$$\rho(\mathbf{r}) = a^{\dagger}(\mathbf{r})a(\mathbf{r}) \tag{1.58}$$

from the usual number-operator property. However, because we are using plane-wave states it follows that

$$a^{\dagger}(\mathbf{r}) = \frac{1}{\sqrt{V}} \sum_{\mathbf{k}} a_{\mathbf{k}}^{\dagger} e^{i\mathbf{k}\cdot\mathbf{r}}, \quad a(\mathbf{r}) = \frac{1}{\sqrt{V}} \sum_{\mathbf{k}'} a_{\mathbf{k}'} e^{-i\mathbf{k}'\cdot\mathbf{r}}, \quad (1.59)$$

which leads to

$$\rho(\mathbf{r}) = \frac{1}{V} \sum_{\mathbf{k},\mathbf{k}'} a_{\mathbf{k}}^{\dagger} a_{\mathbf{k}'} e^{i(\mathbf{k}-\mathbf{k}')\cdot\mathbf{r}}. \quad (1.60)$$

It is also convenient at this stage to define the Fourier components $v(\mathbf{q})$ of the interaction term by

$$w(|\mathbf{r}-\mathbf{r}'|) = \sum_{\mathbf{q}} v(\mathbf{q}) e^{-i\mathbf{q}\cdot(\mathbf{r}-\mathbf{r}')}. \quad (1.61)$$

Substituting all this into the preceding expression for the potential energy leads to the result

$$\hat{W} = \frac{1}{2V^2} \sum_{\mathbf{k}_1,\mathbf{k}_1',\mathbf{k}_2,\mathbf{k}_2',\mathbf{q}} v(\mathbf{q}) a_{\mathbf{k}_1}^{\dagger} a_{\mathbf{k}_1'} a_{\mathbf{k}_2}^{\dagger} a_{\mathbf{k}_2'}$$

$$\times \int\int d^3r \, d^3r' \, e^{-i\mathbf{q}\cdot(\mathbf{r}-\mathbf{r}')} e^{i(\mathbf{k}_1-\mathbf{k}_1')\cdot\mathbf{r}} e^{i(\mathbf{k}_2-\mathbf{k}_2')\cdot\mathbf{r}'}$$

$$- \frac{1}{2V} \sum_{\mathbf{k}_1,\mathbf{k}_1',\mathbf{q}} v(\mathbf{q}) a_{\mathbf{k}_1}^{\dagger} a_{\mathbf{k}_1'} \int d^3r' e^{i(\mathbf{k}_1-\mathbf{k}_1')\cdot\mathbf{r}'}.$$

This can be simplified using the identity in Equation (1.56), which removes the integrations over $\mathbf{r}$ and $\mathbf{r}'$. Also some of the wave vectors are eliminated, giving

$$\hat{W} = \frac{1}{2} \sum_{\mathbf{k}_1,\mathbf{k}_2,\mathbf{q}} v(\mathbf{q}) a_{\mathbf{k}_1}^{\dagger} a_{\mathbf{k}_1-\mathbf{q}} a_{\mathbf{k}_2}^{\dagger} a_{\mathbf{k}_2+\mathbf{q}} - \frac{1}{2} \sum_{\mathbf{k}_1,\mathbf{q}} v(\mathbf{q}) a_{\mathbf{k}_1}^{\dagger} a_{\mathbf{k}_1}$$

$$= \frac{1}{2} \sum_{\mathbf{k}_1,\mathbf{k}_2,\mathbf{q}} v(\mathbf{q}) a_{\mathbf{k}_1}^{\dagger} a_{\mathbf{k}_2}^{\dagger} a_{\mathbf{k}_2+\mathbf{q}} a_{\mathbf{k}_1-\mathbf{q}}. \quad (1.62)$$

We have also used the commutation properties (for bosons) or the anti-commutation properties (for fermions) to get to the final result, which is the same in each case. Putting everything together leads to

$$\mathcal{H} = \sum_{\mathbf{k}} \frac{k^2}{2m} a_{\mathbf{k}}^{\dagger} a_{\mathbf{k}} + \frac{1}{2} \sum_{\mathbf{k}_1,\mathbf{k}_2,\mathbf{q}} v(\mathbf{q}) a_{\mathbf{k}_1}^{\dagger} a_{\mathbf{k}_2}^{\dagger} a_{\mathbf{k}_2+\mathbf{q}} a_{\mathbf{k}_1-\mathbf{q}}. \quad (1.63)$$

We notice that each term conserves the number of particles as well as the overall wave vector. A useful visualization of the two-particle term is shown in Figure 1.3. Later, when we introduce formal diagrammatic perturbation theory in Chapter 8, this kind of diagram will take on a rigorous and quantitative significance.

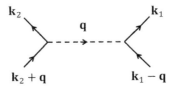

Figure 1.3 Schematic representation of the pairwise potential interaction term in second quantization as in Equation (1.62). The solid lines denote particles being either created or annihilated at the interaction vertex denoted by the dashed line. The labeling of the diagram explicitly shows the overall conservation of the wave vector, and the wave-vector transfer involving $v(\mathbf{q})$.

Electron Gas with Coulomb Interactions

A case of special importance (e.g., in metals and semiconductors) is when the particles are electrons, which are fermions, interacting using the Coulomb repulsion between the charged particles. An additional factor is that we may, in general, need to include the spin degrees of freedom corresponding to the spin quantum number m_s being equal to $\frac{1}{2}$ or $-\frac{1}{2}$. The spin states will be labeled by σ, which takes the possible values ↑ and ↓ for the spin projections "up" and "down," respectively. The Hamiltonian of the system generalizes straightforwardly to become

$$\mathcal{H} = \sum_{\mathbf{k},\sigma} \frac{k^2}{2m} a_{\mathbf{k},\sigma}^\dagger a_{\mathbf{k},\sigma} + \frac{1}{2} \sum_{\mathbf{k}_1,\sigma_1,\mathbf{k}_2,\sigma_2,\mathbf{q}} v(\mathbf{q}) a_{\mathbf{k}_1,\sigma_1}^\dagger a_{\mathbf{k}_2,\sigma_2}^\dagger a_{\mathbf{k}_2+\mathbf{q},\sigma_2} a_{\mathbf{k}_1-\mathbf{q},\sigma_1}. \qquad (1.64)$$

In terms of the spatial variables the Coulomb potential energy (with screening included) is given by

$$w\left(|\mathbf{r} - \mathbf{r}'|\right) = \frac{e^2 \exp(-\lambda|\mathbf{r} - \mathbf{r}'|)}{4\pi\varepsilon_0|\mathbf{r} - \mathbf{r}'|}, \qquad (1.65)$$

where λ is a positive screening parameter. It can easily be shown (see Problem 1.7) that the Fourier transform $v(\mathbf{q})$ of this interaction for a 3D system is

$$v(\mathbf{q}) = \frac{e^2}{\varepsilon_0(q^2 + \lambda^2)}. \qquad (1.66)$$

In the limit of there being no screening ($\lambda \to 0$) we notice that $v(\mathbf{q})$ is proportional to $1/q^2$. This result for $v(\mathbf{q})$ is different for systems with a different dimensionality (e.g., for a 2D charge sheet localized at a semiconductor heterojunction, as discussed also in Problem 1.7).

1.4.3 Hopping Potential and the Hubbard Model

It follows from the introductory discussion given in Subsection 1.4.1 for hopping that we may now write the correponding contribution $\mathcal{H}_h$ to the Hamiltonian in second quantization as

$$\mathcal{H}_h = \sum_{\mathbf{r},\mathbf{r}'} t(\mathbf{r} - \mathbf{r}') a_\mathbf{r}^\dagger a_{\mathbf{r}'} \tag{1.67}$$

when expressed in the position representation. Here we employ the conventional notation $t(\mathbf{r} - \mathbf{r}')$ for the hopping between the two positions. The spatial form of the result is particularly relevant for systems where the tight-binding model of solid state physics (see, e.g., [19, 20]) is a good approximation for obtaining the electronic wave functions. This approach is assumed, for example, in some calculations described for graphene-based structures in later chapters.

The hopping term is also an important ingredient in the treatment of electron correlations in metals using the Hubbard model [21]. The Hubbard Hamiltonian is obtained from Equation (1.67) by adding another term, denoted by $\mathcal{H}_U$, to represent the Coulomb repulsion effects if two electrons are at the same atomic site. As a consequence of the Pauli exclusion principle, the only way in which this can happen is when one electron at the site has spin up and the other has spin down. The Hubbard Hamiltonian in its simplest form can be expressed as

$$\mathcal{H} = \mathcal{H}_h + \mathcal{H}_U = \sum_{\mathbf{r},\mathbf{r}',\sigma} t(\mathbf{r} - \mathbf{r}') a_{\mathbf{r},\sigma}^\dagger a_{\mathbf{r}',\sigma} + U \sum_\mathbf{r} n_{\mathbf{r},\uparrow} n_{\mathbf{r},\downarrow}, \tag{1.68}$$

where $n_{\mathbf{r},\sigma} = a_{\mathbf{r},\sigma}^\dagger a_{\mathbf{r},\sigma}$ is the number operator for electrons with spin projection σ. Because the possible eigenvalues of $n_{\mathbf{r},\sigma}$ are restricted to 0 and 1, we see that the second term in Equation (1.68) is equal to U when $n_{\mathbf{r},\uparrow} = n_{\mathbf{r},\downarrow} = 1$ and zero in all other cases. As we shall see later in Chapter 5, although it is possible to solve exactly for each term of Equation (1.68) on its own, only *approximate* solutions of the Hubbard Hamiltonian in 2D and 3D are so far available.

It is sometimes useful to reexpress the hopping result in Equation (1.67) in terms of wave-vector labels and the corresponding creation and annihilation operators, rather than the position labels. This transformation is discussed, for example, in the context of a hopping model for 2D graphene (see Chapter 2 for complete graphene sheets and Chapter 7 for graphene nanoribbons). A wave-vector transformation of the hopping term in 3D systems is also employed in Chapter 5 for the description of the Hubbard model applied to metal-insulator transitions.

1.5 Hamiltonian Diagonalization Methods

As examples, we now present two physically distinctive cases of more complicated many-body systems for which the Hamiltonian can first be simplified, and then a solution can be found for the approximate Hamiltonian by a "diagonalizing" transformation applied to that Hamiltonian. This approach will be useful in introducing the concept of "quasiparticles" in interacting systems.

1.5.1 Example for an Interacting Boson Gas

We employ the Hamiltonian already derived in Equation (1.63), and we assume now that the interaction is both repulsive and weak, i.e., $v(\mathbf{q})$ is positive and small for all $|\mathbf{q}|$. Specifically, in this calculation we shall assume that $v(\mathbf{q})$ is weak enough that most of the boson particles are in the single-particle ground state (so they have zero momentum and hence zero wave vector). This is the regime of the so-called *Bose–Einstein condensation* (BEC) where the occupation number of the single-particle ground state becomes macroscopically large. The 2001 Nobel Prize in physics was awarded for the experimental detection of BEC in a dilute gas of alkali atoms at ultralow temperatures [22].

We start from a consideration of the total number N of particles, which can be written as

$$N = a_0^\dagger a_0 + {\sum_{\mathbf{k}}}' a_{\mathbf{k}}^\dagger a_{\mathbf{k}}. \tag{1.69}$$

Here the prime symbol following the summation symbol means that the zero wave-vector term ($\mathbf{k} = 0$) in a sum over $\mathbf{k}$ is excluded. Denoting N_0 as the number of particles in the ground state, we have $a_0^\dagger a_0 = N_0$ and, therefore, it follows that $a_0 a_0^\dagger = N_0 + 1 \approx N_0$ because $N_0 \gg 1$. Because the occupation of the ground state is macroscopically large, those creation and annihilation operators at $\mathbf{k} = 0$ behave classically and so essentially $a_0^\dagger \approx a_0 \approx N_0^{1/2}$. With this in mind, we can now rewrite the Hamiltonian by separating out the zero wave-vector terms in the summations and replacing the $a_0^\dagger$ and a_0 operators as mentioned in the preceding text. This gives

$$\mathcal{H} = {\sum_{\mathbf{k}}}' \epsilon_{\mathbf{k}} a_{\mathbf{k}}^\dagger a_{\mathbf{k}} + \frac{1}{2} N_0^2 v(0) + N_0 {\sum_{\mathbf{k}}}' v(\mathbf{k}) a_{\mathbf{k}}^\dagger a_{\mathbf{k}}$$
$$+ \frac{1}{2} N_0 {\sum_{\mathbf{k}}}' v(\mathbf{k}) (a_{\mathbf{k}}^\dagger a_{-\mathbf{k}}^\dagger + a_{\mathbf{k}} a_{-\mathbf{k}})$$
$$+ N_0^{1/2} {\sum_{\mathbf{k},\mathbf{q}}}' v(\mathbf{q}) \left(a_{\mathbf{k}}^\dagger a_{\mathbf{q}}^\dagger a_{\mathbf{k}+\mathbf{q}} + a_{\mathbf{k}}^\dagger a_{\mathbf{k}-\mathbf{q}} a_{\mathbf{q}} \right)$$
$$+ \frac{1}{2} {\sum_{\mathbf{k}_1,\mathbf{k}_2}}' v(0) a_{\mathbf{k}_1}^\dagger a_{\mathbf{k}_2}^\dagger a_{\mathbf{k}_2} a_{\mathbf{k}_1}$$
$$+ \frac{1}{2} {\sum_{\mathbf{k}_1,\mathbf{k}_2,\mathbf{q}}}' v(\mathbf{q}) a_{\mathbf{k}_1}^\dagger a_{\mathbf{k}_2}^\dagger a_{\mathbf{k}_2+\mathbf{q}} a_{\mathbf{k}_1-\mathbf{q}},$$

where $\epsilon_{\mathbf{k}} = k^2/2m$ is the particle energy. By inspection we conclude that the last three terms in the preceding equation that involve the pairwise interaction are negligible compared with the other terms when $N_0 \gg 1$. Next we can approximate by replacing N_0 with N in the remaining terms.

Therefore, apart from an unimportant constant term, we have now arrived at an approximate (or "reduced") Hamiltonian $\mathcal{H}_R$ given by

$$\mathcal{H}_R = {\sum_{\mathbf{k}}}'\left[\epsilon_{\mathbf{k}} + Nv(\mathbf{k})\right]a_{\mathbf{k}}^{\dagger}a_{\mathbf{k}} + \frac{1}{2}{\sum_{\mathbf{k}}}'Nv(\mathbf{k})\left(a_{\mathbf{k}}^{\dagger}a_{-\mathbf{k}}^{\dagger} + a_{\mathbf{k}}a_{-\mathbf{k}}\right). \qquad (1.70)$$

This Hamiltonian still incorporates the leading-order effects of the particle interactions through the appearance of terms like $v(\mathbf{k})$, but it is quadratic in the operators. We will next show that we can solve for $\mathcal{H}_R$ exactly with the following procedure. First we comment that, if we could somehow rewrite the preceding expression as

$$\mathcal{H}_R = {\sum_{\mathbf{k}}}' E_{\mathbf{k}}\alpha_{\mathbf{k}}^{\dagger}\alpha_{\mathbf{k}}, \qquad (1.71)$$

where the $\alpha_{\mathbf{k}}^{\dagger}$ and $\alpha_{\mathbf{k}}$ operators are new boson creation and annihilation operators (undetermined as yet), then this would just represent a Hamiltonian for *noninteracting* particles with a modified energy $E_{\mathbf{k}}$ (also undetermined). From this perspective the interactions would have been taken into account through a changed energy.

While we do not necessarily know in advance if this is possible, we look for an operator transformation to achieve the preceding result and to find $E_{\mathbf{k}}$. The simplest possibility is to try using a linear relationship (known in this context as a *Bogoliubov transformation*) that produces a mixing between the original and new operators with the form

$$a_{\mathbf{k}}^{\dagger} = s_{\mathbf{k}}\alpha_{\mathbf{k}}^{\dagger} + t_{\mathbf{k}}\alpha_{-\mathbf{k}} \quad \text{and} \quad a_{\mathbf{k}} = s_{\mathbf{k}}\alpha_{\mathbf{k}} + t_{\mathbf{k}}\alpha_{-\mathbf{k}}^{\dagger}, \qquad (1.72)$$

where $s_{\mathbf{k}}$ and $t_{\mathbf{k}}$ are unknown scalar functions, which we assume to be real. For the operators to be boson operators they must satisfy

$$[\alpha_{\mathbf{k}}, \alpha_{\mathbf{k}'}^{\dagger}] = \delta_{\mathbf{k},\mathbf{k}'} \quad \text{and} \quad [\alpha_{\mathbf{k}}, \alpha_{\mathbf{k}'}] = [\alpha_{\mathbf{k}}^{\dagger}, \alpha_{\mathbf{k}'}^{\dagger}] = 0. \qquad (1.73)$$

However, we also know that $[a_{\mathbf{k}}, a_{\mathbf{k}'}^{\dagger}] = \delta_{\mathbf{k},\mathbf{k}'}$ and we use this to find conditions on $s_{\mathbf{k}}$ and $t_{\mathbf{k}}$ because

$$[a_{\mathbf{k}}, a_{\mathbf{k}'}^{\dagger}] = [s_{\mathbf{k}}\alpha_{\mathbf{k}} + t_{\mathbf{k}}\alpha_{-\mathbf{k}}^{\dagger}, s_{\mathbf{k}'}\alpha_{\mathbf{k}'}^{\dagger} + t_{\mathbf{k}'}\alpha_{-\mathbf{k}'}]$$
$$= (s_{\mathbf{k}}^2 - t_{\mathbf{k}}^2)\delta_{\mathbf{k},\mathbf{k}'}.$$

Therefore, it is necessary that $s_{\mathbf{k}}^2 - t_{\mathbf{k}}^2 = 1$, which can be automatically satisfied if we denote

$$s_{\mathbf{k}} = \cosh\theta_{\mathbf{k}} \quad \text{and} \quad t_{\mathbf{k}} = \sinh\theta_{\mathbf{k}}. \qquad (1.74)$$

The final step in the calculation is to find the unknown $\theta_{\mathbf{k}}$. This can be done using the Hamiltonian $\mathcal{H}_R$, which when rewritten in terms of the new operators becomes

$$\mathcal{H}_R = \sideset{}{'}\sum_{\mathbf{k}} \left\{ \cosh(2\theta_{\mathbf{k}})\left[\epsilon_{\mathbf{k}} + Nv(\mathbf{k})\right] + \sinh(2\theta_{\mathbf{k}})Nv(\mathbf{k}) \right\} \alpha_{\mathbf{k}}^{\dagger}\alpha_{\mathbf{k}}$$

$$+ \frac{1}{2}\sideset{}{'}\sum_{\mathbf{k}} \left\{ \sinh(2\theta_{\mathbf{k}})\left[\epsilon_{\mathbf{k}} + Nv(\mathbf{k})\right] + \cosh(2\theta_{\mathbf{k}})Nv(\mathbf{k}) \right\} \left(\alpha_{\mathbf{k}}^{\dagger}\alpha_{-\mathbf{k}}^{\dagger} + \alpha_{\mathbf{k}}\alpha_{-\mathbf{k}} \right).$$

$$(1.75)$$

This will have the required form postulated earlier provided the second term vanishes, which is the case if $\theta_{\mathbf{k}}$ is chosen such that

$$\sinh(2\theta_{\mathbf{k}})\left[\epsilon_{\mathbf{k}} + Nv(\mathbf{k})\right] + \cosh(2\theta_{\mathbf{k}})Nv(\mathbf{k}) = 0.$$

The condition for this to be satisfied is

$$\tanh(2\theta_{\mathbf{k}}) = \frac{-Nv(\mathbf{k})}{\epsilon_{\mathbf{k}} + Nv(\mathbf{k})}. \qquad (1.76)$$

From the coefficient of the first term in Equation (1.75) for $\mathcal{H}_R$ we conclude that

$$E_{\mathbf{k}} = \cosh(2\theta_{\mathbf{k}})\left[\epsilon_{\mathbf{k}} + Nv(\mathbf{k})\right] + \sinh(2\theta_{\mathbf{k}})Nv(\mathbf{k})$$

$$= \left\{ \left[\epsilon_{\mathbf{k}} + Nv(\mathbf{k})\right]^2 - \left[Nv(\mathbf{k})\right]^2 \right\}^{1/2}$$

$$= \left\{ \epsilon_{\mathbf{k}}\left[\epsilon_{\mathbf{k}} + 2Nv(\mathbf{k})\right] \right\}^{1/2}. \qquad (1.77)$$

The situation can now be summed up as follows. We know that, if we had a system of *noninteracting* bosons, the exact Hamiltonian is simply

$$\mathcal{H} = \sum_{\mathbf{k}} \epsilon_{\mathbf{k}} a_{\mathbf{k}}^{\dagger}a_{\mathbf{k}}, \qquad (1.78)$$

where $\epsilon_{\mathbf{k}} = k^2/2m$ is the particle kinetic energy. By contrast, for the *weakly interacting* boson system, we have just verified that the approximate Hamiltonian $\mathcal{H}_R$ can be transformed into Equation (1.71). As mentioned earlier, this is like the Hamiltonian for noninteracting particles, but with a modified energy $E_{\mathbf{k}}$ and new boson operators. We refer to these "particles" as *quasiparticles* of energy $E_{\mathbf{k}}$ and the $\alpha_{\mathbf{k}}$ operators are known as the *quasiparticle operators*.

Application to Liquid ⁴He

In simplified terms the low-temperature superfluid phase of liquid ⁴He below its λ-transition at around 2.2 K, otherwise known as liquid He II, is an example of a weakly interacting gas of bosons with a repulsive potential [23]. Therefore, it is of interest to compare the predictions of the preceding theory with the known excitation spectrum for He II, at least in qualitative terms. To do so, we assume the general form for $v(\mathbf{k})$ as a function of $k \equiv |\mathbf{k}|$ to be as sketched in Figure 1.4, i.e., it tends to a constant value at small k and decreases sharply at large k values.

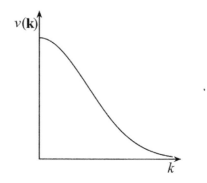

Figure 1.4 Schematic form of a simplified interaction potential $v(\mathbf{k})$ plotted against the magnitude of the wave vector for comparison with liquid He II.

At sufficiently small $\mathbf{k}$ we will have $v(\mathbf{k}) \to v(0)$, and so using $\epsilon_\mathbf{k} = k^2/2m$ in Equation (1.77) we arrive at

$$E_\mathbf{k} \to k \left(\frac{Nv(0)}{m}\right)^{1/2}. \tag{1.79}$$

Hence, the excitation spectrum for small $\mathbf{k}$ is predicted to be similar to that for an acoustic phonon, i.e., its energy is proportional to k. However, for large $\mathbf{k}$ we will be in the regime where $v(\mathbf{k})$ becomes very small, and it follows from Equation (1.77) that

$$\begin{aligned} E_\mathbf{k} &= (k^2/2m) \left\{1 + \left[4mNv(\mathbf{k})/k^2\right]\right\}^{1/2} \\ &\approx (k^2/2m) + Nv(\mathbf{k}) \to k^2/2m. \end{aligned} \tag{1.80}$$

We see that the excitation spectrum for large $\mathbf{k}$ becomes like that for a free particle, as expected. Overall, the predicted behavior for $E_\mathbf{k}$ versus k has a minimum at an intermediate value of k for reasonable values of the parameters, and it has the form sketched in Figure 1.5. Historically, the excitations near to the minimum are known as *rotons*, and they were first postulated for He II by Landau [24] to interpret specific heat measurements. This general form of the phonon-roton spectrum has been verified by inelastic neutron scattering (INS) experiments (see, e.g., [25]) on He II, which provides a more direct technique to probe the excitation. The experimental data reported in [25] at a temperature of 1.12 K are qualitatively very similar to the behavior predicted in Figure 1.5.

1.5.2 Example for Spin Waves in a Ferromagnet

As a second example of diagonalization of a Hamiltonian to obtain an approximation to the excitation spectrum, we consider a Heisenberg ferromagnet with

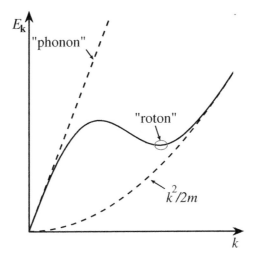

Figure 1.5 Schematic form of dispersion curve from the simplified theory in Subsection 1.5.1 to be compared with the phonon-roton dispersion curve for liquid He II (see text).

nearest-neighbor interactions between the spins (or their localized magnetic moments) at lattice sites in an insulating material.

Suppose we consider a pair of spins for electrons located at different atomic sites. We describe them by the vector operators $\mathbf{S}_i$ and $\mathbf{S}_j$ corresponding to positions denoted by $\mathbf{r}_i$ and $\mathbf{r}_j$ in the ferromagnetic material. It is well known (see, e.g., [19, 26, 27]) that the dominant contribution to the interaction energy is due to the short-range Heisenberg exchange mechanism and is usually written as $-J_{i,j}\mathbf{S}_i \cdot \mathbf{S}_j$. Here $J_{i,j}$ is the QM exchange interaction, which is related to an overlap integral between the electronic wave functions. For a ferromagnet we have each $J_{i,j} > 0$, so that the spins have a tendency to align parallel to each other at low temperatures (below the critical or Curie temperature T_C).

For a lattice of spins forming a crystal, the total Hamiltonian is

$$\mathcal{H} = -\frac{1}{2}\sum_{i,j} J_{i,j}\mathbf{S}_i \cdot \mathbf{S}_j - g\mu_B B_0 \sum_i S_i^z, \tag{1.81}$$

where we have included the Zeeman energy due to an applied magnetic field of magnitude B_0 in the z direction (with g denoting the Landé factor and μ_B the Bohr magneton). The exchange term $J_{i,j}$ depends on the separation of the sites i and j, and typically it is nonzero for nearest neighbors only (otherwise the overlap of wave functions is negligible). The preceding result for a ferromagnet is known as the *Heisenberg Hamiltonian*, and the model ignores other possible contributions such as those from magnetic dipole-dipole interactions and crystal-field anisotropies.

We consider both of these extra terms later. Nevertheless, within these limitations, Equation (1.81) provides us with a useful model for investigating the dynamics of spin systems.

An obvious difference here compared with our previous discussions is that the Hamiltonian contains spin operators, which are neither boson nor fermion operators. Instead they satisfy a set of commutation relations like those for the angular momentum operators in QM [1, 2], namely

$$\left[S_i^x, S_j^y\right] = iS_i^z \delta_{i,j}, \quad \left[S_i^y, S_j^z\right] = iS_i^x \delta_{i,j}, \quad \left[S_i^z, S_j^x\right] = iS_i^y \delta_{i,j}. \tag{1.82}$$

Often it is helpful to define new combinations of these spin operators by introducing

$$S^+ = S^x + iS^y, \qquad S^- = S^x - iS^y, \tag{1.83}$$

and so the basic commutation relations can then be reexpressed as

$$\left[S_i^+, S_j^-\right] = 2S_i^z \delta_{i,j}, \quad \left[S_i^+, S_j^z\right] = -S_i^+ \delta_{i,j}, \quad \left[S_i^-, S_j^z\right] = S_i^- \delta_{i,j}. \tag{1.84}$$

We see that the result of a commutator between two spin operators is another operator (or zero), but not a scalar quantity as it was for bosons.

One way to deal with the preceding difficulty is to look for a transformation to rewrite the set of spin operators (S^+, S^- and S^z) in terms of boson or fermion operators. This can be done in several ways, but one particularly useful form is

$$S_i^+ = \sqrt{2S}\left(1 - \frac{a_i^\dagger a_i}{2S}\right)^{1/2} a_i, \qquad S_i^- = \sqrt{2S} a_i^\dagger \left(1 - \frac{a_i^\dagger a_i}{2S}\right)^{1/2},$$

$$\text{and } S_i^z = S - a_i^\dagger a_i. \tag{1.85}$$

This is the Holstein–Primakoff (HP) transformation [28] for spin quantum number S, where the $a^\dagger$ and a operators are boson operators satisfying the usual relations that $[a_i, a_j^\dagger] = \delta_{i,j}$ and $[a_i, a_j] = [a_i^\dagger, a_j^\dagger] = 0$. It may be verified that the preceding representation correctly reproduces the spin commutation relations (see Problem 1.11).

Next we use the HP transformation to study the excitations in Heisenberg ferromagnets at low temperatures, meaning well below the Curie temperature ($T \ll T_C$). In this case, the spins are well aligned along the direction of the applied magnetic field (taken as defining the z direction), and so $S_i^z \approx S$ for each spin, implying that $a_i^\dagger a_i \ll S$ on average. This allows us to approximate the square root terms in the operator expressions according to the binomial expansion by writing

$$\left(1 - \frac{a_i^\dagger a_i}{2S}\right)^{1/2} \approx 1 - \frac{a_i^\dagger a_i}{4S} + \cdots.$$

Therefore, at $T \ll T_C$ we have the approximate result that

$$S_i^+ \approx \sqrt{2S} a_i, \quad S_i^- \approx \sqrt{2S} a_i^\dagger, \quad S_i^z = S - a_i^\dagger a_i, \tag{1.86}$$

where we are now neglecting terms that are cubic or higher order in the boson operators.

It is necessary to reexpress the Heisenberg Hamiltonian in terms of boson operators using the previously mentioned approximate form and continuing to neglect the higher-order terms. First, we have in component form

$$
\mathcal{H} = -\frac{1}{2}\sum_{i,j} J_{i,j}\left(S_i^x S_j^x + S_i^y S_j^y + S_i^z S_j^z\right) - g\mu_B B_0 \sum_i S_i^z
$$

$$
= -\frac{1}{2}\sum_{i,j} J_{i,j}\left\{\frac{1}{2}\left(S_i^+ S_j^- + S_i^- S_j^+\right) + S_i^z S_j^z\right\} - g\mu_B B_0 \sum_i S_i^z, \tag{1.87}
$$

which leads to

$$
\mathcal{H} \approx -\frac{1}{2}S^2 \sum_{i,j} J_{i,j} - g\mu_B B_0 N S - \frac{1}{2}S \sum_{i,j} J_{i,j}\{a_i a_j^\dagger + a_i^\dagger a_j - a_i^\dagger a_i - a_j^\dagger a_j\}
$$

$$
+ g\mu_B B_0 \sum_i a_i^\dagger a_i. \tag{1.88}
$$

where N is the (macroscopically large) number of spins in the system.

We have arrived at a quadratic expression in terms of the operators, together with some constant terms, and the next step will be to rewrite this in terms of quasiparticles (just as in the previous subsection). It is shown in the text that follows that it can be achieved very simply here by defining a Fourier transform that takes us from position labels to wave-vector labels. We write

$$
a_i^\dagger = \frac{1}{\sqrt{N}}\sum_{\mathbf{k}} a_{\mathbf{k}}^\dagger e^{i\mathbf{k}\cdot\mathbf{r}_i}, \quad a_i = \frac{1}{\sqrt{N}}\sum_{\mathbf{k}} a_{\mathbf{k}} e^{-i\mathbf{k}\cdot\mathbf{r}_i}, \tag{1.89}
$$

and also we introduce Fourier components of the exchange interaction by

$$
J_{i,j} = \frac{1}{N}\sum_{\mathbf{k}} J(\mathbf{k}) e^{-i\mathbf{k}\cdot(\mathbf{r}_i - \mathbf{r}_j)}. \tag{1.90}
$$

Then it may easily be verified by the substitution of Equations (1.89) and (1.90) into (1.88) that the Hamiltonian has the quasiparticle form

$$
\mathcal{H} = \mathcal{E}_0 + \sum_{\mathbf{k}} E_{\mathbf{k}} a_{\mathbf{k}}^\dagger a_{\mathbf{k}}. \tag{1.91}
$$

Here $\mathcal{E}_0$ represents the energy of the ferromagnetic system when all spins are aligned parallel to the applied field (i.e., when the system is in the ground state) and $E_{\mathbf{k}}$ is just the quasiparticle energy. We obtain

$$
\mathcal{E}_0 = -N\left\{g\mu_B B_0 S + \frac{1}{2}S^2 J(0)\right\}, \tag{1.92}
$$

$$
E_{\mathbf{k}} = g\mu_B B_0 + S\{J(0) - J(\mathbf{k})\}. \tag{1.93}
$$

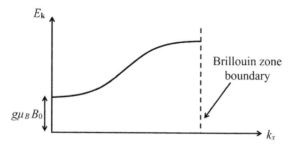

Figure 1.6 Schematic of the spin-wave dispersion relation given by Equation (1.93). The plot shows $E_{\mathbf{k}}$ plotted against the component k_x for $\mathbf{k} = (k_x, 0, 0)$.

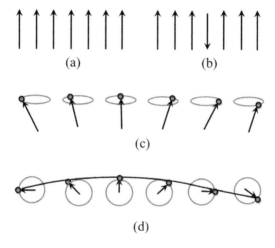

Figure 1.7 Classical interpretation (shown here in 1D) of a spin wave in a ferromagnet: (a) the ground state; (b) a single spin reversal; (c) a spin wave (viewed in perspective); and (d) a spin wave (viewed from above).

These excitations are known as *spin waves* (SWs or *magnons*), and their dispersion relation corresponds to Equation (1.93) for $E_{\mathbf{k}}$ as a function of wave vector $\mathbf{k}$. As an example, it is easy to show (see Problem 1.12) that for a body-centerd cubic (b.c.c.) ferromagnet (such as Ni for one of its phases) the expression for $J(\mathbf{k})$ is

$$J(\mathbf{k}) = 8J \cos\left(\frac{1}{2}k_x a\right) \cos\left(\frac{1}{2}k_y a\right) \cos\left(\frac{1}{2}k_z a\right), \qquad (1.94)$$

where J denotes the exchange coupling to the eight nearest neighbors and a is the lattice constant. A sketch of the dispersion relation is shown in Figure 1.6 for $\mathbf{k} = (k_x, 0, 0)$, so the wave-vector component ranges from zero to the Brillouin zone boundary value at $k_x = \pi/a$. The SWs have a simple classical interpretation as precessing spins with a small phase difference between one spin and its neighbors (see Figure 1.7). The SWs are excitations that have a lower energy than, for example, the energy corresponding to the reversal of a single spin. The overall form of E_k across the Brillouin zone of wave vectors has been well verified in various ferromagnets by inelastic neutron scattering (see, e.g., [19]).

Problems

1.1. Prove the results for the commutators given in the following text. They involve the momentum operator $\mathbf{p}$ (defined as $-i\nabla$ with $\hbar = 1$) for a particle and the position operator $\mathbf{r}$, both in 3D:

$$\left[\exp(-i\mathbf{k}\cdot\mathbf{r}),\mathbf{p}\right] = \mathbf{k}\exp(-i\mathbf{k}\cdot\mathbf{r}),$$
$$\left[\exp(-i\mathbf{k}\cdot\mathbf{r}),\mathbf{p}^2\right] = \exp(-i\mathbf{k}\cdot\mathbf{r})(2\mathbf{k}\cdot\mathbf{p} - k^2).$$

Here $\mathbf{k}$ is a constant vector and we denote $k^2 = \mathbf{k}\cdot\mathbf{k}$.

1.2. Using the discussion given in Subsection 1.1.2, evaluate the following commutators that involve the magnetic vector potential and the EM field variables. These variables are specified at different positions $\mathbf{r}$ and $\mathbf{r}'$ (but the same time t):

$$[A_n(\mathbf{r},t),E_m(\mathbf{r}',t)], \quad [A_n(\mathbf{r},t),A_m(\mathbf{r}',t)], \quad [H_n(\mathbf{r},t),H_m(\mathbf{r}',t)],$$
$$[E_n(\mathbf{r},t),E_m(\mathbf{r}',t)], \quad \text{and} \quad [H_n(\mathbf{r},t),A_m(\mathbf{r}',t)].$$

1.3. Show that the boson operators a and $a^\dagger$ satisfy the commutation relations

$$[a, f(a,a^\dagger)] = \frac{\partial f}{\partial a^\dagger} \quad \text{and} \quad [a^\dagger, f(a,a^\dagger)] = -\frac{\partial f}{\partial a},$$

where f is a differentiable function of the operators. Hence (or otherwise) show that

$$e^{-\alpha a^\dagger a} f(a,a^\dagger) e^{\alpha a^\dagger a} = f(ae^\alpha, a^\dagger e^{-\alpha}).$$

1.4. Prove from first principles that, if the commutator of any two operators A and B is a scalar quantity, then the following identities hold:

(a) $\exp(\lambda A)B\exp(-\lambda A) = B + \lambda[A,B],$

(b) $\dfrac{d}{d\lambda}(\exp(\lambda A)\exp(\lambda B)) = (A + B + \lambda[A,B])$

$$\times \exp(\lambda A)\exp(\lambda B), \text{ and}$$

(c) $\exp(\lambda A)\exp(\lambda B) = \exp\left(\lambda(A + B) + \dfrac{\lambda^2}{2}[A,B]\right).$

Here λ denotes a real scalar quantity. Hence (or otherwise) show that the displacement operator $D(\alpha) = \exp\left(\alpha a^\dagger - \alpha^* a\right)$ for coherent states, which was defined in Equation (1.31), can be rewritten as in Equation (1.33).

1.5. For a coherent state $|\alpha\rangle$, defined as in Section 1.3, show that $|\alpha\rangle\langle\alpha|$ has the properties that

$$a^\dagger |\alpha\rangle\langle\alpha| = \left(\alpha^* + \frac{\partial}{\partial\alpha}\right)|\alpha\rangle\langle\alpha|, \quad \text{and} \quad |\alpha\rangle\langle\alpha|a = \left(\alpha + \frac{\partial}{\partial\alpha^*}\right)|\alpha\rangle\langle\alpha|.$$

1.6. Prove the validity of the representation quoted in Equation (1.56) for the Kronecker delta, where $\mathbf{k}$ and $\mathbf{k}'$ are two discrete wave vectors for plane-wave states in a system of volume V. You may assume the system volume to be a cube with sides of length L (so $V = L^3$) parallel to the x, y, and z coordinate axes and apply periodic boundary conditions over the length L.

1.7. For an interacting electron gas, prove that the interaction term $v(\mathbf{q})$ in the wave-vector representation takes the form quoted in Equation (1.66) when the the spatial interaction in a 3D system has the screened Coulomb form quoted in Equation (1.65). Hence, confirm that $v(\mathbf{q})$ is proportional to $1/q^2$ in the limit of negligible screening. Next repeat this question for an electron gas in the 2D case (a charged sheet), showing that $v(\mathbf{q})$ tends to a constant at small q and for negligible screening.

1.8. Evaluate the commutators $[a_\mathbf{q}^\dagger, \mathcal{H}]$ and $[a_\mathbf{q}^\dagger a_\mathbf{Q}, \mathcal{H}]$, when the $a^\dagger$ and a operators are boson operators and the Hamiltonian $\mathcal{H}$ is given by

$$\mathcal{H} = \sum_\mathbf{k} E_\mathbf{k} a_\mathbf{k}^\dagger a_\mathbf{k},$$

where $E_\mathbf{k}$ is the quasiparticle boson energy.

1.9. By analogy with discussion later in Section 5.6, it is convenient in super-conductivity theory to introduce creation and annihilation operators for *pairs* of electrons. We define $b_\mathbf{k}^\dagger = c_{\mathbf{k},\uparrow}^\dagger c_{-\mathbf{k},\downarrow}^\dagger$ and $b_\mathbf{k} = c_{-\mathbf{k},\downarrow} c_{\mathbf{k},\uparrow}$, where $c^\dagger$ and c are fermion operators and $\uparrow$ and $\downarrow$ denote the spin projections up and down, respectively. Show that $b^\dagger$ and b satisfy the following mixed (anti)commutation relations.

$$[b_\mathbf{k}, b_{\mathbf{k}'}^\dagger] = (1 - n_{\mathbf{k},\uparrow} - n_{-\mathbf{k},\downarrow})\delta_{\mathbf{k},\mathbf{k}'},$$
$$[b_\mathbf{k}, b_{\mathbf{k}'}] = 0,$$
$$\{b_\mathbf{k}, b_{\mathbf{k}'}\} = 2b_\mathbf{k} b_{\mathbf{k}'}(1 - \delta_{\mathbf{k},\mathbf{k}'}),$$

where we denote $n_{\mathbf{k},\sigma} = c_{\mathbf{k},\sigma}^\dagger c_{\mathbf{k},\sigma}$ (with $\sigma = \uparrow$ or $\downarrow$).

1.10. The calculation given in Subsection 1.5.1 for the weakly interacting boson gas with most of the particles in the $\mathbf{k} = 0$ state may be extended to study other properties. For example, the number of particles $F(\mathbf{k})$ with wave vector $\mathbf{k}$ in the quasiparticle ground state, denoted by $|\phi_0\rangle$, can be calculated using

$$F(\mathbf{k}) = \langle \phi_0 | a_{\mathbf{k}}^{\dagger} a_{\mathbf{k}} | \phi_0 \rangle.$$

By reexpressing $a_{\mathbf{k}}^{\dagger} a_{\mathbf{k}}$ in terms of the quasiparticle operators and using the ground-state property that they satisfy $\alpha_{\mathbf{k}} |\phi_0\rangle = 0$ for all $\mathbf{k}$, derive an expression for $F(\mathbf{k})$ in terms of the quantities $k^2/2m$ and $Nv(\mathbf{k})$ as defined in Subsection 1.5.1.

Make a rough sketch of $F(\mathbf{k})$ versus $|\mathbf{k}|$ for the special case when $v(\mathbf{k})$ is a positive constant (independent of $\mathbf{k}$). In this special case show that $F(\mathbf{k})$ has an approximate power-law dependence on $k = |\mathbf{k}|$ in both of the small-k and large-k limits, i.e., show that it behaves asymptotically like k^{λ} where λ is a constant. Deduce the values of λ in each of these limits.

1.11. Evaluate the commutators $[S_i^+, S_j^-]$ and $[S_i^{\pm}, S_j^z]$ for spin one-half operators by making use of the HP transformation in Equation (1.85) and the properties of boson operators. Express your results in terms of spin operators, and check that your results are consistent with Equation (1.84).

1.12. Verify Equation (1.94) for the exchange term $J(\mathbf{k})$ at wave vector $\mathbf{k}$ in a b.c.c. ferromagnetic lattice with nearest-neighbor exchange. Derive also the corresponding expressions for $J(\mathbf{k})$ in the case of a s.c. lattice with nearest-neighbor exchange.

1.13. The coupling between SWs and phonons may be described by the Hamiltonian

$$\mathcal{H} = \sum_{\mathbf{k}} \left[\omega_{\mathbf{k}} a_{\mathbf{k}}^{\dagger} a_{\mathbf{k}} + \Omega_{\mathbf{k}} b_{\mathbf{k}}^{\dagger} b_{\mathbf{k}} + c_{\mathbf{k}} (a_{\mathbf{k}} b_{\mathbf{k}}^{\dagger} + a_{\mathbf{k}}^{\dagger} b_{\mathbf{k}}) \right],$$

where $a_{\mathbf{k}}^{\dagger}$ and $a_{\mathbf{k}}$ are the boson operators to create and annihilate a SW of energy $\omega_{\mathbf{k}}$, while $b_{\mathbf{k}}^{\dagger}$ and $b_{\mathbf{k}}$ are the boson operators to create and annihilate a phonon of energy $\Omega_{\mathbf{k}}$. Also $c_{\mathbf{k}}$ is a real, positive coupling coefficient between the two fields. Show that a transformations to new boson operators, given by

$$a_{\mathbf{k}} = A_{\mathbf{k}} \cos(\phi_{\mathbf{k}}) + B_{\mathbf{k}} \sin(\phi_{\mathbf{k}}) \text{ and } b_{\mathbf{k}} = B_{\mathbf{k}} \cos(\phi_{\mathbf{k}}) - A_{\mathbf{k}} \sin(\phi_{\mathbf{k}})$$

with $\phi_{\mathbf{k}}$ representing a real quantity, can be used to reexpress the Hamiltonian in a quasiparticle (or diagonalized) form as

$$\mathcal{H} = \sum_{\mathbf{k}} \left[\epsilon_{\mathbf{k}} A_{\mathbf{k}}^{\dagger} A_{\mathbf{k}} + E_{\mathbf{k}} B_{\mathbf{k}}^{\dagger} B_{\mathbf{k}} \right].$$

Find the value of $\phi_{\mathbf{k}}$ to achieve this, and deduce the quasiparticle energies $\epsilon_{\mathbf{k}}$ and $E_{\mathbf{k}}$.

2

Time Evolution and Equations of Motion

In this chapter we discuss the time evolution of quantum-mechanical (QM) systems with a focus on the operator properties. This aspect of their behavior can be described in terms of three possible representations (or pictures) that differ in their treatment of the time dependence. The most common divisions are the Schrödinger, Heisenberg, and interaction pictures. In the Schrödinger picture the operators are time independent whereas all the time dependence is put into the wave functions. In the Heisenberg picture, by contrast, all the time dependence is associated with the operators and the wave functions do not change with time. In the interaction picture (sometimes known as the Dirac representation), there is an intermediate situation, with the time dependence being shared between the wave functions and operators, depending on the Hamiltonian. The interaction picture can be particularly useful when the system is acted on by time-dependent external forces or potentials. This last approach provides a pathway to important developments in quantum field theory and diagrammatic perturbation methods that were pioneered by F. J. Dyson, R. P. Feynman, and others.

Starting from general considerations of the time evolution of quantum systems, the operator equation of motion can be established. Then, by solving this equation (either individually for a single operator or sometimes as a set of coupled operator equations), the technique can be applied to examples of many-body systems to deduce properties of the excitation spectrum. This is done here as a preliminary stage before we introduce the more powerful method of Green's functions in Chapter 3. Some of the applications to be considered in this chapter include the forced quantum harmonic oscillator, acoustic and optic phonons in 1D systems, electronic excitations in 2D graphene sheets, density fluctuations in an interacting electron gas (plasmon modes), and a revisiting of the weakly interacting boson gas problem from Section 1.5.

2.1 Operator Methods in Different Quantum Pictures

2.1.1 Schrödinger and Heisenberg Pictures

In the usual description employed in elementary QM (see, e.g., [1, 2]) the wave function is time dependent, whereas the operators do not (normally) have any explicit time dependence. This is referred to as the *Schrödinger picture*. The time-dependent Schrödinger equation is written as

$$i\frac{d}{dt}|\phi_S(t)\rangle = \mathcal{H}|\phi_S(t)\rangle, \tag{2.1}$$

where $\mathcal{H}$ is the time-independent Hamiltonian. This expression can be integrated to give the formal solution for the Schrödinger wave function $|\phi_S(t)\rangle$ as

$$|\phi_S(t)\rangle = e^{-i\mathcal{H}t}|\phi_S(0)\rangle, \tag{2.2}$$

which puts all the time evolution of the wave function into the exponential operator. As mentioned, in the Schrödinger picture we have $dA/dt = 0$ for any operator A, but for an observable, such as the expectation value $\langle A \rangle = \langle \phi_S(t)|A_S|\phi_S(t)\rangle$ corresponding to the operator $A_S = A_S(0) \equiv A$, this time independence may not be applicable. In particular, it follows that

$$i\frac{d}{dt}\langle A \rangle = i\left\{\left(\frac{d}{dt}\langle \phi_S(t)|\right)A|\phi_S(t)\rangle + \langle \phi_S(t)|\frac{d}{dt}A|\phi_S(t)\rangle + \langle \phi_S(t)|A\frac{d}{dt}|\phi_S(t)\rangle\right\}$$

$$= -\langle \phi_S(t)|\mathcal{H}A|\phi_S(t)\rangle + \langle \phi_S(t)|A\mathcal{H}|\phi_S(t)\rangle$$

$$= \langle \phi_S(t)|[A, \mathcal{H}]|\phi_S(t)\rangle = \langle [A, \mathcal{H}]\rangle,$$

where the Schrödinger Equation (2.1) was used to simplify the preceding expression. Thus we see that, if the operator A commutes with the Hamiltonian $\mathcal{H}$, the expectation value $\langle A \rangle$ is a constant of motion; otherwise it will be time dependent.

By contrast, in the *Heisenberg picture* a transformation is defined to make the operators time dependent, while the wave functions become time independent. This is achieved by defining the new wave function in terms of the time-independent $\mathcal{H}$ as

$$|\phi_H(t)\rangle = e^{i\mathcal{H}t}|\phi_S(t)\rangle = e^{i\mathcal{H}t}e^{-i\mathcal{H}t}|\phi_S(0)\rangle = |\phi_S(0)\rangle, \tag{2.3}$$

The result is obviously independent of t, and we denote it simply by $|\phi_H\rangle$. We suppose that any operator A (in the Schrödinger picture) becomes $A_H(t)$ in the Heisenberg picture, and we need to find this quantity. The connection is that any QM matrix element must be the same in each picture, and so

$$\langle \phi'_H|A_H(t)|\phi_H\rangle = \langle \phi'_S(t)|A|\phi_S(t)\rangle.$$

However, we know that $\langle \phi'_S(t)|A|\phi_S(t)\rangle = \langle \phi'_H|e^{i\mathcal{H}t}Ae^{-i\mathcal{H}t}|\phi_H\rangle$ from Equation (2.3), and therefore the required result is

$$A_H(t) = e^{i\mathcal{H}t}Ae^{-i\mathcal{H}t}. \tag{2.4}$$

It is seen that the time derivatives of the wave function and operator in the Heisenberg picture are

$$\frac{d}{dt}|\phi_H\rangle = \frac{d}{dt}|\phi_S(0)\rangle = 0,$$

and

$$\frac{d}{dt}A_H(t) = i\mathcal{H}e^{i\mathcal{H}t}Ae^{-i\mathcal{H}t} - ie^{i\mathcal{H}t}Ae^{-i\mathcal{H}t}\mathcal{H} = i\mathcal{H}A_H(t) - iA_H(t)\mathcal{H}.$$

Hence the equation of motion for the operator is

$$\frac{d}{dt}A_H(t) = i[\mathcal{H}, A_H(t)]. \tag{2.5}$$

It follows that the Hamiltonian $\mathcal{H}$ remains time independent in the Heisenberg picture (because it commutes with itself). More generally, an additional term $e^{i\mathcal{H}t}(\partial A/\partial t)e^{-i\mathcal{H}t}$ could be included on the right-hand side [17].

Later in this chapter we will present several examples in which Equation (2.5) is employed to study the frequencies (or energies) of the excitations in many-body systems. This will include a second look at some of the systems that were treated in Section 1.5 using diagonalization transformations.

We note that in the Heisenberg picture the time evolution of the system is determined either by integrating the operator equation of motion or by using frequency Fourier transforms. In the Schrödinger picture the system evolution is found by integrating the Schrödinger equation. In both cases we get the same answers, and thus we may use whichever specific picture is more convenient. For example, the magnetic and electric fields in Equation (1.16) of Subsection 1.1.2 are in the Heisenberg picture. However, in some calculations such as for the interaction between fields and atoms, it may be more convenient to use the Schrödinger picture.

The interaction picture, which we consider next, is usually convenient if we want to study the effect of a small interaction term $\mathcal{H}_{int}(t)$ when it is added to the Hamiltonian of a solved system $\mathcal{H}_0$ (see, e.g., [29, 30]). Another application of the interaction picture, which also illustrates the use of coherent states, is given later in Section 2.2.

2.1.2 Interaction Picture and the Time Evolution Operator

To solve problems where there is a time-dependent term in the Hamiltonian, it is often convenient to work in the interaction picture, which is defined in the following text. It is helpful first to write the Hamiltonian in the form $\mathcal{H} = \mathcal{H}_0 + \mathcal{H}_{int}(t)$, where $\mathcal{H}_0$ denotes the "free" (or noninteracting) Hamiltonian of the system and is generally time independent. The extra term $\mathcal{H}_{int}(t)$ represents an external perturbation part, which is typically assumed to be weak compared to $\mathcal{H}_0$.

We start by considering the Schrödinger Equation (2.1). Bearing in mind that the total Hamiltonian $\mathcal{H}$ is now time dependent, the equation can be formally integrated to give

$$|\phi_S(t)\rangle = U(t)|\phi_S(0)\rangle. \tag{2.6}$$

This provides us with the definition of the unitary *time-evolution operator* $U(t)$, which gives us $|\phi_S(t)\rangle$ at any time $t > 0$ in terms of the initial value $|\phi_S(0)\rangle$ at $t = 0$. This operator satisfies the differential equation

$$i\frac{d}{dt}U(t) = \mathcal{H}U(t), \tag{2.7}$$

subject to the initial condition that $dU(t)/dt = 0$ at $t = 0$.

First, in the absence of $\mathcal{H}_{int}(t)$ the time evolution of the noninteracting Hamiltonian $\mathcal{H}_0$ is given by

$$i\frac{d}{dt}U_0(t) = \mathcal{H}_0U_0(t), \tag{2.8}$$

which may be integrated to yield the solution

$$U_0(t) = \exp\left[-i\int_0^t \mathcal{H}_0(\tau)d\tau\right]. \tag{2.9}$$

If $\mathcal{H}_0$ is independent of time (which might typically be the case) we have simply $U_0(t) = \exp(-i\mathcal{H}_0 t)$, as in Subsection 2.1.1.

Now we introduce another form of the wave function, denoted by $|\phi_{int}(t)\rangle$, which will represent the wave function in the *interaction picture*:

$$|\phi_{int}(t)\rangle = U_0^{\dagger}(t)|\phi_S(t)\rangle. \tag{2.10}$$

This relation can also be rearranged using the unitary property to give

$$|\phi_S(t)\rangle = U_0(t)|\phi_{int}(t)\rangle. \tag{2.11}$$

To study the time evolution of the wave function in the interaction picture we will find an equation of motion for the operators. We first substitute Equation (2.11) into the Schrödinger equation to obtain

$$i\frac{d}{dt}U_0(t)|\phi_{int}(t)\rangle = \mathcal{H}U_0(t)|\phi_{int}(t)\rangle. \tag{2.12}$$

After some straightforward algebraic manipulation (as in Problem 2.1), this equation gives the time dependence more explicitly in the form

$$i\frac{d}{dt}|\phi_{int}(t)\rangle = V_{int}(t)|\phi_{int}(t)\rangle, \tag{2.13}$$

where we define the form of the Hamiltonian expressed in the interaction picture as

$$V_{int}(t) = U_0^\dagger(t)\mathcal{H}_{int}(t)U_0(t). \tag{2.14}$$

This result is just a unitary transformation of $\mathcal{H}_{int}(t)$. Our conclusion is that the wave function $|\phi_{int}(t)\rangle$ satisfies a Schrödinger equation with the interaction picture Hamiltonian $V_{int}(t)$ as defined in Equation (2.14).

The time-evolution operator in the interaction picture, denoted by $U_{int}(t)$, is defined by analogy with Equation (2.6) as having the basic property

$$|\phi_{int}(t)\rangle = U_{int}(t)|\phi_{int}(0)\rangle. \tag{2.15}$$

From Equations (2.11) and (2.15) it follows that

$$|\phi_S(t)\rangle = U_0(t)U_{int}(t)|\phi_{int}(0)\rangle = U_0(t)U_{int}(t)|\phi_S(0)\rangle.$$

Then, by comparing the preceding result with Equation (2.6), we see that

$$U(t) = U_0(t)U_{int}(t).$$

We require the equation of motion in the interaction picture. If we start from $\langle A \rangle = \langle \phi_S(t)|A|\phi_S(t)\rangle$ in the Schrödinger picture, the expectation value of an operator A can be reexpressed as

$$\begin{aligned}\langle A \rangle &= \langle \phi_S(0)|U^\dagger(t)AU(t)|\phi_S(0)\rangle \\ &= \langle \phi_S(0)|U_{int}^\dagger U_0^\dagger(t)AU_0(t)U_{int}(t)|\phi_S(0)\rangle \\ &= \langle \phi_{int}(t)|A_{int}|\phi_{int}(t)\rangle.\end{aligned}$$

Here we have used Equation (2.15) to arrive at the definition for the operator A_{int} that

$$A_{int} = U_0^\dagger(t)AU_0(t). \tag{2.16}$$

To summarize, the important results obtained here relating to the interaction picture are the wave equation in (2.13) and the equation of motion that

$$\frac{d}{dt}A_{int} = i[\mathcal{H}_0, A_{int}], \tag{2.17}$$

which is deduced directly from Equation (2.16).

As a simple check on some limiting cases, we note that for $\mathcal{H}_0 = 0$ we have $\mathcal{H} = \mathcal{H}_{int}(t)$, and, therefore, $dA_{int}/dt = 0$ and $i(d/dt)|\phi_S(t)\rangle = \mathcal{H}|\phi_S(t)\rangle$, yielding the

Schrödinger picture. By contrast, for $\mathcal{H}_{int}(t) = 0$ we have $\mathcal{H} = \mathcal{H}_0$, and, therefore, $id A/dt = [A, \mathcal{H}_0]$ and $(d/dt)|\phi_S(t)\rangle = 0$, yielding the Heisenberg picture.

The Dyson Series

In our discussion of the interaction picture we introduced the time evolution operator $U_{int}(t)$ by Equation (2.15). We now show how a solution for $U_{int}(t)$ in the form of a perturbation expansion, known as the *Dyson series*, can be obtained.

First, by differentiating Equation (2.15) we find that

$$\frac{d}{dt}|\phi_{int}(t)\rangle = \frac{d}{dt}U_{int}(t)|\phi_{int}(0)\rangle.$$

Then, substituting Equation (2.13) into the preceding equation leads to

$$i\frac{d}{dt}U_{int}(t) = V_{int}(t)U_{int}(t),$$

which is subject to the initial condition $U_{int}(0) = 1$. Integrating both sides of the preceding equation with respect to time over the time interval from 0 to t, we find

$$U_{int}(t) = 1 - i\int_0^t dt_1 V_{int}(t_1)U_{int}(t_1), \tag{2.18}$$

where we have used the initial condition. Note that this is only a formal solution (as an integral equation) because U appears on the left-hand side and then again within the integral on the right-hand side.

We may now proceed by substituting the formal solution for U into the right-hand side, giving

$$U_{int}(t) = 1 - i\int_0^t dt_1 V_{int}(t_1) + (-i)^2 \int_0^t dt_1 V_{int}(t_1) \int_0^{t_1} dt_2 U_{int}(t_2).$$

This process may be continued iteratively on the right-hand side to give a series expansion, which is the Dyson series in the form

$$U_{int}(t) = 1 - i\int_0^t dt_1 V_{int}(t_1) + (-i)^2 \int_0^t dt_1 V_{int}(t_1) \int_0^{t_1} dt_2 V_{int}(t_2)$$
$$\cdots + (-i)^n \int_0^t dt_1 V_{int}(t_1) \int_0^{t_1} dt_2 V_{int}(t_2) \cdots \int_0^{t_{n-1}} dt_n V_{int}(t_n) + \cdots .$$

$$\tag{2.19}$$

We note that the general term in the preceding Dyson series is a time-ordered expression because $t > t_1 > t_2 > \cdots > t_{n-1} > 0$ and the operators V_{int} at different times do not commute with one another in general. It is possible to use symmetry properties to rewrite the limits of the integral in Equation (2.19) in a more convenient form. To help in doing this, we introduce an ordering operator for

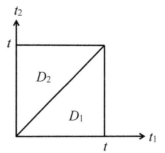

Figure 2.1 Domains of integration for $U_{int}(t)$ in the 2D space defined by the time variables t_1 and t_2.

the t-labels as follows. The operator $\hat{T}_W$ is defined so that it rearranges the order of any product of operators in such a way that the associated t-labels decrease on reading from left to right.

Suppose we examine the effect of this operator for the second-order term only in the Dyson series (the term with a product of two V_{int} terms). In the 2D space defined by t_1 and t_2 we see from Figure 2.1 that an integration is required over the domain labeled as D_1. Suppose we consider, however, the related integral over a square, which can be split into two parts as

$$(-i)^2 \int_0^t dt_1 \int_0^t dt_2 \hat{T}_W \{V_{int}(t_1)V_{int}(t_2)\} = (-i)^2 \int \int_{D_1} dt_1 dt_2 V_{int}(t_1)V_{int}(t_2)$$

$$+ (-i)^2 \int \int_{D_2} dt_1 dt_2 V_{int}(t_2)V_{int}(t_1)$$

$$= 2(-i)^2 \int \int_{D_1} dt_1 dt_2 V_{int}(t_1)V_{int}(t_2).$$

The last step follows because the time labels involved are just interchangeable (dummy) integration labels. Hence the second-order term in the expansion in Equation (2.19) is equivalent to

$$\frac{(-i)^2}{2} \int_0^t dt_1 \int_0^t dt_2 \hat{T}_W \{V_{int}(t_1)V_{int}(t_2)\}.$$

Generalizing the preceding argument to the higher-order terms in Equation (2.19), we arrive at the Dyson series in an alternative form as

$$U_{int}(t) = 1 + \sum_{n=1}^{\infty} \frac{(-i)^n}{n!} \int_0^t dt_1 \int_0^t dt_2 \cdots \int_0^t dt_n$$

$$\times \hat{T}_W \{V_{int}(t_1)V_{int}(t_2) \cdots V_{int}(t_n)\}. \tag{2.20}$$

A shorthand for the preceding result, which is sometimes convenient to employ, involves rewriting it in an exponential form as

$$U_{int}(t) = \hat{T}_W \left[\exp \left\{ -i \int_0^t dt_1 \, V_{int}(t_1) \right\} \right]. \tag{2.21}$$

As a word of caution, we must keep in mind that the exponential expansion in the preceding expression must always be made *before* doing the integration, otherwise the role of the $\hat{T}_W$ operator is not meaningful.

It will be seen later, in Chapter 8, that a formalism analogous to the Dyson series described here is used for the S-matrix expansion as an important step in establishing a diagrammatic perturbation theory using Green's functions.

2.2 Forced Quantum Harmonic Oscillator

In this section, we use the interaction picture in QM to study the behavior of a harmonic oscillator acted on by a time-dependent external force. This system will also serve as an example in which coherent states play an important role. A more general account of the time evolution of coherent states is given afterward in Section 2.3.

The forced harmonic oscillator (FHO) presents a problem of some significance in QM because it is one of the relatively few nontrivial problems that can be solved exactly in 1D. It allows us to know exactly how the states of the FHO evolve with time. As background the FHO was first examined using the path integral technique by Feynman [31]. Later Husimi [32] and Kerner [33] found an exact solution of the Schrödinger equation for this system. Interestingly it was shown [34] that the evolution of the quantum FHO leads to coherent states.

The Hamiltonian for the forced system in 1D (with position coordinate x and momentum operator p, as before) can be written as

$$\mathcal{H} = \frac{p^2}{2m} + \frac{1}{2}m\omega^2 x^2 - xf(t), \tag{2.22}$$

where $f(t)$ is a time-dependent external force that is switched on at time $t = 0$, so that $f(t) \to 0$ as $t \to 0$. We assume the system to be initially in the oscillator ground state $|0\rangle$ at time $t = 0$, and we are interested here in obtaining the state of the forced system at any later time t. It follows from the previous section that in the interaction picture this state can be found from $|\phi_{int}(t)\rangle = U_{int}(t)|0\rangle$. The Hamiltonian may be expressed as $\mathcal{H} = \mathcal{H}_0 + V_{int}(t)$, where $\mathcal{H}_0$ refers to the unperturbed system and is given by Equation (1.4) in second-quantized form, while $V_{int}(t) = -x_{int} f(t)$ is the potential energy in the interaction picture. On employing Equation (2.16) we obtain $x_{int}(t)$ in operator form as

$$x_{int} = \left[a \exp(-i\omega t) + a^\dagger \exp(i\omega t)\right]/\sqrt{2m\omega}, \tag{2.23}$$

where we have also used Equation (1.2). Therefore, we find

$$V_{int}(t) = F(t)[a \exp(-i\omega t) + a^\dagger \exp(i\omega t)], \tag{2.24}$$

where we have defined $F(t) = -f(t)/\sqrt{2m\omega}$ as a shorthand.

We now refer to the Dyson series in Equation (2.21) to obtain U_{int} for the system. For this purpose it is very convenient to use the Magnus expansion [18, 35], which allows a time-dependent interaction (V_{int} in this case) to be expanded in a series of integral terms involving nested commutators of increasing order. Specifically, its application to U_{int} allows us to write

$$U_{int}(t) = \exp\left[V_{int}^{(1)}(t) + V_{int}^{(2)}(t) + V_{int}^{(3)}(t) + \cdots\right], \tag{2.25}$$

where the lowest order terms in the preceding exponent are

$$V_{int}^{(1)}(t) = (-i) \int_0^t dt_1 V_{int}(t_1),$$

$$V_{int}^{(2)}(t) = \frac{(-i)^2}{2!} \int_0^t dt_1 \int_0^{t_1} dt_2 [V_{int}(t_1), V_{int}(t_2)],$$

$$V_{int}^{(3)}(t) = \frac{(-i)^3}{3!} \int_0^t dt_1 \int_0^{t_1} dt_2 \int_0^{t_2} dt_3 \{[V_{int}(t_1), [V_{int}(t_2), V_{int}(t_3)]]$$
$$+ [V_{int}(t_3), [V_{int}(t_2), V_{int}(t_1)]]\}. \tag{2.26}$$

The higher-order terms have increasing products of V_{int} in nested commutators. The validity of the expansion requires V_{int} to be small; in fact, convergence is ensured only if t is sufficiently small that $|V_{int}|t \ll 1$.

The relevance of the preceding identity in the context of the FHO is that, in this case, we have the commutator property

$$[V_{int}(t_1), V_{int}(t_2)] = -2i F(t_1) F(t_2) \sin[\omega(t_1 - t_2)] \tag{2.27}$$

at any two time labels t_1 and t_2. This is easily proved (see Problem 2.3) by using Equation (2.24) and the commutation properties of the boson operators a and $a^\dagger$. The useful result is that the right-hand side of the preceding equation is a scalar quantity and *not* an operator. Consequently, the higher-order commutators like $[V_{int}(t), [V_{int}(t_1), V_{int}(t_2)]]$ are zero. This means that $V_{int}^{(3)}(t)$ and all higher-order terms in Equation (2.25) are identically zero and the expansion terminates. Using this simplification we obtain an explicit solution for the Dyson series as

$$U_{int}(t) = \exp\left\{-i \int_0^t dt_1 V_{int}(t_1) - \frac{1}{2} \int_0^t dt_1 \int_0^{t_1} dt_2 [V_{int}(t_1), V_{int}(t_2)]\right\}.$$
$$(2.28)$$

To explore further the consequences of this result we define two scalar quantities $\xi(t)$ and $\zeta(t)$ by

$$\xi(t) = -i \int_0^t dt_1 F(t_1) e^{i\omega t_1},$$

$$\zeta(t) = \int_0^t dt_1 F(t_1) \int_0^{t_1} dt_2 F(t_2) \sin[\omega(t_1 - t_2)].$$

The insertion of these expressions into the time evolution operator in Equation (2.28) then leads to

$$U_{int}(t) = \exp\{\xi(t)a^\dagger - \xi^*(t)a + i\zeta(t)\}$$
$$= e^{i\zeta(t)} \exp\{\xi(t)a^\dagger - \xi^*(t)a\} = e^{i\zeta(t)} D[\xi(t)]. \qquad (2.29)$$

In the last step we have employed the definition of the coherent-state displacement D from Equation (1.31).

Now that we have obtained the time evolution operator for the FHO we can examine the state of the system in the interaction picture. We note that, if the initial state at $t = 0$ is the ground state $|0\rangle$, the state of the system at a later time t is given by

$$|\phi_{int}(t)\rangle = U_{int}(t)|0\rangle$$
$$= e^{i\zeta(t)} D[\xi(t)]|0\rangle = e^{i\zeta(t)} |\xi(t)\rangle, \qquad (2.30)$$

where Equation (1.35) has been employed. We see that the ground state in the FHO evolves into a coherent state, which has the property of minimum uncertainty.

Next we consider the situation in which the initial state is a coherent state $|\phi_{int}(0)\rangle = |\alpha\rangle$ and we examine the time evolution of this coherent state in the FHO. By employing Equations (1.35) and (2.29) we obtain

$$|\phi_{int}(t)\rangle = U_{int}(t)|\alpha\rangle = U_{int}(t)D(\alpha)|0\rangle$$
$$= e^{i\zeta(t)} D[\xi(t)]D(\alpha)|0\rangle.$$

We may now substitute Equation (1.31) into the preceding equation to find

$$|\phi_{int}(t)\rangle = e^{i\zeta(t)} \exp\{\xi(t)a^\dagger - \xi^*(t)a\} \exp\{\alpha a^\dagger - \alpha^* a\}|0\rangle.$$

Next, if we use the expansion given by Equation (1.32), we obtain

$$|\phi_{int}(t)\rangle = e^{i\zeta(t)} \exp\left\{\frac{1}{2}(\xi(t)\alpha^* - \xi^*(t)\alpha)\right\}$$
$$\times \exp\{(\alpha + \xi(t))a^\dagger - (\alpha^* + \xi^*(t))a\}|0\rangle.$$

Finally, employing Equation (1.35) one more time, we arrive at the important result that

$$|\phi_{int}(t)\rangle = e^{i\zeta(t)} \exp\left\{\frac{1}{2}(\xi(t)\alpha^* - \xi^*(t)\alpha)\right\} D[\alpha + \xi(t)]|0\rangle$$
$$= e^{i\zeta(t)} \exp\left\{\frac{1}{2}(\xi(t)\alpha^* - \xi^*(t)\alpha)\right\} |\alpha + \xi(t)\rangle$$
$$= e^{i\gamma(t)} |\alpha + \xi(t)\rangle. \tag{2.31}$$

Here the phase term appearing in the last line has been shown to correspond to $\gamma(t) = \zeta(t) - (i/2)[\xi(t)\alpha^* - \xi^*(t)\alpha]$. We may conclude, therefore, that when the FHO starts in a coherent state $|\alpha\rangle$ at $t = 0$ it will evolve with time into other coherent states that are specified by $|\alpha + \xi(t)\rangle$. Other aspects of the time evolution of the FHO will be examined later (see Subsection 6.4.2).

2.3 Time Evolution of Coherent States

Continuing with the discussion of the basic properties of coherent states given in Section 1.3, we now address the time evolution of a coherent state $|\alpha(t)\rangle$. As in Chapter 1, this will be done in terms of the simple harmonic oscillator (SHO) Hamiltonian. In addition to the general references given previously for coherent states, we mention that another useful reference for the time evolution is the book by Bongaarts [36].

By analogy with Equation (2.2) we have the formal result that $|\alpha(t)\rangle = e^{-i\mathcal{H}t}|\alpha(0)\rangle$, where $\mathcal{H}$ is the time-independent SHO Hamiltonian. Therefore, using the expansion in terms of harmonic oscillator states obtained in Equation (1.38) we find

$$|\alpha(t)\rangle = e^{-i\mathcal{H}t} e^{-|\alpha(0)|^2/2} \sum_{n=0}^{\infty} \frac{(\alpha(0))^n}{\sqrt{n!}} |n\rangle$$
$$= e^{-|\alpha(0)|^2/2} \sum_{n=0}^{\infty} \frac{(\alpha(0))^n}{\sqrt{n!}} e^{-i(n+1/2)\omega t} |n\rangle.$$

We have used Equation (1.5) in the second line of the preceding text. Then, with the aid of Equation (1.7) it follows that

$$|\alpha(t)\rangle = e^{-|\alpha(0)|^2/2} e^{-i\omega t/2} \sum_{n=0}^{\infty} \frac{\left(\alpha(0)a^\dagger e^{-i\omega t}\right)^n}{n!} |0\rangle$$

$$= \exp\left[-|\alpha(0)|^2/2 - i\omega t/2 + \alpha(0)a^\dagger e^{-i\omega t}\right] |0\rangle$$

$$= e^{-i\omega t/2} \left|e^{-i\omega t}\alpha(0)\right\rangle. \tag{2.32}$$

We have now established the important property that a coherent state remains coherent as it evolves with time.

Furthermore, from the preceding expression we see that $\alpha(t) = e^{-i\omega t}\alpha(0)$ for the evolving coherent state, and so it follows that

$$\frac{d}{dt}\alpha(t) = -i\omega\alpha(t) = -i\omega\,\mathrm{Re}[\alpha(t)] + \omega\,\mathrm{Im}[\alpha(t)]. \tag{2.33}$$

Thus, equating real and imaginary parts on both sides of the preceding equation, we have

$$\frac{d}{dt}\mathrm{Re}[\alpha(t)] = \omega\,\mathrm{Im}[\alpha(t)], \qquad \frac{d}{dt}\mathrm{Im}[\alpha(t)] = -\omega\,\mathrm{Re}[\alpha(t)]. \tag{2.34}$$

Finally, we may examine the time dependence of the expectation values for the position and momentum operators x and p by using

$$\langle x(t)\rangle = \langle\alpha(t)|x|\alpha(t)\rangle, \qquad \langle p(t)\rangle = \langle\alpha(t)|p|\alpha(t)\rangle. \tag{2.35}$$

From Equation (1.2) we note that

$$x(t) = \sqrt{\frac{1}{2m\omega}}(a^\dagger + a), \qquad p(t) = i\sqrt{\frac{m\omega}{2}}(a^\dagger - a) \tag{2.36}$$

in terms of the boson creation and annihilation operators. Therefore, we have

$$\langle x(t)\rangle = \sqrt{\frac{1}{2m\omega}}\langle\alpha(t)|(a^\dagger + a)|\alpha(t)\rangle$$

$$= \sqrt{\frac{1}{2m\omega}}\left(\alpha(t) + \alpha^*(t)\right) = \sqrt{\frac{2}{m\omega}}\,\mathrm{Re}[\alpha(t)]. \tag{2.37}$$

In a similar way, it is found that

$$\langle p(t)\rangle = \sqrt{2m\omega}\,\mathrm{Im}[\alpha(t)]. \tag{2.38}$$

From the previously mentioned equations we may conclude

$$\frac{d}{dt}\langle x(t)\rangle = \sqrt{\frac{2}{m\omega}}\frac{d}{dt}\mathrm{Re}[\alpha(t)] = \sqrt{\frac{2\omega}{m}}\,\mathrm{Im}[(\alpha(t)] = \frac{\langle p(t)\rangle}{m}, \tag{2.39}$$

and similarly

$$\frac{d}{dt}\langle p(t)\rangle = -m\omega^2\,\langle x(t)\rangle. \tag{2.40}$$

This analysis, culminating in Equations (2.39) and (2.40), has shown that the expectation values of x and p satisfy the classical equations of motion. The significance is that the classical limit of QM can be studied by using coherent states.

In concluding this section we remark that, due to the connection with the oscillator model, the applicability of coherent states extends beyond the original concept for photons in quantum optics to other wavelike excitations. For example, the applicability to magnons (spin waves) was developed in [37, 38].

2.4 Lattice Dynamics for Phonons

We now turn from the rather formal results considered so far in this chapter for the QM time evolution to make some applications to excitations in condensed matter systems. We start by considering phonons.

The simplest treatment of lattice dynamics, as a description of the quantized vibrational waves or phonons in solids, can be developed in terms of mass-and-spring arrangements in 1D chains of atoms, as in [9, 19, 20, 39] for example. For a fuller discussion of lattice dynamics the 3D character of the crystal structure has to be included (see, e.g., [40, 41]). In an alternative approach, calculations for phonons as waves in continuous elastic media are presented later in Chapters 6 and 7.

2.4.1 Monatomic Chain

Here we consider the simplest case of the monatomic lattice shown in Figure 2.2, where an infinite number of identical masses m are joined by identical springs (with spring constant C). We assume u_n is the longitudinal displacement of the mass labeled n from its equilibrium position denoted as x_{0n}. Then the instantaneous position is denoted by $x_n = x_{0n} + u_n$. Therefore, the Hamiltonian for small (elastic) displacements in the system is given by

$$\mathcal{H} = \sum_n \frac{p_n^2}{2m} + \frac{C}{2} \sum_n (u_n - u_{n+1})^2,$$ (2.41)

where p_n denotes the momentum for mass n.

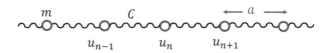

Figure 2.2 A simple 1D model for vibrational waves in a monatomic lattice with identical masses m coupled by identical springs with constant C. The longitudinal displacement of the nth mass is denoted by u_n.

Quantization is introduced into the theory by the usual commutator condition between operators that $[u_n, p_m] = i\delta_{n,m}$. The next step involves making a transformation to wavenumber coordinates (i.e., taking a 1D Fourier transform) for the momentum and the displacement as

$$p_n = \frac{1}{\sqrt{N}} \sum_k p_k e^{ikx_n}, \qquad u_n = \frac{1}{\sqrt{N}} \sum_k u_k e^{ikx_n}, \qquad (2.42)$$

where k is the wavenumber. Substitution of Equation (2.42) into the Hamiltonian (2.41) leads to

$$\mathcal{H} = \sum_{n=1}^N \frac{1}{2m} \sum_{k,k'} \frac{1}{N} p_k p_{k'} e^{i(k+k')x_n} + \frac{C}{2} \sum_n \frac{1}{N} \left[\sum_k u_k \{ e^{ikx_n} - e^{ik(x_n+a)} \} \right]^2.$$

Here a denotes the equilibrium distance between adjacent masses. Further simplification comes about by noting the Kronecker-delta relationship (the analogue of the property stated in Equation (1.56) for 3D wave vectors) that

$$\frac{1}{N} \sum_{n=1}^N e^{i(k+k')x_n} = \delta_{k,-k'}. \qquad (2.43)$$

We find after further manipulation that the Hamiltonian is

$$\mathcal{H} = \sum_k \left(\frac{1}{2m} p_k p_{-k} + \frac{1}{2} m\omega_k^2 u_k u_{-k} \right), \qquad (2.44)$$

where

$$\omega_k = \sqrt{\frac{2C}{m} \{1 - \cos(ka)\}} = \sqrt{\frac{4C}{m}} |\sin(ka/2)|. \qquad (2.45)$$

Some limiting cases for small and large wavenumbers are

$$\omega_k^2 = \begin{cases} (Ca^2/m)k^2, & \text{for } ka \ll 1, \\ 4C/m, & \text{for } ka = \pi. \end{cases} \qquad (2.46)$$

From the first line of the previously mentioned expression we see that the mode behaves like a sound wave with $\omega_k = vk$, where $v^2 = Ca^2/m$, in the long-wavelength limit. The second line in the equation provides the maximum frequency, which occurs at the Brillouin zone boundary $k = \pi/a$ in 1D.

We note that the other convenient way to derive the dispersion relation in Equation (2.45) is to use the equation-of-motion method [39]. The Newtonian equation of motion for any mass in the chain is given by

$$m\frac{d^2 u_n}{dt^2} = -C(u_n - u_{n-1}) + C(u_{n+1} - u_n)$$

$$= -C(2u_n - u_{n-1} - u_{n+1}). \tag{2.47}$$

The set of coupled equations (for different n) is solved by using Bloch's theorem for the 1D system (see, e.g., [19, 20, 39]). This theorem is a symmetry requirement following from the translational invariance of the 1D lattice of atoms under displacements by a. In the present case, it tells us to seek wavelike solutions of the form

$$u_n = u \exp[i(kan - \omega_k t)], \tag{2.48}$$

where ω_k is the angular frequency. Substituting Equation (2.48) into (2.47) just leads to the same expression as in Equation (2.45).

The conclusion from Equation (2.44) is that the Hamiltonian has the form of a simple harmonic oscillation for each wavenumber. By analogy with the procedure in Chapter 1 we may define creation and annihilation operators as

$$a_k = \frac{1}{\sqrt{2}}\left(\frac{u_k}{l_k} + ip_{-k}l_k\right),$$

$$a_k^\dagger = \frac{1}{\sqrt{2}}\left(\frac{u_k}{l_k} - ip_{-k}l_k\right), \tag{2.49}$$

with the shorthand notation that $l_k = 1/\sqrt{m\omega_k}$. These operators obey the usual boson commutation relations in terms of the 1D wavenumbers:

$$[a_k, a_{k'}^\dagger] = \delta_{k,k'}, \qquad [a_k, a_{k'}] = [a_k^\dagger, a_{k'}^\dagger] = 0.$$

Now, by using Equation (2.49) and the commutation relations that $[u_n, \mathcal{H}] = i\, du_n/dt$ and $[u_k, p_{k'}] = i\delta_{k,k'}$, the Hamiltonian in Equation (2.44) may be reexpressed as

$$\mathcal{H} = \sum_k \omega_k \left(a_k^\dagger a_k + 1/2\right). \tag{2.50}$$

A typical dispersion relation is shown in Figure 2.3 for ω_k versus ka. We see, as anticipated, that when $k \to 0$ the relationship becomes linear and also ω_k vanishes at zero wavenumber, i.e. there is a soft-mode behavior at zero wavenumber (as indicated by the dotted circle in the figure).

2.4.2 Diatomic Chain

As an extension we next consider the 1D diatomic lattice shown in Figure 2.4, where all the connecting springs are identical (and of length a), but there are

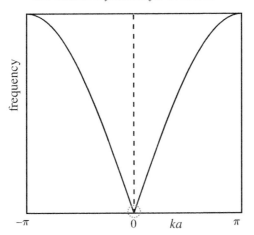

Figure 2.3 Dispersion relation showing a plot of the angular frequency ω_k versus dimensionless wavenumber ka for the acoustic phonon mode in a monatomic 1D lattice (see the text).

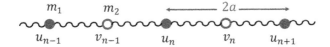

Figure 2.4 The 1D chain model for vibrational waves in a diatomic lattice with alternating masses m_1 and m_2 coupled by springs C. The periodicity length is $2a$.

alternating masses m_1 and m_2 along the chain. We will assume that $m_1 > m_2$, for convenience. We may visualize the system as consisting of a repeated pattern of "cells" (labeled by n) with each cell of length $2a$ consisting of a mass m_1 and a mass m_2. The Hamiltonian in this case can be written as

$$\mathcal{H} = \sum_n \left[\frac{\dot{u}_n^2}{2m_1} + \frac{\dot{v}_n^2}{2m_2} \right] + \frac{C}{2} \sum_n \left[(u_n - v_n)^2 + (u_n - v_{n-1})^2 \right], \qquad (2.51)$$

where u_n (or v_n) denotes the longitudinal displacement from its equilibrium position of the mass m_1 (or m_2) in cell n. Here the dot denotes a time derivative d/dt.

We may proceed as in the case of a monatomic chain to find the dispersion relation for any mode. Here it is convenient to follow the equation-of-motion method. Newton's equations of motion for the masses m_1 and m_2, respectively, are

$$m_1 \ddot{u}_n + C\,(2u_n - v_n - v_{n-1}) = 0,$$
$$m_2 \ddot{v}_n + C\,(2v_n - u_n - u_{n+1}) = 0. \qquad (2.52)$$

Following Bloch's theorem as before, we now look for solutions of the form

$$u_n = u\,\exp[i\,(kan - \omega_k t)] \quad \text{and} \quad v_n = v\,\exp[i\,(kan - \omega_k t)],$$

where u and v are amplitude factors. Substituting for u_n and v_n in Equation (2.52) we obtain

$$m_1\omega_k^2 = \frac{C}{u}\left(2u - ve^{ika} - ve^{-ika}\right) = 2C\left(1 - \frac{v}{u}\cos(ka)\right),$$

$$m_2\omega_k^2 = \frac{C}{v}\left(2v - ue^{-ika} - ue^{ika}\right) = 2C\left(1 - \frac{u}{v}\cos(ka)\right). \tag{2.53}$$

Then, by eliminating u/v, the implicit dispersion relation for the diatomic chain is found to be

$$\left(1 - \frac{m_1\omega_k^2}{2C}\right)\left(1 - \frac{m_2\omega_k^2}{2C}\right) = \cos^2(ka),$$

which can be rearranged to give the result

$$\omega_k^2 = C\left(\frac{1}{m_1} + \frac{1}{m_2}\right) \pm C\left[\left(\frac{1}{m_1} + \frac{1}{m_2}\right)^2 - \frac{4\sin^2(ka)}{m_1 m_2}\right]^{1/2}. \tag{2.54}$$

We see now that there are two modes in the excitation spectrum, corresponding to the upper and lower signs. This is to be expected on general principles because there are two atoms in each unit cell. As a check, if we put $m_1 = m_2 = m$ we find that one solution becomes the same as in Equation (2.45) for the monatomic case. The interpretation of the other branch is left for consideration in Problem 2.5.

The dispersion relations of the two modes in the general case have the form shown in Figure 2.5. The lower (or acoustic) phonon branch occupies the frequency range $\omega \le (2C/m_1)^{1/2}$, while the higher (or optic) phonon branch occupies $(2C/m_2)^{1/2} \le \omega \le [2C(m_1 + m_2)/m_1 m_2]^{1/2}$. It is useful to examine the relative phases of the atomic displacements for the two modes, which we will do for wavenumber $k = 0$. In this case, it is found from Equations (2.53) and (2.54) that

$$u/v = \begin{cases} 1, & \text{for the acoustic phonon,} \\ -m_2/m_1, & \text{for the optic phonon.} \end{cases} \tag{2.55}$$

Thus, for the acoustic mode, the direction and the amplitude of the oscillations of both masses are the same and they are in phase. By contrast, however, for the optic mode the two masses oscillate in opposite directions and the heavier one has the smaller amplitude.

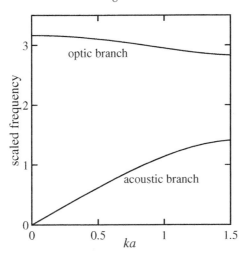

Figure 2.5 Illustration of the dispersion relations for acoustic and optic phonons in a 1D diatomic chain. The scaled angular frequency $\omega_k m_1^{1/2}/C^{1/2}$ is plotted versus the dimensionless wavenumber ka. We employed Equation (2.54) with parameter values corresponding to $m_1 = 2m_2$. The Brillouin zone boundary is at $ka = \pi/2 \simeq 1.57$.

2.5 The Interacting Boson Gas Revisited

For the next application we briefly reconsider the earlier calculation given in Subsection 1.5.1 for the excitation spectrum of a weakly interacting boson gas at sufficiently low temperatures that most of the particles are in the single-particle ground state, as in a Bose–Einstein condensation (BEC). We recall that an approximate, or reduced, Hamiltonian $\mathcal{H}_R$ was obtained in Equation (1.70), which was solved by making a Bogoliubov transformation to a new set of operators. It will be shown here that the use of the operator equation of motion provides an alternative approach without requiring an explicit transformation to be found.

We begin by rewriting the expression for $\mathcal{H}_R$ more conveniently as

$$\mathcal{H}_R = {\sum_{\mathbf{k}}}' \left[A_{\mathbf{k}} a_{\mathbf{k}}^\dagger a_{\mathbf{k}} + B_{\mathbf{k}}(a_{\mathbf{k}}^\dagger a_{-\mathbf{k}}^\dagger + a_{\mathbf{k}} a_{-\mathbf{k}}) \right], \tag{2.56}$$

where $A_{\mathbf{k}}$ and $B_{\mathbf{k}}$ are simply shorthands for $[(k^2/2m) + Nv(\mathbf{k})]$ and $Nv(\mathbf{k})/2$, respectively. From the operator equation of motion in Equation (2.5) it follows that $a_{\mathbf{k}}$ satisfies

$$\frac{d}{dt} a_{\mathbf{k}} = i[\mathcal{H}_R, a_{\mathbf{k}}].$$

The commutator on the right-hand side of the preceding expression can easily be evaluated, keeping in mind that the only contributions come when the wave-vector labels in the commutators match. The result leads to

$$\frac{d}{dt}a_{\mathbf{k}} = -i A_{\mathbf{k}} a_{\mathbf{k}} - 2i B_{\mathbf{k}} a^{\dagger}_{-\mathbf{k}}, \tag{2.57}$$

so we see that this equation provides a coupling between the operators $a_{\mathbf{k}}$ and $a^{\dagger}_{-\mathbf{k}}$. Another similar equation of motion can be formed for $a^{\dagger}_{-\mathbf{k}}$, giving

$$\frac{d}{dt}a^{\dagger}_{-\mathbf{k}} = i A_{\mathbf{k}} a^{\dagger}_{-\mathbf{k}} + 2i B_{\mathbf{k}} a_{\mathbf{k}}. \tag{2.58}$$

The equation of motion for the $a^{\dagger}_{-\mathbf{k}}$ operator couples back to the $a_{\mathbf{k}}$ operator.

To find the excitations we seek solutions of the coupled Equations (2.57) and (2.58) that have a normal-mode time variation like $\exp(-i\omega t)$, where ω is the angular frequency of an excitation. This means that we can make the replacement $d/dt \rightarrow -i\omega$, which yields the following pair of equations:

$$(\omega - A_{\mathbf{k}})a_{\mathbf{k}} = 2B_{\mathbf{k}}a^{\dagger}_{-\mathbf{k}} \quad \text{and} \quad (\omega + A_{\mathbf{k}})a^{\dagger}_{-\mathbf{k}} = 2B_{\mathbf{k}}a_{\mathbf{k}}. \tag{2.59}$$

The solutions for the frequency (and hence for the energy $\hbar\omega$ because we use units such that $\hbar = 1$) are seen to be $\omega = \pm E_{\mathbf{k}}$, where

$$E_{\mathbf{k}} = \sqrt{A^2_{\mathbf{k}} - 4B^2_{\mathbf{k}}}. \tag{2.60}$$

Comparison of Equation (2.60) with the expression obtained for the quasiparticle energy in Subsection 1.5.1 verifies that the same dispersion relation has indeed been obtained by the equation-of-motion method.

2.6 Exchange and Dipole-Exchange Spin Waves

A simple introduction to SWs at low temperatures $T \ll T_C$ was given in Subsection 1.5.2 for a ferromagnet described by the Heisenberg exchange Hamiltonian. Here we will supplement that calculation of the dispersion relation by adding the effects of magnetic dipole-dipole interactions, which turn out to be of significance for the SWs at sufficiently small wave vectors in the Brillouin zone. Here we will for convenience employ the operator equation-of-motion method.

The total spin Hamiltonian can now be written as $\mathcal{H} = \mathcal{H}_{ex} + \mathcal{H}_{dip}$, where $\mathcal{H}_{ex}$ is just the previous expression in Equation (1.81) that included the exchange interactions plus an applied magnetic field term, whereas $\mathcal{H}_{dip}$ is the extra term describing the dipolar interactions between all sites (see, e.g., [7, 26, 39]):

$$\mathcal{H}_{dip} = \frac{1}{2}\left(\frac{\mu_0 g^2 \mu_B^2}{4\pi}\right) \sum_{i,j}\left\{\frac{\mathbf{S}_i \cdot \mathbf{S}_j}{r^3_{ij}} - \frac{3(\mathbf{r}_{ij} \cdot \mathbf{S}_i)(\mathbf{r}_{ij} \cdot \mathbf{S}_j)}{r^5_{ij}}\right\}. \tag{2.61}$$

Here $\mathbf{r}_{ij}$ is the vector joining the sites i and j. By contrast with the short-range exchange interactions between nearest neighbors, the dipolar interactions fall off more slowly with distance apart of the atomic sites.

Despite the more complicated form of the preceding expression compared with the exchange term, we may follow the same approach as in Subsection 1.5.2 by employing the Holstein–Primakoff (HP) transformation from the spin operators to boson operators. Also we will approximate the HP transformation for low temperatures as in Equation (1.85). After some straightforward but lengthy algebra (see Problem 2.7) it is found that the previous Equation (1.91) for the Hamiltonian in second quantization and in the wave-vector representation becomes generalized to

$$\mathcal{H} = \sum_{\mathbf{k}} \left\{ P(\mathbf{k}) a_{\mathbf{k}}^{\dagger} a_{\mathbf{k}} + Q(\mathbf{k}) a_{\mathbf{k}}^{\dagger} a_{-\mathbf{k}}^{\dagger} + Q^{*}(\mathbf{k}) a_{\mathbf{k}} a_{-\mathbf{k}} \right\}, \tag{2.62}$$

where we have dropped an unimportant constant term to focus on the dynamics. Here the coefficients $P(\mathbf{k})$ and $Q(\mathbf{k})$ are given by

$$P(\mathbf{k}) = g\mu_B B_0 + S[J(0) - J(\mathbf{k})] - S\left[D^{zz}(0) + \frac{1}{2}D^{zz}(\mathbf{k}) \right],$$

$$Q(\mathbf{k}) = \frac{1}{4}S[D^{xx}(\mathbf{k}) - D^{yy}(\mathbf{k}) + 2iD^{xy}(\mathbf{k})].$$

The Fourier-transformed exchange terms $J(\mathbf{k})$ were defined previously in Subsection 1.5.2, and the $D^{\alpha\beta}(\mathbf{k})$ are corresponding terms for the dipole-dipole interactions (with α and β denoting Cartesian components):

$$D^{\alpha\beta}(\mathbf{k}) = \frac{\mu_0 g^2 \mu_B^2}{4\pi} \sum_{\mathbf{r}} \frac{3r^{\alpha}r^{\beta} - |\mathbf{r}|^2 \delta_{\alpha,\beta}}{|\mathbf{r}|^5} \exp(-i\mathbf{k}\cdot\mathbf{r}). \tag{2.63}$$

Here the sum over $\mathbf{r}$ is over all lattice translation vectors (but excluding the $\mathbf{r} = 0$ term).

When we form the operator equation of motion for $a_{\mathbf{k}}$ using Equations (2.5) and (2.62) we find that $a_{\mathbf{k}}$ is coupled to the $a_{-\mathbf{k}}^{\dagger}$ operator through the dipolar terms. The equation of motion is

$$\frac{d}{dt}a_{\mathbf{k}} = -iP(\mathbf{k})a_{\mathbf{k}} - 2iQ(\mathbf{k})a_{-\mathbf{k}}^{\dagger}, \tag{2.64}$$

while the corresponding equation for $a_{-\mathbf{k}}^{\dagger}$ is found to be

$$\frac{d}{dt}a_{-\mathbf{k}}^{\dagger} = iP(\mathbf{k})a_{-\mathbf{k}}^{\dagger} + 2iQ^{*}(\mathbf{k})a_{\mathbf{k}}. \tag{2.65}$$

Then we may follow the same steps to find the excitation energy or frequency as used earlier in Section 2.5. These steps include making the replacement $d/dt \rightarrow -i\omega$ in Equations (2.64) and (2.65) leading to $\omega = \pm E_{\mathbf{k}}$ as a consistency condition, where

$$E_{\mathbf{k}} = \sqrt{|P(\mathbf{k})|^2 - 4|Q(\mathbf{k})|^2}. \tag{2.66}$$

This result for the *dipole-exchange* SW regime shows how the previous SW dispersion relation for a Heisenberg ferromagnet in Equation (1.93) becomes generalized when the dipole-dipole interactions are included. It is known that another physical effect due to the dipolar terms (see, e.g., [26]), apart from modifying the dispersion relation for the SWs, is that the precession of the spin vectors becomes *elliptical*, rather than the circular precession depicted in Figure 1.7 for the Heisenberg ferromagnet case.

An approximate analytic evaluation of the dipole-dipole sums in Equation (2.63) was given in [42] by assuming a s.c. lattice structure and small wave vectors such that $L^{-1} \ll |\mathbf{k}| \ll a^{-1}$, where L is of the order of the sample dimensions and a is the lattice parameter. This allows the dipole-dipole sums to be related to the sample magnetization M_0. On using these results it is found that the SW dispersion corresponding to Equation (2.66) approximates at small $|\mathbf{k}|$ to

$$E_{\mathbf{k}} = \left[\left\{ g\mu_B B_0 + SJ(ak)^2 + g\mu_B \mu_0 M_0 \sin^2\theta \right\} \left\{ g\mu_B B_0 + SJ(ak)^2 \right\} \right]^{1/2}.$$

(2.67)

Here the notation is that the wave vector $\mathbf{k}$ has magnitude k and is at an angle θ relative to the z direction, which is the magnetization direction. The sample demagnetizing factor has been taken as zero for simplicity. We notice that the SW energies (or frequencies) now depend on the direction, as well as the magnitude, of the propagation wave vector $\mathbf{k}$. Equation (2.67), which is sometimes known as the Kittel formula, generalizes the result obtained in Subsection 1.5.2 in the absence of the dipole-dipole interactions.

In the limit of very small k, such that $SJa^2k^2 \ll g\mu_B\mu_0 M_0$ when exchange effects are negligible, there is a further simplification of Equation (2.67) to

$$E_{\mathbf{k}} = g\mu_B \left[B_0(B_0 + \mu_0 M_0 \sin^2\theta) \right]^{1/2}.$$

(2.68)

This dipole-dominated SW, known as the magnetostatic mode, ranges between $g\mu_B B_0$ and $g\mu_B[B_0(B_0 + \mu_0 M_0)]^{1/2}$ as the angle θ is varied. This behavior has been well verified by ferromagnetic resonance (FMR) experiments [26].

2.7 Electronic Bands of Graphene

Graphene is a 2D (monolayer) form of carbon that has been the subject of intense investigations since its experimental fabrication [43] was demonstrated in 2004. Graphene provides a remarkable example of a 2D elemental crystal lattice, and it has been shown to display special electronic properties including zero-gap features in its dispersion relation for the electronic excitations. It has amazing versatility, whether in pristine condition as a single sheet or when defects, impurities or adatoms are present. The introduction of edges (as in graphene nanoribbons) has

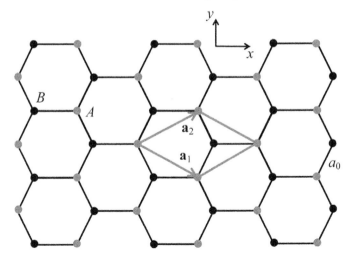

Figure 2.6 Lattice structure for a single sheet of graphene, where each carbon atom lies on one of two interpenetrating sublattices A and B (shown as gray and black circles, respectively). A possible choice of unit cell is shown with the lattice vectors $\mathbf{a}_1$ and $\mathbf{a}_2$.

led to the study of novel types of localized edge modes known as topological edge modes (to be discussed in Chapter 7) and to speculations about associated magnetic properties. Reviews of graphene are given, for example, in [44, 45].

In this section we describe the graphene structure and then we employ a hopping model for the electrons (in a tight-binding approximation) to derive the band structure of electronic excitations in this material using the operator equation-of-motion method. The graphene lattice structure for a single sheet is shown in Figure 2.6. It has a 2D hexagonal or honeycomb structure with two interpenetrating sublattices of carbon atoms denoted by A and B. The A and B sites are characterized by having different bond directions (with a relative rotation of $60°$). There are two carbon atoms per unit cell, and a possible choice for the basic lattice vectors $\mathbf{a}_1$ and $\mathbf{a}_2$ is shown in Figure 2.6 corresponding to

$$\mathbf{a}_1 = \frac{\sqrt{3}a_0}{2}(\sqrt{3}, -1), \quad \mathbf{a}_2 = \frac{\sqrt{3}a_0}{2}(\sqrt{3}, 1). \qquad (2.69)$$

Here $a_0 \simeq 0.14$ nm denotes the nearest-neighbor carbon-carbon distance. It is easily shown that the vectors of the 2D reciprocal lattice are (see Problem 2.9)

$$\mathbf{b}_1 = \frac{2\pi}{3a_0}(1, -\sqrt{3}), \quad \mathbf{b}_2 = \frac{2\pi}{3a_0}(1, \sqrt{3}). \qquad (2.70)$$

Next we need a Hamiltonian to describe the electronic states in graphene, where each carbon atom contributes three σ-bonded electron and one π-bonded electron in a sp^2-hybridized scheme. The wave functions for the π electrons are only nonzero out of the plane and form much weaker bonds than the in-plane σ electrons. As a consequence, it turns out to be the π electrons that control the density of electronic states near the Fermi level, and to a good approximation it is assumed that they can be treated within the tight-binding model (see, e.g., [20]). In the notation of second quantization this means that the Hamiltonian for the electrons may be expressed in terms of a hopping Hamiltonian as in Subsection 1.4.3. In the present case, we assume for simplicity that the hopping is restricted to the nearest neighbors (along the carbon-carbon bonds in Figure 2.6). We write

$$\mathcal{H} = -\sum_{i,j} t_{ij}(a_i^\dagger b_j + \text{H.c.}), \tag{2.71}$$

where H.c. denotes Hermitian conjugate. Also i and j refer to sites on the A and B sublattices, respectively, $a_i^\dagger$ (and a_i) and $b_j^\dagger$ (and b_j) are the corresponding creation (and annihilation) operators for the electrons, and t_{ij} is the hopping parameter (assumed to have the value t between all nearest neighbors and zero otherwise). A negative sign has been included on the right-hand side of Equation (2.71) so that $t > 0$ for graphene.

By analogy with the examples in the previous two sections, we may proceed by forming the operator equations of motion for a_i and b_j (which turn out to be coupled to one another) using Equations (2.5) and (2.71). After the operators have been Fourier transformed from the site representation to a 2D wave-vector representation with $\mathbf{k} = (k_x, k_y)$, we find

$$\frac{d}{dt}a_{\mathbf{k}} = i[\mathcal{H}, a_{\mathbf{k}}]$$

$$= -it\left\{2\exp\left(\frac{1}{2}ik_x a_0\right)\cos\left(\frac{\sqrt{3}}{2}k_y a_0\right) + \exp\left(-ik_x a_0\right)\right\}b_{\mathbf{k}} \tag{2.72}$$

and

$$\frac{d}{dt}b_{\mathbf{k}} = i[\mathcal{H}, b_{\mathbf{k}}]$$

$$= it\left\{2\exp\left(-\frac{1}{2}ik_x a_0\right)\cos\left(\frac{\sqrt{3}}{2}k_y a_0\right) + \exp\left(ik_x a_0\right)\right\}a_{\mathbf{k}}. \tag{2.73}$$

The trigonometric factors here are deduced using Figure 2.6. Next we find the normal mode solutions by making the replacement $d/dt \to -i\omega$ in Equations (2.72) and (2.73). After some straightforward algebra we find that the solutions are $\omega = \pm E_{\mathbf{k}}$ where

$$E_{\mathbf{k}} = t\left[4\cos^2\left(\frac{\sqrt{3}}{2}k_y a_0\right) + 4\cos\left(\frac{\sqrt{3}}{2}k_y a_0\right)\cos\left(\frac{3}{2}k_x a_0\right) + 1\right]^{1/2}. \tag{2.74}$$

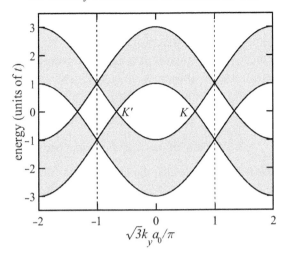

Figure 2.7 Dispersion relation for a 2D graphene sheet, showing the scaled energy $E_{\mathbf{k}}/t$ plotted against the dimensionless wavenumber component $\sqrt{3}k_y a_0/\pi$. The bands of electronic modes, arising as $k_x a_0$ varies, appear as the shaded regions. The vertical dashed lines correspond to the edges of the first Brillouin zone.

This dispersion relation for a complete graphene sheet has several remarkable features, as reviewed in [44]. We note that the excitation energy is symmetric about zero energy (as measured relative to the Fermi energy), although this is no longer the case if the small effects of next-nearest-neighbor hopping are included (see Problem 2.10). In Figure 2.7 we show a plot of the energy versus the scaled wavenumber k_y, which ranges across the Brillouin zone from $-\pi/\sqrt{3}a_0$ to $\pi/\sqrt{3}a_0$. Because k_x may also vary, we obtain energy bands that are shown as shaded regions with boundaries in this type of plot. It may be noted, in particular, that there are special points labeled as K and K' where $E_{\mathbf{K}} = 0$. These are known as the Dirac points, and it is their existence that gives rise to many of the special electronic properties of graphene, as we will see in later chapters.

For an electronic excitation close to the K point we can easily deduce from Equation (2.74) that the approximate dispersion relation is

$$E_{\mathbf{k}} \simeq t\sqrt{\delta_x^2 + 3\delta_y^2} \qquad (|\delta_x| \ll 1, \ |\delta_y| \ll 1), \qquad (2.75)$$

where $\delta_x = \pi - \frac{3}{2}k_x a$ and $\delta_y = \frac{\pi}{3} - \frac{\sqrt{3}}{2}k_y a$ are the rescaled dimensionless wave-vector components used in making the previously mentioned expansions.

2.8 Density Fluctuations in an Electron Gas

We now study some more properties of fermion systems, this time in 3D, by considering the density fluctuations in an electron gas. This may be realized, for example, in a metal or semiconductor where an electron gas plasma moves in a background of positively charged quasistatic ions that provide the overall charge neutrality. This is often referred to as the "jellium" model. It is helpful to start with the noninteracting gas in 3D as a special case, and then we introduce the interactions afterward to obtain a description of the plasma oscillations or "plasmons."

2.8.1 Noninteracting Fermion Gas

In the absence of any interaction terms, the Hamiltonian is given simply by

$$\mathcal{H} = \sum_{\mathbf{k}} \frac{k^2}{2m} a_{\mathbf{k}}^{\dagger} a_{\mathbf{k}}, \tag{2.76}$$

where m plays the role of an effective mass and $\mathbf{k}$ is the wave vector for the electrons. We now consider the effect of the operator $\rho_{\mathbf{q}}^{\dagger}(\mathbf{k})$ defined by

$$\rho_{\mathbf{q}}^{\dagger}(\mathbf{k}) = a_{\mathbf{k+q}}^{\dagger} a_{\mathbf{k}}. \tag{2.77}$$

This creates an electron of wave vector $\mathbf{k} + \mathbf{q}$ and also destroys an electron (thereby creating a "hole," or the absence of an electron) of wave vector $\mathbf{k}$. We will first examine its effect on the ground state of the noninteracting system, which is just a Fermi sphere in wave-vector space with all the electronic states filled up to a maximum value k_F, which is related to the Fermi energy by $\epsilon_F = k_F^2/2m$. We will ignore the effects of temperature here, noting that the Fermi–Dirac (FD) distribution function at temperature $T = 0$ is simply

$$n_{\mathbf{k}}^0 = \begin{cases} 1, & k \le k_F \\ 0, & k > k_F \end{cases}. \tag{2.78}$$

The effect of the operator $\rho_{\mathbf{q}}^{\dagger}(\mathbf{k})$ is illustrated schematically in Figure 2.8. By the Pauli exclusion principle, we must have $|\mathbf{k} + \mathbf{q}| > k_F$ and $|\mathbf{k}| < k_F$. Therefore, the overall effect of the operator is to create a particle-hole pair with total wave vector equal to $\mathbf{q}$. The energy of the excitation (for the electron-hole pair) is simply

$$\omega_0(\mathbf{k}, \mathbf{q}) = \frac{(\mathbf{k} + \mathbf{q})^2}{2m} - \frac{k^2}{2m} = \frac{q^2 + 2\mathbf{k} \cdot \mathbf{q}}{2m}. \tag{2.79}$$

For a given magnitude q of the wave vector $\mathbf{q}$ it follows that the maximum value of $\omega_0(\mathbf{k}, \mathbf{q})$ is $(q^2 + 2k_F q)/2m$, whereas the minimum value is either $(q^2 - 2k_F q)/2m$ if $q > 2k_F$ or 0 if $q \le 2k_F$. Hence the prediction is for a continuum of particle-hole states with energies lying between the maximum and minimum values just stated. This behavior is sketched as the shaded region in Figure 2.9.

Next, by including the Coulomb-type interactions, we will show that a new type of electron-hole excitation may occur with its energy lying outside the continuum region.

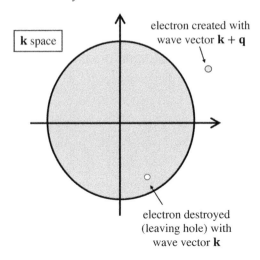

Figure 2.8 The effect of the operator $\rho_{\mathbf{q}}^{\dagger}(\mathbf{k})$ in creating an electron outside the filled Fermi sphere and leaving a hole (or absence of an electron) inside the sphere.

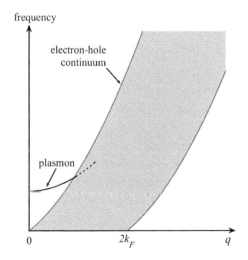

Figure 2.9 Schematic dispersion relations for the electron-hole excitations in an interacting electron gas, showing the continuum of states (shaded region), as in Equation (2.79), and the additional plasmon mode when Coulomb interactions are included (see Subsection 2.8.2).

2.8.2 Interacting Fermion Gas

We now study the effect of the particle-hole operator $\rho_{\mathbf{q}}^{\dagger}(\mathbf{k})$ defined in Equation (2.77) by using the full form of the Hamiltonian as given in Equation (1.63). The first step is to form the operator equation of motion for $\rho_{\mathbf{q}}^{\dagger}(\mathbf{k})$, which is found from the expression

$$\frac{d}{dt}\rho_{\mathbf{q}}^{\dagger}(\mathbf{k}) = i[\mathcal{H}, \rho_{\mathbf{q}}^{\dagger}(\mathbf{k})]. \tag{2.80}$$

Using the anticommutation properties of the fermion operators to simplify the preceding commutator, it may be verified that

$$[\mathcal{H}, \rho_{\mathbf{q}}^{\dagger}(\mathbf{k})] = \omega_0(\mathbf{k}, \mathbf{q})\rho_{\mathbf{q}}^{\dagger}(\mathbf{k})$$

$$-\frac{1}{2}\sum_{\mathbf{k}'} v(\mathbf{q})\left(a_{\mathbf{k}+\mathbf{q}}^{\dagger}a_{\mathbf{k}+\mathbf{k}'} - a_{\mathbf{k}+\mathbf{q}-\mathbf{k}'}^{\dagger}a_{\mathbf{k}}\right)\left(\sum_{\mathbf{k}''}\rho_{\mathbf{k}'}^{\dagger}(\mathbf{k}'')\right)$$

$$-\frac{1}{2}\sum_{\mathbf{k}'} v(\mathbf{q})\left(\sum_{\mathbf{k}''}\rho_{\mathbf{k}'}^{\dagger}(\mathbf{k}'')\right)\left(a_{\mathbf{k}+\mathbf{q}}^{\dagger}a_{\mathbf{k}+\mathbf{k}'} - a_{\mathbf{k}+\mathbf{q}-\mathbf{k}'}^{\dagger}a_{\mathbf{k}}\right). \tag{2.81}$$

This result is very complicated to interpret, but we remark first that if the interactions are neglected the equation reduces to

$$[\mathcal{H}, \rho_{\mathbf{q}}^{\dagger}(\mathbf{k})] = \omega_0(\mathbf{k}, \mathbf{q})\rho_{\mathbf{q}}^{\dagger}(\mathbf{k}), \tag{2.82}$$

where $\omega_0(\mathbf{k}, \mathbf{q})$ was defined in Equation (2.79). The equation of motion (2.80) then becomes

$$\frac{d}{dt}\rho_{\mathbf{q}}^{\dagger}(\mathbf{k}) = i\omega_0(\mathbf{k}, \mathbf{q})\rho_{\mathbf{q}}^{\dagger}(\mathbf{k}). \tag{2.83}$$

This is just an oscillator equation with a time dependence like $\rho_{\mathbf{q}}^{\dagger}(\mathbf{k}) \propto \exp[i\omega_0(\mathbf{k}, \mathbf{q})t]$, which corresponds to the excitation frequency found before, as might be expected.

In the case in which there are interactions, a complete solution of Equation (2.81) is not possible, but we can look for a simplification as an approximation for the right-hand side. To express everything in terms of the particle-hole $\rho_{\mathbf{q}}^{\dagger}(\mathbf{k})$ operator we consider making the replacement

$$\left(a_{\mathbf{k}+\mathbf{q}}^{\dagger}a_{\mathbf{k}+\mathbf{k}'} - a_{\mathbf{k}+\mathbf{q}-\mathbf{k}'}^{\dagger}a_{\mathbf{k}}\right) \rightarrow \langle a_{\mathbf{k}+\mathbf{q}}^{\dagger}a_{\mathbf{k}+\mathbf{k}'}\rangle - \langle a_{\mathbf{k}+\mathbf{q}-\mathbf{k}'}^{\dagger}a_{\mathbf{k}}\rangle$$

$$= \langle a_{\mathbf{k}+\mathbf{q}}^{\dagger}a_{\mathbf{k}+\mathbf{q}}\rangle\delta_{\mathbf{q},\mathbf{k}'} - \langle a_{\mathbf{k}}^{\dagger}a_{\mathbf{k}}\rangle\delta_{\mathbf{q},\mathbf{k}'}$$

$$= (n_{\mathbf{k}+\mathbf{q}}^0 - n_{\mathbf{k}}^0)\delta_{\mathbf{q},\mathbf{k}'}.$$

The approximation is in the first line and consists of replacing the chosen combination of operators, specifically those that do not directly involve $\rho_{\mathbf{q}}^{\dagger}(\mathbf{k})$, by their *average* value. This corresponds to ignoring some fluctuation effects in the system in a mean-field sense, and it is an example of a *decoupling approximation*. In the second line we have made use the wave-vector conservation property of the system as represented by the Kronecker deltas, and in the last line $n_{\mathbf{k}}^0$ is the Fermi–Dirac distribution function at zero temperature, as mentioned previously.

The type of approximation proposed here is based on the simplest qualitative reasons, and a justification comes later through any success in the predictions that follow. We shall be discussing decoupling approximations more generally in Chapter 5. Also, in Chapter 9 as an example of the diagrammatic perturbation technique, we shall consider the density fluctuations of an electron gas once more. It will be shown there that the decoupling approximation made here is consistent with choosing a particular set of Feynman diagrams in a low order of perturbation.

On using the decoupling approximation we find that Equation (2.81) takes the form

$$[\mathcal{H}, \rho_{\mathbf{q}}^{\dagger}(\mathbf{k})] = \omega_0(\mathbf{k}, \mathbf{q}) \rho_{\mathbf{q}}^{\dagger}(\mathbf{k}) - v(\mathbf{q}) \left(n_{\mathbf{k}+\mathbf{q}}^0 - n_{\mathbf{k}}^0 \right) \sum_{\mathbf{k}''} \rho_{\mathbf{q}}^{\dagger}(\mathbf{k}''). \tag{2.84}$$

We will solve this by determining what is the quasiparticle operator. Therefore, by analogy with Equation (2.82), we would like to rewrite Equation (2.84) in the simple form

$$[\mathcal{H}, P_{\mathbf{q}}^{\dagger}] = \Omega(\mathbf{q}) P_{\mathbf{q}}^{\dagger}, \tag{2.85}$$

where $\Omega(\mathbf{q})$ denotes the quasiparticle energy, and $P_{\mathbf{q}}^{\dagger}$ is the quasiparticle operator to create an interacting electron-hole pair with total wave vector $\mathbf{q}$. The solution is far from being obvious, but using linear superposition we may try forming it as

$$P_{\mathbf{q}}^{\dagger} = \sum_{\mathbf{k}} A(\mathbf{k}, \mathbf{q}) \rho_{\mathbf{q}}^{\dagger}(\mathbf{k}), \tag{2.86}$$

where the scalar coefficients A will need to be determined self-consistently later. On substituting for $P_{\mathbf{q}}^{\dagger}$ in Equation (2.85) we obtain

$$\sum_{\mathbf{k}} A(\mathbf{k}, \mathbf{q})[\mathcal{H}, \rho_{\mathbf{q}}^{\dagger}(\mathbf{k})] = \sum_{\mathbf{k}} \Omega(\mathbf{q}) A(\mathbf{k}, \mathbf{q}) \rho_{\mathbf{q}}^{\dagger}(\mathbf{k}).$$

Now by using Equation (2.84) to replace the commutator, we find

$$\sum_{\mathbf{k}} A(\mathbf{k}, \mathbf{q}) \omega_0(\mathbf{k}, \mathbf{q}) \rho_{\mathbf{q}}^{\dagger}(\mathbf{k}) - \sum_{\mathbf{k}} A(\mathbf{k}, \mathbf{q}) v(\mathbf{q}) \left(n_{\mathbf{k}+\mathbf{q}}^0 - n_{\mathbf{k}}^0 \right) \sum_{\mathbf{k}''} \rho_{\mathbf{q}}^{\dagger}(\mathbf{k}'')$$

$$= \sum_{\mathbf{k}} \Omega(\mathbf{q}) A(\mathbf{k}, \mathbf{q}) \rho_{\mathbf{q}}^{\dagger}(\mathbf{k}).$$

Rearranging this expression gives

$$\sum_{\mathbf{k}} A(\mathbf{k}, \mathbf{q}) \left\{ \omega_0(\mathbf{k}, \mathbf{q}) - \Omega(\mathbf{q}) \right\} \rho_{\mathbf{q}}^{\dagger}(\mathbf{k}) = \sum_{\mathbf{k}} A(\mathbf{k}, \mathbf{q}) v(\mathbf{q}) \left(n_{\mathbf{k}+\mathbf{q}}^0 - n_{\mathbf{k}}^0 \right) \sum_{\mathbf{k}''} \rho_{\mathbf{q}}^{\dagger}(\mathbf{k}'').$$

$$\tag{2.87}$$

This result still looks fairly formidable, but we can make some progress by examining the functional form of the preceding equation for $A(\mathbf{k}, \mathbf{q})$. We note that the right-hand side has the form of a scalar function multiplied by

$$\sum_{\mathbf{k}''} \rho_{\mathbf{q}}^{\dagger}(\mathbf{k}'').$$

The left side will have a similar form if $A(\mathbf{k}, \mathbf{q}) \{\omega_0(\mathbf{k}, \mathbf{q}) - \Omega(\mathbf{q})\}$ is a function that is overall independent of $\mathbf{k}$. We denote it by $B(\mathbf{q})$, and therefore we have shown that

$$A(\mathbf{k}, \mathbf{q}) = \frac{B(\mathbf{q})}{\{\omega_0(\mathbf{k}, \mathbf{q}) - \Omega(\mathbf{q})\}}.$$

If we now choose to denote

$$\rho_{\mathbf{q}}^{\dagger} = \sum_{\mathbf{k}} \rho_{\mathbf{q}}^{\dagger}(\mathbf{k}) = \sum_{\mathbf{k}''} \rho_{\mathbf{q}}^{\dagger}(\mathbf{k}''), \tag{2.88}$$

then Equation (2.87) becomes

$$B(\mathbf{q})\rho_{\mathbf{q}}^{\dagger} = \sum_{\mathbf{k}} B(\mathbf{q})v(\mathbf{q})\frac{\left(n_{\mathbf{k}+\mathbf{q}}^{0} - n_{\mathbf{k}}^{0}\right)}{\omega_0(\mathbf{k}, \mathbf{q}) - \Omega(\mathbf{q})}\rho_{\mathbf{q}}^{\dagger}. \tag{2.89}$$

Some factors cancel out (including the unknown function B), and as a consistency condition we have now arrived at

$$\sum_{\mathbf{k}} \frac{n_{\mathbf{k}}^{0} - n_{\mathbf{k}+\mathbf{q}}^{0}}{\Omega(\mathbf{q}) - \omega_0(\mathbf{k}, \mathbf{q})} = \frac{1}{v(\mathbf{q})}. \tag{2.90}$$

This is our main result in this section because it represents an implicit dispersion relation from which the quasiparticle energy $\Omega(\mathbf{q})$ can be deduced.

To analyze this result we first rewrite the expression by splitting the left-hand side into two terms and then make a change of variable $\mathbf{k} + \mathbf{q} \to \mathbf{k}'$ in the second term, yielding

$$\sum_{\mathbf{k}} \frac{n_{\mathbf{k}}^{0}}{\Omega(\mathbf{q}) - \omega_0(\mathbf{k}, \mathbf{q})} - \sum_{\mathbf{k}'} \frac{n_{\mathbf{k}'}^{0}}{\Omega(\mathbf{q}) - \omega_0(\mathbf{k}' - \mathbf{q}, \mathbf{q})} = \frac{1}{v(\mathbf{q})}$$

or

$$\sum_{\mathbf{k}} n_{\mathbf{k}}^{0} \left[\frac{1}{\Omega(\mathbf{q}) - \omega_0(\mathbf{k}, \mathbf{q})} - \frac{1}{\Omega(\mathbf{q}) - \omega_0(\mathbf{k} - \mathbf{q}, \mathbf{q})} \right] = \frac{1}{v(\mathbf{q})}.$$

Next, by substituting for $\omega_0(\mathbf{k}, \mathbf{q})$ from Equation (2.79) and rearranging, we have

$$\sum_{k<k_F} \left[\left(\Omega(\mathbf{q}) - \frac{\mathbf{k} \cdot \mathbf{q}}{m} \right)^2 - \left(\frac{q^2}{2m} \right)^2 \right]^{-1} = \frac{m}{q^2 v(\mathbf{q})}. \tag{2.91}$$

This alternative form of the result in Equation (2.90) is particularly useful. We shall examine the solution for the case of Coulomb interactions in 3D, for which the expression for $v(\mathbf{q})$ is quite simple (see Subsection 1.4.2). In the absence of screening we have $v(\mathbf{q}) \propto 1/q^2$, and so we obtain

$$\sum_{k<k_F} \left[\left(\Omega(\mathbf{q}) - \frac{\mathbf{k} \cdot \mathbf{q}}{m} \right)^2 - \left(\frac{q^2}{2m} \right)^2 \right]^{-1} = \frac{\varepsilon_0 m}{e^2}, \qquad (2.92)$$

with no $\mathbf{q}$ dependence on the right-hand side. In the case of $\mathbf{q} = 0$ this expression gives

$$\sum_{k<k_F} \left(\frac{1}{\Omega(0)} \right)^2 = \frac{\varepsilon_0 m}{e^2}.$$

The left-hand side of the preceding equation simply reduces to $n_0/[\Omega(0)]^2$, where $n_0 = N/V$ is the number of electrons per unit volume. Therefore, we have deduced that

$$\Omega(0) = \left(\frac{n_0 e^2}{\varepsilon_0 m} \right)^{1/2}. \qquad (2.93)$$

This is known as the *plasma frequency*, and it corresponds to the natural frequency of oscillations in a classical electron gas plasma (see, e.g., [20]).

For small nonzero $\mathbf{q}$ we may find solutions of Equation (2.92) by inserting a power-series expansion for $\Omega(\mathbf{q})$ into the left-hand side. Then by equating coefficients of powers of q on both sides of the equation, we may deduce (see Problem 2.13) that

$$\Omega(\mathbf{q}) = \Omega(0) \left[1 + \left(\frac{3\varepsilon_0 k_F^2}{10 m n_0 e^2} \right) q^2 + \mathcal{O}(q^4) \right]. \qquad (2.94)$$

At sufficiently small q this energy can be much larger than the maximum of the electron-hole continuum band, which was discussed earlier. The quasiparticle excitation corresponding to $\Omega(\mathbf{q})$ is known as a *plasmon* and it is included schematically as the additional excitation branch in Figure 2.9.

Problems

2.1. By using the operator equation-of-motion method as described in Subsection 2.1.2, verify that the time evolution of the wave function in the interaction picture proceeds in accordance with Equation (2.14).

2.2. The Hamiltonian of the forced harmonic oscillator (FHO) was specified in Equation (2.22) in terms of momentum and position variables. Show that this Hamiltonian can be rewritten in the notation of second quantization as

$$\mathcal{H} = \omega(a^\dagger a + 1/2) - h(t)(a^\dagger + a),$$

where $a^\dagger$ and a are the creation and annihilation operators. Find the relationship between $h(t)$ and the force $f(t)$ in Equation (2.22).

2.3. The Hamiltonian for a FHO may be expressed as $\mathcal{H} = \mathcal{H}_0 + V_{int}(t)$, where $\mathcal{H}_0$ refers to the unperturbed system and $V_{int}(t)$ is the time-dependent potential energy in the interaction picture. Using the Baker–Campbell–Hausdorff identity stated in the text that follows (in terms of any two operators X and Y)

$$e^X Y e^{-X} = Y + [X, Y] + \frac{1}{2!}[X, [X, Y]] + \frac{1}{3!}[X, [X, [X, Y]]] + \cdots,$$

verify that $V_{int}(t)$ is as quoted in Equation (2.24). Next show that the commutation relation $[V_{int}(t_1), V_{int}(t_2)]$ at any two time t_1 and t_2 is a scalar quantity given by Equation (2.27).

2.4. The following Hamiltonian $\mathcal{H}$ describes the coupling between two sets of bosons, one described by operators a and $a^\dagger$ and the other by operators b and $b^\dagger$:

$$\mathcal{H} = \lambda a^\dagger a + \varepsilon \left(ab^\dagger + a^\dagger b \right).$$

Here λ and ε are positive constants and the two sets of operators are independent (i.e., they commute with one another). Evaluate the commutators $[a, \mathcal{H}]$ and $[b, \mathcal{H}]$, and hence use the operator equations of motion to find da/dt and db/dt. Assuming time dependences like $\exp(-i\omega t)$, deduce the value or values for the frequency ω of the coupled mode(s). Sketch the form of your results for ω plotted as a function of the coupling strength ε.

2.5. Consider the diatomic chain where the mode frequencies for longitudinal lattice vibrations (phonons) are given by Equation (2.54). Now take the special case of $m_1 = m_2 \equiv m$ in Subsection 2.4.2 and show that

$$\omega_k^2 = \frac{2C}{m}\left(1 \pm \cos(ka)\right).$$

One of the preceding solutions is the same as for the monatomic chain (as expected). What is the interpretation of the other solution? As a hint, consider what has happened in this limit regarding the Brillouin zone boundary wave vector.

2.6. Consider a system of bosons for which $a_\mathbf{k}^\dagger$ and $a_\mathbf{k}$ are the creation and annihilation operators at wave vector $\mathbf{k}$ and the Hamiltonian is

$$\mathcal{H} = \sum_\mathbf{k} \left[F(\mathbf{k})a_\mathbf{k}^\dagger a_\mathbf{k} + \frac{1}{2}G(\mathbf{k})(a_\mathbf{k}^\dagger a_{-\mathbf{k}}^\dagger + a_\mathbf{k}a_{-\mathbf{k}}) \right].$$

Here $F(\mathbf{k})$ and $G(\mathbf{k})$, which are real functions of the wave vector $\mathbf{k}$, satisfy $F(\mathbf{k}) > G(\mathbf{k}) > 0$. Use the operator equation of motion to calculate the time derivative $da_{\mathbf{k}}^{\dagger}/dt$. Show that this equation couples $a_{\mathbf{k}}^{\dagger}$ to another operator, whose time derivative you should also calculate. Now assume time dependences like $\exp(-i\omega t)$ in the coupled equations to solve for the positive frequency ω of the excitations. Sketch the form of the dispersion relation for ω versus $k = |\mathbf{k}|$ when

$$F(\mathbf{k}) = \alpha k^2 + \exp(-k^2) \quad \text{and} \quad G(\mathbf{k}) = \exp(-k^2),$$

assuming α is a positive constant.

2.7. Consider a ferromagnet described by the dipole-exchange Hamiltonian, as obtained from Equations (1.81) and (2.61). Apply the HP transformation from spin operators to boson operators at low temperatures to verify that the previous Heisenberg Hamiltonian in Equation (1.91) expressed in second-quantized form generalizes to the Hamiltonian given by Equation (2.62).

2.8. The Hamiltonian for a single spin with quantum number $S = \frac{1}{2}$ placed in a static magnetic field B_0 along the z direction is given by $\mathcal{H} = -\gamma B_0 S^z \equiv -\omega_L S^z$, where ω_L is known as the Larmor frequency of the system. Assume that the system is initially (at $t = 0$) in a superposition state such that $|\phi(0)\rangle = \alpha_0|u\rangle + \beta_0|d\rangle$, and it evolves at later time t into

$$|\phi(t)\rangle = \alpha(t)|u\rangle + \beta(t)|d\rangle,$$

where $|u\rangle$ and $|d\rangle$ denote, respectively, the normalized spin "up" and "down" eigenstates of $\mathcal{H}$. Find (a) the functions $\alpha(t)$ and $\beta(t)$ and then (b) the time evolution of the longitudinal spin average $\langle S^z(t)\rangle$, as well as $\langle S^x(t)\rangle$ and $\langle S^y(t)\rangle$ in the transverse directions.

2.9. Consider the graphene lattice structure for a single sheet, as depicted in Figure 2.6, and verify that the vectors of the 2D reciprocal lattice are given by Equation (2.70), as stated.

2.10. Generalize the operator calculation given in Section 2.7 for a single sheet of graphene to include the extra effects of next-nearest-neighbor hopping t', in addition to the nearest-neighbor hopping t considered previously. Show that the electronic dispersion relation, which was previously given by Equation (2.74), is modified and is no longer symmetric about the zero of energy.

2.11. Use the Heisenberg Hamiltonian of Subsection 1.5.2, together with Equations (1.81) and (2.5), to find the operator equations of motion for each of the Cartesian components $S_{\mathbf{r}}^x$, $S_{\mathbf{r}}^y$, and $S_{\mathbf{r}}^z$ of the spin vector $\mathbf{S_r}$ at any site $\mathbf{r}$ in the crystal. Next, by interpreting the spins on the right-hand side of these equations as being classical vectors, prove that the result for $d\mathbf{S_r}/dt$ may be written in terms of vector (or cross) products as

$$i\frac{d\mathbf{S_r}}{dt} = J\sum_{\delta}(\mathbf{S_{r-\delta}} \times \mathbf{S_r} - \mathbf{S_r} \times \mathbf{S_{r+\delta}}) + g\mu_B\mathbf{S_r} \times \mathbf{B_0}.$$

Here $\mathbf{B_0}$ is a vector of magnitude B_0 along the z direction and the vector δ connects site $\mathbf{r}$ to its nearest neighbors. The preceding result is equivalent to the expression obtained directly from the torque equation of motion for exchange-coupled classical spin vectors [9].

2.12. Consider a free particle with mass m and charge e placed in a uniform magnetic field B_0 along the z direction. Its Hamiltonian is

$$\mathcal{H} = \frac{1}{2m}(\mathbf{p} - \frac{e}{c}\mathbf{A})^2,$$

where $\mathbf{p}$ is the particle momentum and $\mathbf{A}$ is the magnetic vector potential corresponding to the applied field. By using the EM gauge in which $\mathbf{A} = (-\frac{1}{2}B_0\,y, \frac{1}{2}B_0\,x, 0)$ obtain the operator equations of motion for the position and momentum of the particle and deduce the general form of the trajectory.

2.13. Use Equation (2.92) derived for the implicit dispersion relation of a plasmon to verify the power series expansion quoted in Equation (2.94). Specifically, confirm that the q^2 term has the coefficient quoted and that terms proportional to q and q^3 vanish.

3

Formal Properties of Green's Functions

The operator methods developed in the previous two chapters, including the equation-of-motion formalism, have been useful tools for calculating certain properties of the excitations in many-body systems, especially the frequencies (or energies) of these excitations. Some other properties, however, such as the statistical weighting or relative intensities that depend on the amplitudes of the excitations, are not so readily obtained. The statistical weighting may be different for the different excitations existing in a system, and this quantity will typically depend on parameters such as the wave vector and temperature. For interpreting the results of an experiment using inelastic light scattering, for example, it would be important to know the intensities as well as the frequencies corresponding to the excitation peaks in the measured spectrum. Green's functions provide us with a powerful technique for making such an analysis: they will be introduced formally in this chapter and then utilized throughout the rest of this book.

The Green's functions defined in the following sections will be expressible in terms of statistical-mechanical thermal averages of the products between two quantum-mechanical (QM) operators. They are, however, the analogues of the classical Green's functions introduced by G. Green in the 1820s that are now extensively used in mathematics for solving the differential equations in boundary-value problems (see, e.g., [46]). For completeness, the classical functions are briefly surveyed later in this chapter. We remark that the connection between the classical and QM counterparts will not be at all obvious to begin with, but it should become apparent later, particularly when we describe applications using linear response theory in Chapter 6. The operators appearing in the equilibrium thermal averages will have time labels associated with them (through a modified Heisenberg picture) because we want to study the dynamical behavior. As a generalization of the operators (and the Green's functions) having *real*-time labels, we shall see that it is also formally of interest to investigate the analogous quantities with *imaginary* time labels. This may seem surprising at first, but there are useful properties in both cases when Fourier transforms are made from a time to a frequency domain. In particular, the imaginary-time Green's functions are required in the development of Feynman diagrammatic techniques, as discussed in later chapters.

General references for Green's functions in condensed matter physics are to be found in the books by Mahan [47], Economou [48], Rickayzen [49], and Coleman [50]. Some other texts that have a particular focus on diagrammatic perturbation techniques are by Abrikosov et al. [51] and Mattuck [52].

3.1 Real-Time Green's Functions

We start by introducing the Green's functions (GFs) with a dependence on *real-time* labels because these quantities can be more straightforwardly related to physical properties and there are analogies with the equation-of-motion methods in Chapter 2. As mentioned earlier, the definitions involve equilibrium thermal averages of products of operators, so it is helpful to review some basic results from statistical mechanics.

3.1.1 Equilibrium Thermal Averages

Many-body systems typically have a very large number of interacting particles, for which it is possible that the energy eigenvalues and the number of particles (or quasiparticles in some cases) may vary. For this reason it is appropriate to work in terms of a grand canonical ensemble when discussing equilibrium thermal averages. We will assume here an undergraduate background in statistical physics, as covered for example in the standard text books by Reichl [3] and Pathria [4].

We start by recalling the formal result that the equilibrium thermal average $\langle A \rangle$ corresponding to any operator A in a grand canonical ensemble is given by

$$\langle A \rangle = \frac{1}{Q} \sum_i \langle i|A|i \rangle e^{-\beta(E_i - \mu N_i)}, \tag{3.1}$$

where Q is the grand partition function defined as

$$Q = \sum_i e^{-\beta(E_i - \mu N_i)}. \tag{3.2}$$

Here $|i\rangle$ represents a state of the system with energy E_i and number of particles N_i. We denote $\beta = 1/k_B T$ as before, and μ is the chemical potential.

For our purposes it is usually convenient to rewrite these results in a compact form using the trace (Tr) notation, defining as usual

$$\mathrm{Tr}(A) = \sum_i A_{ii} \quad \text{or} \quad \sum_i \langle i|A|i \rangle. \tag{3.3}$$

When the previous results from statistical physics are expressed in the new notation we have

$$\langle A \rangle = \frac{1}{Q} \mathrm{Tr} \left\{ A e^{-\beta(\mathcal{H} - \mu \mathcal{N})} \right\}, \tag{3.4}$$

$$Q = \mathrm{Tr} \left\{ e^{-\beta(\mathcal{H} - \mu \mathcal{N})} \right\}. \tag{3.5}$$

Here $\mathcal{H}$ denotes the total Hamiltonian and $\mathcal{N}$ is the number operator for the system.

One of the properties of the trace that will be useful later is that it is invariant under cyclic permutations of a product of any number of operators. This means, for example, in the case of a product of three operators that $\mathrm{Tr}(ABC) = \mathrm{Tr}(BCA) = \mathrm{Tr}(CAB)$. The stated property can easily be proved first for a product of just two operators, using the definition in Equation (3.3) and the rule for matrix multiplication. Then it can be generalized to a product of any number of operators by induction.

3.1.2 Definitions of the Green's Functions

Following the methods used in the pioneering work by Zubarev [53], we introduce Green's functions (GFs) mathematically as functions of any two QM operators A and B that are associated with real-time labels t and t', respectively. The physical significance of these quantities will gradually emerge later after their properties have been established. We first define the so-called retarded and advanced GFs by

Retarded Green's functions:

$$g_r(A; B \mid t - t') = -i\theta(t - t')\langle [A(t), B(t')]_\varepsilon \rangle. \tag{3.6}$$

Advanced Green's functions:

$$g_a(A; B \mid t - t') = i\theta(t' - t)\langle [A(t), B(t')]_\varepsilon \rangle. \tag{3.7}$$

Several explanations are needed to take in the preceding notations. First, $\theta(t)$ is the unit step function defined by

$$\theta(t) = \begin{cases} 1 & (t > 0) \\ 0 & (t < 0) \end{cases}. \tag{3.8}$$

Then the time dependences of the operators are defined as

$$A(t) = e^{i(\mathcal{H} - \mu \mathcal{N})t} A e^{-i(\mathcal{H} - \mu \mathcal{N})t}. \tag{3.9}$$

This expression is similar to the one given in Equation (2.4) for the operator A transformed to the Heisenberg picture, except that $\mathcal{H}$ has been replaced by $(\mathcal{H} - \mu \mathcal{N})$. Finally, we define

$$[A(t), B(t')]_\varepsilon = A(t)B(t') - \varepsilon B(t')A(t), \tag{3.10}$$

where ε is a constant that can take the values 1 or -1. Therefore, the preceding quantity represents either a commutator if $\varepsilon = 1$ or an anticommutator if $\varepsilon = -1$. It is important to emphasize that the actual value of ε can be chosen according to what is convenient in any particular case. Normally, however, we would take $\varepsilon = 1$ if A and B are boson operators and $\varepsilon = -1$ if A and B are fermion operators because we would then have commutation or anticommutation relations, respectively, in the definitions. If A and/or B are neither boson nor fermion operators (e.g., they could be spin or angular momentum operators), there is no clear indication of the choice for ε.

Also we note that the retarded GF is nonzero only when $t > t'$, whereas the reverse is true for the advanced GF. This property will lead to implications regarding causality, which we will discuss later. For completeness, although it is redundant to do so, we will now define a third type of real-time GF by

   Causal Green's functions:

$$g_c(A; B \mid t - t') = -i\langle \hat{T}_W A(t) B(t') \rangle, \tag{3.11}$$

where $\hat{T}_W$ is known as the Wick time-ordering operator, and it is defined by

$$\hat{T}_W A(t) B(t') = \begin{cases} A(t) B(t') & \text{if } t > t' \\ \varepsilon B(t') A(t) & \text{if } t < t' \end{cases}$$
$$= \theta\left(t - t'\right) A(t) B(t') + \theta\left(t' - t\right) \varepsilon B(t') A(t). \tag{3.12}$$

In other words, when $\hat{T}_W$ acts on the product $A(t) B(t')$, it leaves it unchanged if $t > t'$, but otherwise it inverts the order of the operators and multiplies them by ε if $t < t'$.

We notice that none of the three types of GFs is defined at $t = t'$ because of the discontinuity for the unit step function when this occurs. It will be shown later that, in general, the GFs have a discontinuity at $t = t'$. It is evident that all three of the GFs provide information about correlations in the system, i.e., they give information about the time-dependent averages $\langle A(t) B(t') \rangle$ and $\langle B(t') A(t) \rangle$ that describe correlations between one operator at time t and another operator at time t'. Because two operators are involved in the GFs, these quantities are sometimes referred to as two-time GFs.

One of the simple properties of GFs, which we will next establish, is that they depend on the time labels t and t' only through the time difference $(t - t')$. It is clear from the GF definitions that the preceding statement will be true if the constituent correlation functions $\langle A(t) B(t') \rangle$ and $\langle B(t') A(t) \rangle$ have the property of depending only on $(t - t')$. To show that this is the case we first consider $\langle A(t) B(t') \rangle$. From the definition of the thermal average in Equation (3.4) and by introducing the shorthand notation that $\overline{\mathcal{H}} = \mathcal{H} - \mu\mathcal{N}$, we have

$$\langle A(t)B(t')\rangle = Q^{-1}\mathrm{Tr}(e^{i\mathcal{H}t}Ae^{-i\mathcal{H}t}e^{i\mathcal{H}t'}Be^{-i\mathcal{H}t'}e^{-\beta\mathcal{H}})$$

$$= Q^{-1}\mathrm{Tr}(e^{i\mathcal{H}(t-t')}Ae^{-i\mathcal{H}(t-t')}Be^{-\beta\mathcal{H}}).$$

In the preceding equation we have employed the invariance property for cyclic permutation of the operators within the trace, as well as the result that $e^{-i\mathcal{H}t'}$ and $e^{-\beta\mathcal{H}}$ commute, to advance from the first to second line of the equation. The second line is evidently a function of $(t - t')$. A similar conclusion can be reached for the other correlation function $\langle B(t')A(t)\rangle$, and so the stated result holds for the GFs.

Regarding an alternative notation we mention that a double angular bracket form, as in $\ll A(t); B(t') \gg_r$, is sometimes used in the literature for the retarded GF denoted here as $g_r(A; B \,|\, t - t')$. Whenever it is convenient we shall abbreviate our notation to just $g_r(t - t')$ in cases in which the choice of operators is implicit.

3.1.3 Equations of Motion

We may start with the definition of any one of the preceding GFs, say the retarded GF in Equation (3.6), and differentiate it with respect to one of its time labels, say t. The result is

$$\frac{d}{dt}g_r(A; B \,|\, t - t') = \frac{d}{dt}\left\{-i\theta(t - t')\langle[A(t), B(t')]_\varepsilon\rangle\right\}$$

$$= -i\frac{d\theta(t - t')}{dt}\langle[A(t), B(t')]_\varepsilon\rangle - i\theta(t - t')\left\langle\left[\frac{dA(t)}{dt}, B(t')\right]_\varepsilon\right\rangle.$$

$$(3.13)$$

Next we can simplify the terms appearing in the last line. For the first term we may use the result

$$\frac{d\theta\left(t - t'\right)}{dt} = \delta\left(t - t'\right).$$

$$(3.14)$$

This can easily be proved by noting the delta-function property that

$$\int_{-\infty}^{t}\delta(t'' - t')dt'' = \begin{cases} 0 & \text{if } t < t' \\ 1 & \text{if } t > t' \end{cases}.$$

The right-hand side is just the definition of $\theta(t - t')$. Then we differentiate both sides with respect to t to obtain the required result. Hence the first term in Equation (3.13) becomes

$$-i\delta(t - t')\langle[A(t), B(t')]_\varepsilon\rangle = -i\delta(t - t')\langle[A(t), B(t)]_\varepsilon\rangle$$

$$= -i\delta(t - t')\langle[A(0), B(0)]_\varepsilon\rangle = -i\delta(t - t')\langle[A, B]_\varepsilon\rangle,$$

where we have used the property that the correlation functions depend only on the time difference. Next we observe that the second term in the last line of Equation (3.13) can be rewritten using the GF definition as

$$g_r(dA/dt; B \mid t - t').$$

On employing the operator equation of motion (2.5) to replace dA/dt, this GF is seen to be the same as $ig_r([\overline{\mathcal{H}}, A]; B \mid t - t')$. Putting all these results together, it follows that the retarded GF equation of motion becomes

$$\frac{d}{dt} g_r(A; B \mid t - t') = -i\delta(t - t')\langle [A, B]_\varepsilon \rangle + ig_r([\overline{\mathcal{H}}, A]; B \mid t - t').$$

It may easily be verified that exactly the same final result holds for the advanced and causal GFs, even though some of the intermediate steps are different. Thus we can write quite generally (dropping the r, a, and c subscripts) that

$$\frac{d}{dt} g(A; B \mid t - t') = -i\delta(t - t')\langle [A, B]_\varepsilon \rangle + ig([\overline{\mathcal{H}}, A]; B \mid t - t'). \qquad (3.15)$$

We note that the last term in the preceding expression, which is just another GF, always involves a commutator, whereas the other term on the right-hand side may be a commutator or anticommutator (according to the choice made for ε).

Later we shall come back to Equation (3.15) as a differential equation that we can, in principle, solve to obtain the required GF $g(A; B \mid t - t')$ with which we started. However, the feasibility of obtaining a solution depends on the form of the last term in Equation (3.15), and usually we may only be able to obtain an *approximate* solution except in simple cases.

3.2 Time Correlation Functions

Here we explore further some of the properties of the correlation functions. We have already shown that the correlation functions appear in the definitions of all the GFs and that they depend on the time only through the difference $(t - t')$. We, therefore, introduce the convenient shorthand notations that

$$F_{BA}(t - t') = \langle B(t')A(t) \rangle, \qquad F_{AB}(t - t') = \langle A(t)B(t') \rangle, \qquad (3.16)$$

according to the order of the operators. We can next define a frequency Fourier transform for one of these by

$$F_{BA}(t - t') = \int_{-\infty}^{\infty} J(\omega)e^{-i\omega(t-t')}d\omega, \qquad (3.17)$$

where ω is a real angular frequency and the transformed quantity $J(\omega)$ is usually called the *spectral function* or *spectral intensity* of the correlation function. In broad terms, it provides us with a measure of the strength associated with each frequency in the Fourier spectrum. In a similar manner we may choose to define another spectral function $J'(\omega)$ for the other correlation function, giving

$$F_{AB}(t - t') = \int_{-\infty}^{\infty} J'(\omega) e^{-i\omega(t-t')} d\omega. \tag{3.18}$$

Because the same two operators A and B are involved, we might speculate that there could be a connection between the two correlation functions $F_{BA}(t - t')$ and $F_{AB}(t - t')$, and hence between their respective spectral functions. This is indeed the case, as we show in the following text.

Starting with the time correlation function F_{AB}, we can make use of the definition of the equilibrium thermal average and the property of invariance of the trace under cyclic permutation of the operators to show that

$$
\begin{aligned}
\langle A(t)B(t') \rangle &= Q^{-1}\mathrm{Tr}\left\{ e^{i\overline{\mathcal{H}}t} A e^{-i\overline{\mathcal{H}}t} e^{i\overline{\mathcal{H}}t'} B e^{-i\overline{\mathcal{H}}t'} e^{-\beta\overline{\mathcal{H}}} \right\} \\
&= Q^{-1}\mathrm{Tr}\left\{ e^{i\overline{\mathcal{H}}t'} B e^{-i\overline{\mathcal{H}}t'} e^{-\beta\overline{\mathcal{H}}} e^{i\overline{\mathcal{H}}t} A e^{-i\overline{\mathcal{H}}t} \right\}.
\end{aligned}
$$

With further algebraic rearrangement and then using the definition in Equation (3.9), we obtain

$$
\begin{aligned}
\langle A(t)B(t') \rangle &= Q^{-1}\mathrm{Tr}\left\{ B(t') e^{i\overline{\mathcal{H}}(t+i\beta)} A e^{-i\overline{\mathcal{H}}(t+i\beta)} e^{-\beta\overline{\mathcal{H}}} \right\} \\
&= Q^{-1}\mathrm{Tr}\left\{ B(t') A(t + i\beta) e^{-\beta\overline{\mathcal{H}}} \right\} \\
&= \langle B(t') A(t + i\beta) \rangle. \tag{3.19}
\end{aligned}
$$

Therefore, we have established a connection between the two correlation functions as

$$F_{AB}(t - t') = F_{BA}(t - t' + i\beta). \tag{3.20}$$

This result formally relates the two correlation functions in the time domain through a shift by the imaginary interval $i\beta$.

The implications of the preceding result for the spectral functions $J(\omega)$ and $J'(\omega)$ can readily be seen. We substitute Equations (3.17) and (3.18) into (3.20) to obtain

$$\int_{-\infty}^{\infty} J'(\omega) e^{-i\omega(t-t')} d\omega = \int_{-\infty}^{\infty} J(\omega) e^{-i\omega(t-t'+i\beta)} d\omega.$$

Then, equating the integrands on each side of the preceding equation and cancelling out common factors, we deduce that there is a simple relation between $J'(\omega)$ and $J(\omega)$, which is given by

$$J'(\omega) = J(\omega) e^{\beta\omega} \qquad (\beta = 1/k_B T). \tag{3.21}$$

An important consequence of the preceding result is that we can work in terms of a single spectral function, say $J(\omega)$, and this provides information about *both* correlation functions. It should not be a surprising outcome that there is a close connection between the spectral functions, bearing in mind that the same two operators are involved.

Eventually we want to be able to calculate correlation functions because they are the quantities that have the most direct connection with what is measured in an experiment. We demonstrate this connection by examples in later chapters. Typically, however, the correlation functions are difficult to evaluate from their definitions or through an equation of motion. Instead, the usual approach is first to evaluate the corresponding GF, which turns out to be an easier task, and afterward we deduce the correlation functions from the GF. For this, we need to examine more closely the relationships between GFs and correlation functions, which is conveniently accomplished using frequency Fourier transforms.

3.3 Spectral Representations

By analogy with the expressions for the Fourier transforms of correlation functions, we may define a frequency Fourier transform for any of the three GFs by

$$g(t - t') = \int_{-\infty}^{\infty} G(\omega) e^{-i\omega(t-t')} d\omega. \tag{3.22}$$

Here ω is an angular frequency as before, and $G(\omega)$ denotes a frequency Fourier component for the GF. The inverse relation of the preceding transform is easily shown to be

$$G(\omega) = \frac{1}{2\pi} \int_{-\infty}^{\infty} g(t) e^{i\omega t} dt. \tag{3.23}$$

We now explore some results for $G(\omega)$ in the different cases, while working toward a connection with the spectral intensity $J(\omega)$ of the correlation functions.

3.3.1 Retarded Green's Functions

From Equations (3.6) and (3.23), and with the subscript r for retarded included, we have

$$\begin{aligned}
G_r(\omega) &= \frac{1}{2\pi} \int_{-\infty}^{\infty} (-i)\theta(t) \left\{ \langle A(t)B(0) \rangle - \varepsilon \langle B(0)A(t) \rangle \right\} e^{i\omega t} dt \\
&= \frac{1}{2\pi i} \int_{-\infty}^{\infty} \theta(t) \left\{ F_{AB}(t) - \varepsilon F_{BA}(t) \right\} e^{i\omega t} dt.
\end{aligned} \tag{3.24}$$

This may be expressed, using Equations (3.17) and (3.20), in terms of the spectral intensity $J(\omega)$ as

$$G_r(\omega) = \int_{-\infty}^{\infty} d\omega' \, J(\omega') \left(e^{\beta\omega'} - \varepsilon\right) \frac{1}{2\pi i} \int_{-\infty}^{\infty} \theta(t) \, e^{i(\omega - \omega')t} \, dt. \qquad (3.25)$$

We next show how to simplify this by employing the following integral representation of the step function:

$$\theta(t) = \frac{i}{2\pi} \int_{-\infty}^{\infty} \frac{e^{-ixt}}{(x + i\eta)} \, dx, \qquad (3.26)$$

where η denotes a positive infinitesimal quantity (with the understanding that the limit $\eta \to 0$ is taken). We detour briefly to provide a proof for this result.

Proof Here we use complex analysis in terms of the contour integration method and the theorem of residues (see textbooks on mathematical methods, e.g., [54, 55]) to evaluate the right-hand side of the preceding integral. Thus we consider the related contour integral

$$\frac{i}{2\pi} \oint_C \frac{e^{-izt}}{(z + i\eta)} \, dz$$

for a complex variable z, where the contour C goes all the way along the real axis and is closed at infinity in the complex plane to form a loop. By convention, it is anticlockwise. The integrand has just one simple pole at $z = -i\eta$, and there are two cases to consider, depending on whether $t > 0$ or $t < 0$.

First, taking $t > 0$, we choose the contour as $C = C_1$ in the lower half-plane, as shown in Figure 3.1. With this choice the contribution from the part of the semicircle at infinity is negligible, which follows because we can write

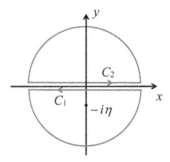

Figure 3.1 The two choices of contours (C_1 and C_2) used in proving the integral representation for the step function $\theta(t)$ in Equation (3.26).

$e^{-izt} = e^{-i(x+iy)t} = e^{-ixt}e^{yt}$. Thus, with $t > 0$ and $y < 0$, it is seen that $e^{-izt} \to 0$ as $|z| \to \infty$. The simple pole at $z = -i\eta$ is enclosed by the contour, and so

$$\frac{i}{2\pi} \int_{-\infty}^{\infty} \frac{e^{-ixt}}{(x+i\eta)} dx = \frac{-i}{2\pi} \oint_{C_1} \frac{e^{-izt}}{(z+i\eta)} dz$$

$$= \frac{-i}{2\pi} \times 2\pi i \times (\text{residue at } z = -i\eta)$$

$$= e^{-\eta t} \to 1 \quad (\text{as } \eta \to 0).$$

Therefore, the integral is 1 when $t > 0$. However, when $t < 0$ we must choose the contour C_2 in the upper half plane (see Figure 3.1). The contribution from the semicircle is again negligible as $|z| \to \infty$, but this time no pole is enclosed. Therefore, the integral is 0 for $t < 0$. Hence we have proved that the integral is equal to the step function $\theta(t)$.

We now return to the problem of simplifying Equation (3.25). As part of the right-hand side of this expression we have an integral involving $\theta(t)$, which we can rewrite using Equation (3.26) as

$$\frac{1}{2\pi i} \int_{-\infty}^{\infty} \theta(t) e^{i(\omega-\omega')t} dt = \frac{1}{2\pi i} \int_{-\infty}^{\infty} e^{i(\omega-\omega')t} dt \frac{i}{2\pi} \int_{-\infty}^{\infty} \frac{e^{-ixt}}{(x+i\eta)} dx$$

$$= \frac{1}{4\pi^2} \int_{-\infty}^{\infty} \frac{dx}{(x+i\eta)} \int_{-\infty}^{\infty} e^{i(\omega-\omega'-x)t} dt.$$

We next employ the well-known integral representation for the Dirac delta function that

$$\delta(y) = \frac{1}{2\pi} \int_{-\infty}^{\infty} e^{iyt} dt \tag{3.27}$$

for any real variable y. This leads to

$$\frac{1}{2\pi i} \int_{-\infty}^{\infty} \theta(t) e^{i(\omega-\omega')t} dt = \frac{1}{2\pi} \int_{-\infty}^{\infty} \frac{dx}{(x+i\eta)} \delta(\omega - \omega' - x)$$

$$= \frac{1}{2\pi} \left(\frac{1}{\omega - \omega' + i\eta} \right).$$

Substituting this result back into Equation (3.25) gives the following useful relationship between $G_r(\omega)$ and $J(\omega)$:

$$G_r(\omega) = \frac{1}{2\pi} \int_{-\infty}^{\infty} \frac{J(\omega')(e^{\beta\omega'} - \varepsilon)}{\omega - \omega' + i\eta} d\omega'. \tag{3.28}$$

3.3.2 Advanced and Causal Green's Functions

The derivation for the advanced GF follows in the same general way as the case mentioned previously, starting from Equation (3.7). The final result, quoted here for reference, is

$$G_a(\omega) = \frac{1}{2\pi} \int_{-\infty}^{\infty} \frac{J(\omega')(e^{\beta\omega'} - \varepsilon)}{\omega - \omega' - i\eta} d\omega'. \tag{3.29}$$

We note that the only difference compared with the retarded case is a change in sign before the imaginary term in the denominator.

Again, in the case of the causal GF the derivation follows in the same general way as before, starting from Equation (3.11). The final result is

$$G_c(\omega) = \frac{1}{2\pi} \int_{-\infty}^{\infty} J(\omega') \left\{ \frac{e^{\beta\omega'}}{\omega - \omega' + i\eta} - \frac{\varepsilon}{\omega - \omega' - i\eta} \right\} d\omega', \tag{3.30}$$

which is slightly more complicated than the previous cases because both types of denominator terms now appear.

3.4 Real and Imaginary Parts of Green's Functions

We have seen in Section 3.3 that the expressions for the Fourier transforms of the GFs all involve denominators like

$$\frac{1}{\omega - \omega' \pm i\eta}$$

with the understanding that $\eta \to 0$. Hence the GFs are complex quantities with real and imaginary parts. We shall now deduce expressions for these real and imaginary parts, showing that they are *not* independent of one another but are intricately connected.

We shall need to make use of the following symbolic identity:

$$\frac{1}{x \pm i\eta} = \mathcal{P}\left(\frac{1}{x}\right) \mp i\pi\delta(x), \tag{3.31}$$

where x denotes any real variable, η is a positive infinitesimal, and $\mathcal{P}$ denotes that the Cauchy principal value is taken in any integration over x.

Proof We will show the proof of Equation (3.31), by taking the case of the lower set of signs, namely

$$\frac{1}{x - i\eta} = \mathcal{P}\left(\frac{1}{x}\right) + i\pi\delta(x).$$

This relation is equivalent to stating that, for any real analytic function $f(x)$ in the integrand, we must have

$$\lim_{\eta \to 0} \int_{-\infty}^{\infty} \frac{f(x)dx}{x - i\eta} = \mathcal{P}\int_{-\infty}^{\infty} \frac{f(x)dx}{x} + i\pi \int_{-\infty}^{\infty} f(x)\delta(x)dx$$

$$= \mathcal{P}\int_{-\infty}^{\infty} \frac{f(x)dx}{x} + i\pi f(0). \tag{3.32}$$

Starting from the integral on the left-hand side of the preceding result, we will split the range of integration (which goes from $-\infty$ to ∞) into three separate segments by writing

$$\lim_{\eta \to 0} \int_{-\infty}^{\infty} \frac{f(x)dx}{x - i\eta} = \lim_{\eta \to 0, \tau \to 0} \left\{ \int_{-\infty}^{-\tau} \frac{f(x)dx}{x - i\eta} + \int_{-\tau}^{\tau} \frac{f(x)dx}{x - i\eta} + \int_{\tau}^{\infty} \frac{f(x)dx}{x - i\eta} \right\}.$$

The preceding is true for τ taking any positive value (because we are free to split up the range of integration in any way), and so there is no loss of generality in assuming that $\eta \ll \tau$. In other words, we are taking the double limit in the order that $\eta \to 0$ first and then $\tau \to 0$. Proceeding now with the evaluation, we find that the right-hand side of the preceding equation can be rearranged and expanded as

$$\lim_{\tau \to 0} \left\{ \int_{-\infty}^{-\tau} \frac{f(x)dx}{x} + \int_{\tau}^{\infty} \frac{f(x)dx}{x} \right\} + \lim_{\eta \to 0, \tau \to 0} \int_{-\tau}^{\tau} \frac{\left[f(0) + xf'(0) + \mathcal{O}(x^2) \right] dx}{x - i\eta}.$$

With some further rewriting it becomes equal to

$$\mathcal{P} \int_{-\infty}^{\infty} \frac{f(x)dx}{x} + f(0) \lim_{\eta \to 0, \tau \to 0} \int_{-\tau}^{\tau} \frac{dx}{x - i\eta}$$

$$= \mathcal{P} \int_{-\infty}^{\infty} \frac{f(x)dx}{x} + f(0) \lim_{\eta \to 0, \tau \to 0} \left\{ 2i \tan^{-1}(\tau/\eta) \right\}$$

$$= \mathcal{P} \int_{-\infty}^{\infty} \frac{f(x)dx}{x} + i\pi f(0).$$

This is seen to be the required result, and we may develop the proof for the other set of signs in Equation (3.31) in a very similar manner.

It is now straightforward to make use of Equation (3.31) to obtain the expressions for the real and imaginary parts of the various GFs. From Equations (3.28), (3.29), and (3.30) the results are found to be

$$\mathrm{Re}\, G_r(\omega) = \mathrm{Re}\, G_a(\omega) = \mathrm{Re}\, G_c(\omega)$$

$$= \frac{1}{2\pi} \mathcal{P} \int_{-\infty}^{\infty} \frac{J(\omega')\left(e^{\beta\omega'} - \varepsilon \right) d\omega'}{\omega - \omega'} \tag{3.33}$$

for the real parts, while the imaginary parts satisfy

$$\mathrm{Im}\, G_r(\omega) = -\mathrm{Im}\, G_a(\omega) = -\frac{1}{2}\left(e^{\beta\omega} - \varepsilon \right) J(\omega), \tag{3.34}$$

$$\mathrm{Im}\, G_c(\omega) = -\frac{1}{2}\left(e^{\beta\omega} + \varepsilon \right) J(\omega). \tag{3.35}$$

It is striking that all three GFs have the same real part, and that their differences arise only from their imaginary parts. Some important consequences of the previously mentioned results are the Kramers–Kronig relations and the fluctuation-dissipation theorem, and each of these is described in the following subsections.

3.4.1 Kramers–Kronig Relations

We may eliminate the $J(\omega')$ factor within the integrand of Equation (3.33) to reexpress the right-hand side in terms of the imaginary parts of any of the GFs by using Equations (3.34) and (3.35). The results for the three types of GFs are obtained as

$$\operatorname{Re} G_r(\omega) = -\frac{1}{\pi} \mathcal{P} \int_{-\infty}^{\infty} \frac{\operatorname{Im} G_r(\omega')\, d\omega'}{\omega - \omega'}, \tag{3.36}$$

$$\operatorname{Re} G_a(\omega) = \frac{1}{\pi} \mathcal{P} \int_{-\infty}^{\infty} \frac{\operatorname{Im} G_a(\omega')\, d\omega'}{\omega - \omega'}, \tag{3.37}$$

$$\operatorname{Re} G_c(\omega) = -\frac{1}{\pi} \mathcal{P} \int_{-\infty}^{\infty} \frac{\operatorname{Im} G_c(\omega')\left(e^{\beta\omega} - \varepsilon\right) d\omega'}{\left(\omega - \omega'\right)\left(e^{\beta\omega} + \varepsilon\right)}. \tag{3.38}$$

These results show that the real and imaginary parts of the GFs are intricately related through integral expressions.

We see that the connection between the real and imaginary parts is particularly simple for the retarded and advanced GFs, and the results that we have just obtained in Equations (3.36) and (3.37) represent examples of *Kramers–Kronig relations*, which have a more general validity for a class of complex functions that are analytic in either the upper or lower half of the complex plane (see, e.g., the discussion by Landau and Lifshitz [56]). The significance of the results may be viewed as a consequence in the frequency (or energy) domain of the fact that the retarded and advanced GFs in the time domain are nonzero only when $t > t'$ or vice versa. They are related, therefore, to ideas of causality in these cases. We shall come back to this topic to explore some applications of the Kramers–Kronig relations in later chapters, particularly in Chapter 6 in the context of linear response theory.

As a further comment, it can be shown that the relationships obtained here are bidirectional in the sense that the imaginary part of $G_r(\omega)$, for example, can be expressed as an integral over an expression that contains the real part of $G_r(\omega)$. Thus the "inverse" relation to Equation (3.36) for the retarded GF is found to be

$$\operatorname{Im} G_r(\omega) = \frac{1}{\pi} \mathcal{P} \int_{-\infty}^{\infty} \frac{\operatorname{Re} G_r(\omega')\, d\omega'}{\omega - \omega'}. \tag{3.39}$$

The proof involves further use of the method of contour integration in the complex frequency plane, and is left as a problem (see Problem 3.4).

The Kramers–Kronig relations have allowed us to express the real (or imaginary) part of the GF in terms of its imaginary (or real) part. It follows that the complete GF can be rewritten in an integral form in terms of either its real part only or imaginary part only. For example, from the preceding results we may take

$$G_r(\omega) = \mathrm{Re}\,G_r(\omega) + i\,\mathrm{Im}\,G_r(\omega)$$

$$= \mathrm{Re}\,G_r(\omega) + \frac{i}{\pi}P\int_{-\infty}^{\infty}\frac{\mathrm{Re}\,G_r(\omega')\,d\omega'}{\omega - \omega'}$$

$$= \frac{i}{\pi}\int_{-\infty}^{\infty}\frac{\mathrm{Re}\,G_r(\omega')\,d\omega'}{\omega - \omega' + i\eta},$$

where η is a positive infinitesimal quantity ($\eta \to 0$) as before. Alternatively, we can write

$$G_r(\omega) = \mathrm{Re}\,G_r(\omega) + i\,\mathrm{Im}\,G_r(\omega)$$

$$= -\frac{1}{\pi}P\int_{-\infty}^{\infty}\frac{\mathrm{Im}\,G_r(\omega')\,d\omega'}{\omega - \omega'} + i\,\mathrm{Im}\,G_r(\omega)$$

$$= -\frac{1}{\pi}\int_{-\infty}^{\infty}\frac{\mathrm{Im}\,G_r(\omega')\,d\omega'}{\omega - \omega' + i\eta}.$$

3.4.2 Fluctuation-Dissipation Theorem

We can rearrange the results obtained in Equations (3.34) and (3.35) for the imaginary parts of the GFs so that they become expressions for the spectral intensity, giving in the case of the retarded GF the result that

$$J(\omega) = \frac{-2}{\left(e^{\beta\omega} - \varepsilon\right)}\,\mathrm{Im}\,G_r(\omega). \tag{3.40}$$

The alternative expressions using the other two GFs are

$$J(\omega) = \frac{2}{\left(e^{\beta\omega} - \varepsilon\right)}\,\mathrm{Im}\,G_a(\omega), \quad J(\omega) = \frac{-2}{\left(e^{\beta\omega} + \varepsilon\right)}\,\mathrm{Im}\,G_c(\omega). \tag{3.41}$$

The preceding results are particularly important, because they provide us with a direct way of deducing the spectral function $J(\omega)$, and hence the time correlation functions, once we have calculated any of the GFs, either $G_r(\omega)$, $G_a(\omega)$, or $G_c(\omega)$. The results are usually known collectively as the *fluctuation-dissipation theorem* because in many physical applications (as we will see later) the imaginary part of the GF may be related to the dissipative (or "frictional") effects in a system, while the spectral function contains information about the excitations (or "fluctuations") in a related property of the system.

A case of special interest for the preceding results, usually for a retarded GF when we take $\varepsilon = 1$ (as would typically be the situation with bosons or with spin operators), occurs when $\beta\omega \ll 1$ (or more explicitly when $\hbar\omega \ll k_B T$) for the relevant frequencies. This is the "high-temperature" or "classical" regime for which Equation (3.40) in the retarded GF case simplifies to

$$J(\omega) = -\left(\frac{2}{\beta\omega}\right)\mathrm{Im}\,G_r(\omega) = -\left(\frac{2k_B T}{\hbar\omega}\right)\mathrm{Im}\,G_r(\omega). \tag{3.42}$$

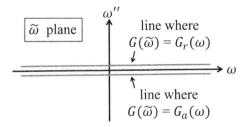

Figure 3.2 The complex $\tilde{\omega} = \omega + i\omega''$ plane for frequency (or energy). The gray lines just above and below the real axis for the analytically continued GF represent the retarded and advanced GFs, respectively, at a real frequency ω.

The preceding form of the fluctuation-dissipation theorem is often derived directly (e.g., for a damped oscillator model) assuming classical properties for the spectrum of fluctuations and the equipartition theorem [39, 56, 57].

3.4.3 Analytic Continuation for Green's Functions

In the previous expressions, we have taken ω to be a real variable representing the frequency (or it could alternatively have been described in terms of energy E, through the relation $E = \hbar\omega$ recalling that we employ units such that $\hbar = 1$).

As a matter of convenience we can now try to extend the previous results (in a formal or mathematical sense) to complex frequencies. This process is generally known as *analytic continuation* and it is useful for the retarded and advanced GFs, $G_r(\omega)$ and $G_a(\omega)$. The essential idea behind the method is that we introduce a complex frequency $\tilde{\omega} = \omega + i\omega''$. The complex $\tilde{\omega}$ plane is, therefore, as depicted in Figure 3.2 with real and imaginary axes.

The previous definitions given for $G_r(\omega)$ and $G_a(\omega)$ apply only along the real (or ω) axis. Suppose, however, we consider defining a new GF as a function of this complex $\tilde{\omega}$, by the expression

$$G(\tilde{\omega}) = \frac{1}{2\pi} \int_{-\infty}^{\infty} \frac{J(\omega')\left(e^{\beta\omega'} - \varepsilon\right)d\omega'}{\tilde{\omega} - \omega'}, \qquad (3.43)$$

where the integration variable ω' is real. This is known as the *analytically continued* GF, and it is a function defined in the complex plane. From the form of the integrand we see that it is an analytic function everywhere, except along the real axis where there is a singularity.

We now would like to see what this means and whether it represents more than just a mathematical curiosity. Suppose we put $\tilde{\omega} = \omega + i\eta$ (which holds at all points along the upper gray line in Figure 3.2) in Equation (3.43), where η is a positive infinitesimal as before. Then it follows that we have

$$G(\omega + i\eta) = \frac{1}{2\pi} \int_{-\infty}^{\infty} \frac{J(\omega')\left(e^{\beta\omega'} - \varepsilon\right)d\omega'}{\omega - \omega' + i\eta} = G_r(\omega), \qquad (3.44)$$

where the last step makes use of Equation (3.28). Similarly, if we now suppose $\tilde{\omega} = \omega - i\eta$ (which holds at all points along the lower gray line in Figure 3.2), we obtain

$$G(\omega - i\eta) = \frac{1}{2\pi} \int_{-\infty}^{\infty} \frac{J(\omega')\left(e^{\beta\omega'} - \varepsilon\right)d\omega'}{\omega - \omega' - i\eta} = G_a(\omega), \qquad (3.45)$$

where the last step follows from Equation (3.29).

To summarize, we have found here that immediately above the real axis the analytically continued GF $G(\tilde{\omega})$ corresponds to the retarded GF, whereas immediately below the real axis $G(\tilde{\omega})$ corresponds to the advanced GF. Also it follows that there is a discontinuity across the real frequency axis, because

$$G(\omega + i\eta) - G(\omega - i\eta) = G_r(\omega) - G_a(\omega)$$
$$= -i\left(e^{\beta\omega} - \varepsilon\right)J(\omega), \qquad (3.46)$$

where we have used the preceding two equations, together with Equations (3.33) and (3.34). By rearranging the preceding result, we see that an alternative statement of the fluctuation-dissipation theorem for $J(\omega)$ is

$$J(\omega) = \frac{-1}{\left(e^{\beta\omega} - \varepsilon\right)} \operatorname{Im}\left\{G(\omega + i\eta) - G(\omega - i\eta)\right\}. \qquad (3.47)$$

For the present there is no particular meaning that we attach to the analytically continued GF at other points in the complex frequency plane, apart from just below and just above the real axis. Later, when we study GFs in the context of diagrammatic perturbation methods, we will examine the behavior along the imaginary-frequency (or ω'') axis. This is the case with the Matsubara GFs that we introduce in the next section.

3.5 Imaginary-Time Green's Functions

The typical situation in interacting many-body systems is that there may be no systematic or rigorous procedures for calculating the real-time GFs, except for special cases. Nevertheless, there are various approximation methods that have been developed for the real-time GFs, and these will be covered in the next few chapters of this book. By contrast, however, perturbation methods (usually expressed in terms of a diagrammatic representation) are applicable for another type of GF that is defined with *imaginary*-time labels.

As a bridge to these perturbation methods, which are the main topics for Chapters 8 and 9, we give here a brief introduction to the imaginary-time GFs. These were proposed by T. Matsubara and are sometimes known as *Matsubara GFs* [58]. The connection between the imaginary-time GFs and the real-time (specifically the retarded and advanced GFs described earlier) will eventually be established here by employing the concept of analytic continuation.

The so-called imaginary-time (or Matsubara) GFs, which we shall denote as $g_M(\tau - \tau')$ to distinguish them from the previous real-time GFs, can be defined formally by

$$g_M(\tau - \tau') = -\langle \hat{T}_W \check{A}(\tau)\check{B}(\tau')\rangle. \tag{3.48}$$

As before, A and B are any two QM operators, the angular brackets $\langle\cdots\rangle$ denote an equilibrium thermal average (taken with respect to the Hamiltonian $\mathcal{H}$), and $\hat{T}_W$ is the Wick time-ordering operator defined in Equation (3.12). The labels τ and τ' that are associated with the operators take real values in this case. In fact, we will see later that they may be restricted to a particular finite range, which depends on the temperature T for the system (assuming $T \neq 0$). The definitions for the τ (and τ') dependences of the operators in this formalism are that

$$\check{A}(\tau) = e^{\overline{\mathcal{H}}\tau} A e^{-\overline{\mathcal{H}}\tau}, \tag{3.49}$$

and similarly for $\check{B}(\tau')$. We notice that these transformations are like those used for the modified Heisenberg picture in Equation (3.9), and they again involve the combination $\overline{\mathcal{H}} = \mathcal{H} - \mu\mathcal{N}$. The difference is that in Equation (3.9) the exponents were pure imaginary, whereas they are real in the previously mentioned definition for $\check{A}(\tau)$. It is in this sense that τ (and τ'), although they are specified here as real parameters, play a role analogous to that for an "imaginary" time.

We note here that care must be taken in forming the Hermitian conjugates of the τ-dependent operators. For example, using $\check{A}^{\dagger}(\tau) = e^{\overline{\mathcal{H}}\tau} A^{\dagger} e^{-\overline{\mathcal{H}}\tau}$, it can easily be checked that $\check{A}^{\dagger}(\tau) \neq [\check{A}(\tau)]^{\dagger}$. In fact the Hermitian conjugate of $\check{A}(\tau)$ is $\check{A}^{\dagger}(-\tau)$.

Another comment regarding the imaginary-time GFs is that they depend on τ and τ' only through the difference term $\tau - \tau'$, as implicitly assumed in the definition in Equation (3.48). This result follows by analogy with the result found for the GFs in the real-time formalism (and the proof using properties of the trace and the definition for a thermal average is very similar). Without loss of generality, we can therefore limit our consideration to the function

$$g_M(\tau) \equiv -\langle \hat{T}_W \check{A}(\tau)\check{B}(0)\rangle. \tag{3.50}$$

3.5.1 Periodicity and Antiperiodocity in Imaginary Time

An important property of the imaginary-time GF, written as in Equation (3.50) is that it obeys either a periodicity or antiperiodicity condition over an interval β for the τ variable, where we denote $\beta = 1/k_B T$ as before and consider $T \neq 0$. The property may be stated as follows:

$$g_M(\tau) = \varepsilon\, g_M(\tau + \beta), \tag{3.51}$$

where ε was defined earlier as having the possible values ± 1, so with $\varepsilon = 1$ (as typically for boson) we have periodicity whereas with $\varepsilon = -1$ (as typically for fermions) we have antiperiodicity.

Proof There are two cases to consider, depending on the sign of τ. Taking the case of $\tau < 0$, we have

$$g_M(\tau) = -\langle \hat{T}_W \check{A}(\tau)\check{B}(0)\rangle = -\varepsilon\langle \check{B}(0)\check{A}(\tau)\rangle$$
$$= -\varepsilon \frac{1}{Q}\mathrm{Tr}\left[e^{-\beta\overline{\mathcal{H}}}Be^{\overline{\mathcal{H}}\tau}Ae^{-\overline{\mathcal{H}}\tau}\right] = -\varepsilon\frac{1}{Q}\mathrm{Tr}\left[e^{\overline{\mathcal{H}}\tau}Ae^{-\overline{\mathcal{H}}\tau}e^{-\beta\overline{\mathcal{H}}}B\right],$$

where we have employed properties of the trace and the definition of thermal average in Equation (3.4). With further rearrangement this yields

$$g_M(\tau) = -\varepsilon\frac{1}{Q}\mathrm{Tr}\left[e^{-\beta\overline{\mathcal{H}}}e^{(\tau+\beta)\overline{\mathcal{H}}}Ae^{-(\tau+\beta)\overline{\mathcal{H}}}B\right]$$
$$= -\varepsilon\frac{1}{Q}\mathrm{Tr}\left[e^{-\beta\overline{\mathcal{H}}}\check{A}(\tau+\beta)\check{B}(0)\right]$$
$$= -\varepsilon\,\langle\check{A}(\tau+\beta)\check{B}(0)\rangle = \varepsilon\,g_M(\tau+\beta).$$

This proves the required result when $\tau < 0$. The case of $\tau > 0$ is treated similarly.

If Equation (3.51) is applied twice in succession to the GF we obtain

$$g_M(\tau) = \varepsilon\,g_M(\tau+\beta) = \varepsilon^2\,g_M(\tau+2\beta) = g_M(\tau+2\beta). \tag{3.52}$$

This result means that the imaginary-time GF is always periodic with a period equal to 2β. Consequently, without loss of generality we are free to choose τ to satisfy $-\beta < \tau < \beta$. Also, because of this periodicity, it follows that we may expand $g_M(\tau)$ as a Fourier series in this chosen interval. The expansion can be written in the form

$$g_M(\tau) = \frac{1}{\beta}\sum_{m=-\infty}^{\infty} e^{-i\omega_m\tau}\mathcal{G}(i\omega_m). \tag{3.53}$$

The overall $1/\beta$ factor appearing in the preceding expansion is optional, but it is included for later convenience. The quantities $\mathcal{G}(i\omega_m)$ denote the Fourier components for the imaginary-time GF. The summation in Equation (3.53) is over the full spectrum of frequencies denoted by ω_m. It is now straightforward to show that these frequencies are discrete, which is why we have a summation in Equation (3.53) rather than an integration. In fact, there is an infinite number of frequency values, and they take a different form for bosons and for fermions. To see this, we substitute Equation (3.53) into (3.51) giving

$$\frac{1}{\beta} \sum_{m=-\infty}^{\infty} e^{-i\omega_m \tau} \mathcal{G}(i\omega_m) = \varepsilon \frac{1}{\beta} \sum_{m=-\infty}^{+\infty} e^{-i\omega_m(\tau+\beta)} \mathcal{G}(i\omega_m).$$

This leads to the consistency condition that we must have $\exp(-i\omega_m\beta) = \varepsilon$, implying for ω_m that

$$\omega_m = \begin{cases} (2m+1)\pi/\beta & \text{for fermions } (\varepsilon = -1) \\ \\ 2m\pi/\beta & \text{for bosons } (\varepsilon = 1) \end{cases}, \tag{3.54}$$

where m can take all integer values from $-\infty$ to ∞. The frequencies ω_m are sometimes known as *Matsubara frequencies*. They depend on temperature through $\beta = 1/k_B T$ and form a discrete (rather than continuous) set provided $T \neq 0$. The frequencies are always nonzero for fermions but can be zero (when $m = 0$) for bosons. In the zero-temperature limit (when $\beta \to \infty$) the interval between adjacent frequencies, given by $2\pi/\beta$ tends to zero and the frequency spectrum becomes continuous in this limit.

It is of interest for us to obtain the inverse Fourier transformation to Equation (3.53). We start by noting that this will have the form

$$\mathcal{G}(i\omega_m) = \beta \frac{1}{2\beta} \int_{-\beta}^{\beta} d\tau\, e^{i\omega_m \tau} g_M(\tau)$$

$$= \frac{1}{2} \int_{-\beta}^{0} d\tau\, e^{i\omega_m \tau} g_M(\tau) + \frac{1}{2} \int_{0}^{\beta} d\tau\, e^{i\omega_m \tau} g_M(\tau).$$

In the second line in the preceding equation we have split the range of integration into two parts. Then, by introducing a change of variable to $\tilde{\tau} = \tau + \beta$ in the first term in this line and subsequently using the property stated in Equation (3.51), it may be shown that the two terms are equal to one another. The details are left for Problem 3.6. The outcome is that we can now write the inverse Fourier transformation more concisely as

$$\mathcal{G}(i\omega_m) = \int_{0}^{\beta} d\tau\, e^{i\omega_m \tau} g_M(\tau). \tag{3.55}$$

The "frequency" labels represented by $i\omega_m$ are, of course, pure imaginary numbers.

3.5.2 *The Lehmann Representation*

Here we will show that there is a simple connection between the imaginary-time GF and its corresponding real-time retarded GF. This convenient result is achieved in terms of their frequency Fourier transforms using what is known as the Lehmann representation.

The starting point is the definition in Equation (3.50) for the imaginary-time GF. We choose to employ a representation of the complete set of quantum states of $\overline{\mathcal{H}}$, and so we write $\overline{\mathcal{H}}|j\rangle = E_j|j\rangle$. This means that the single-particle energy eigenvalues E_j are measured here with respect to the chemical potential μ. When $\tau > 0$ it follows that we have

$$g_M(\tau) = -\langle \check{A}(\tau)\check{B}(0)\rangle = -\frac{1}{Q}\text{Tr}\left[e^{-\beta\overline{\mathcal{H}}}e^{\overline{\mathcal{H}}\tau}Ae^{-\overline{\mathcal{H}}\tau}B\right]$$

$$= -\frac{1}{Q}\sum_j \langle j|e^{-\beta\overline{\mathcal{H}}}e^{\overline{\mathcal{H}}\tau}Ae^{-\overline{\mathcal{H}}\tau}B|j\rangle.$$

Then we can employ the completeness property of the states that

$$\sum_{j'}|j'\rangle\langle j'| = 1 \tag{3.56}$$

to obtain a decomposition in terms of the matrix elements of A and B individually as

$$g_M(\tau) = -\frac{1}{Q}\sum_{j,j'}\langle j|e^{-\beta\overline{\mathcal{H}}}e^{\overline{\mathcal{H}}\tau}Ae^{-\overline{\mathcal{H}}\tau}|j'\rangle\langle j'|B|j\rangle$$

$$= -\frac{1}{Q}\sum_{j,j'}e^{-\beta E_j}e^{\tau(E_j-E_{j'})}A_{j,j'}B_{j',j}.$$

Here we have denoted $A_{j,j'} = \langle j|A|j'\rangle$ and $B_{j',j} = \langle j'|B|j\rangle$ as a shorthand for the matrix elements.

Now we substitute the preceding result into the right-hand side of Equation (3.55) to find initially that

$$\mathcal{G}(i\omega_m) = -\int_0^\beta d\tau \frac{1}{Q}\sum_{j,j'}e^{-\beta E_j}e^{\tau(i\omega_m+E_j-E_{j'})}A_{j,j'}B_{j',j}$$

$$= -\frac{1}{Q}\sum_{j,j'}e^{-\beta E_j}\frac{e^{i\omega_m\beta}e^{(E_j-E_{j'})\beta}-1}{i\omega_m+E_j-E_{j'}}A_{j,j'}B_{j',j}.$$

Then, after using the property that $\exp(i\omega_m\beta) = \varepsilon$ for the boson and fermion Matsubara frequencies, the preceding result can be expressed as

$$\mathcal{G}(i\omega_m) = \frac{1}{Q} \sum_{j,j'} \frac{e^{-\beta E_j} - \varepsilon e^{-\beta E_{j'}}}{i\omega_m + E_j - E_{j'}} A_{j,j'} B_{j',j}. \tag{3.57}$$

Following an analogous derivation by H. Lehmann in [59], the preceding type of result for the frequency Fourier transform of a GF in terms of a summation over matrix elements of the operators is often referred to as the *Lehmann representation*.

We may next go through a very similar procedure to obtain the Lehmann representation for the case of the real-time retarded GF $G_r(\omega)$ in terms of the same operators. The main results are quoted in the following text (with the details being left to Problem 3.8). Using Equation (3.6) and setting $t' = 0$, because only the time difference is relevant, we find

$$g_r(t) = -i\theta(t)\left\{\langle A(t)B(0)\rangle - \varepsilon\langle B(0)A(t)\rangle\right\}$$

$$= -i\theta(t)\frac{1}{Q}\sum_j \left\{e^{-\beta E_j}\langle j|e^{i\mathcal{H}t}Ae^{-i\mathcal{H}t}B|j\rangle - \varepsilon\, e^{-\beta E_j}\langle j|Be^{i\mathcal{H}t}Ae^{-i\mathcal{H}t}|j\rangle\right\}$$

$$= -i\theta(t)\frac{1}{Q}\sum_{j,j'} \left\{e^{-\beta E_j}e^{i(E_j - E_{j'})t}A_{j,j'}B_{j',j} - \varepsilon\, e^{-\beta E_{j'}}e^{i(E_j - E_{j'})t}A_{j,j'}B_{j',j}\right\}.$$

$$\tag{3.58}$$

The Fourier transform of the preceding retarded GF at frequency ω, as defined in Equation (3.23), can then be obtained in the form

$$G_r(\omega) = \frac{1}{2\pi Q} \sum_{j,j'} \frac{e^{-\beta E_j} - \varepsilon e^{-\beta E_{j'}}}{\omega + E_j - E_{j'} + i\eta} A_{j,j'} B_{j',j}, \tag{3.59}$$

which provides us with the required Lehmann representation. Again η denotes a positive infinitesimal quantity.

By now comparing the right-hand sides of Equations (3.57) and (3.59) we see that the only difference regarding the summation parts is in the denominator term. The important result is that, if we have already calculated the imaginary-time GF $\mathcal{G}(i\omega_m)$, we may directly obtain the corresponding retarded GF $G_r(\omega)$ by making the simple replacement $i\omega_m \rightarrow \omega + i\eta$. There is also the overall factor of $(1/2\pi)$ to take account of when comparing $\mathcal{G}(i\omega_m)$ and $G_r(\omega)$: this is a consequence of the conventions adopted for the respective Fourier transforms. We will make use of this connection between the GFs when developing the diagrammatic perturbation method in Chapter 8. We note that the quantities $i\omega_m$ lie along the imaginary axis in the complex $\tilde{\omega}$ plane, as seen in Figure 3.3.

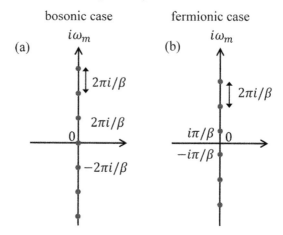

Figure 3.3 Representation of the imaginary quantities $i\omega_m$ in the complex frequency (or $\tilde{\omega}$) plane, where ω_m denote the Matsubara frequencies, for (a) bosons and (b) fermions.

3.6 Methods of Evaluating Green's Functions

There are three main methods that we shall be following throughout this book to calculate the GFs and to deduce other information from them (such as the excitation frequencies, the spectral intensities, and correlation functions). Briefly, these methods are

- *The equation-of-motion method.* This employs the equation of motion for the real-time (usually the retarded) GFs as derived earlier in this chapter. Because exact solutions are possible only in a few special cases, the method is typically employed in conjunction with a so-called decoupling approximation.
- *The linear response function method.* This involves finding the linearized response of an unperturbed system to a small (either real or fictitious) perturbation that couples to the excitations of the system. It will be shown that the response functions are related to the retarded GFs in a straightforward fashion.
- *The diagrammatic perturbation method.* This makes use of the imaginary-time GFs, involving a formal perturbation expansion. The technique employs an important result known as Wick's theorem (to simplify the process of taking thermal averages) and a Feynman diagrammatic representation (to simplify the algebra).

Here and in the next few chapters we lay the foundations for the first of these methods, which is summarized in Subsection 3.6.2. The other two methods are left until later (see Chapter 6 for the linear response function method and Chapters 8 and 9 for the diagrammatic perturbation method).

It is worthwhile at this stage, however, to give a brief survey in Subsection 3.6.1 of the "classical" GFs used in mathematics. In part, this is done for completeness, but it will also enable certain comparisons and analogies with the quantum-mechanical GFs to be highlighted later.

3.6.1 Survey of Classical Green's Functions

Suppose we consider the problem in mathematics of solving a linear differential equation for y as a function of the real variable x, having the form

$$\mathcal{L}\, y(x) - \lambda\, y(x) = f(x), \tag{3.60}$$

where $\mathcal{L}$ denotes a Hermitian differential operator (usually a linear combination of terms involving powers of d/dx), λ is a constant, and $f(x)$ on the right-hand side is an inhomogeneous term. In general, initial-value boundary conditions would also be specified. Note that in the absence of $f(x)$ we would have a homogeneous eigenvalue equation.

A common procedure used to solve Equation (3.60) involves expanding both $y(x)$ and $f(x)$ in terms of the eigenfunctions $y_n(x)$, with $n = 1, 2, 3, \ldots$, of the operator $\mathcal{L}$, giving

$$y(x) = \sum_n a_n\, y_n(x), \qquad f(x) = \sum_n b_n\, y_n(x). \tag{3.61}$$

A connection between the coefficients a_n and b_n can simply be obtained by substituting the previously mentioned expressions into Equation (3.60) and using the orthogonality properties of the eigenfunction. In this way it can be deduced that

$$y(x) = \sum_n \frac{b_n\, y_n(x)}{\lambda_n - \lambda},$$

denoting λ_n as the eigenvalue corresponding to eigenfunction $y_n(x)$. Then, by utilizing the inverse expansions to those in Equation (3.61), the preceding result can be rewritten as

$$y(x) = \int G(x, x')\, f(x')\, dx'. \tag{3.62}$$

The quantity $G(x, x')$ is just the classical GF, and it can be expressed formally as

$$G(x, x') = \sum_n \frac{y_n(x)\, y_n^*(x')}{\lambda_n - \lambda}. \tag{3.63}$$

The preceding results hold quite generally for any $f(x)$, but suppose we now take the inhomogeneous term to consist of a delta function at the origin, i.e., we choose $f(x) = \delta(x)$. Equation (3.62) then gives

$$y(x) = \int G(x,x')\,\delta(x')\,dx' = G(x,0).$$

As a minor extension it follows that the classical GF $G(x,x')$ can be obtained as the solution of

$$(\mathcal{L} - \lambda)G(x,x') = \delta(x - x'). \tag{3.64}$$

The preceding result nicely illustrates a physical interpretation of the classical GF. Specifically, $G(x,x')$ represents the solution (as a function of x) for the differential equation when the source term has the form of a "spike" at x', i.e., $f(x) = \delta(x - x')$. We shall later refer to analogies with Equations (3.63) and (3.64) when we discuss the linear response methods in Chapter 6.

3.6.2 The Green's Function Equation-of-Motion Method

We have already obtained a formal equation of motion given in Equation (3.15) for any real-time GF $g(A; B \,|\, t - t')$ of the operators A and B. This equation relates the time derivative of the GF to a static correlation function, multiplied by $\delta(t - t')$, plus another GF that depends on the form of the Hamiltonian of the system.

This equation will now be reexpressed more conveniently in the frequency representation. We define a Fourier transform from time labels to frequency by analogy with Equation (3.22), so that

$$g(A; B \,|\, t - t') = \int_{-\infty}^{\infty} G(A; B \,|\, \omega)e^{-i\omega(t-t')}d\omega. \tag{3.65}$$

The quantities $G(A; B \,|\, \omega)$ are the Fourier components of the original GF. It then follows that we have for the time derivative

$$\frac{d}{dt} g(A; B \,|\, t - t') = \frac{d}{dt} \int_{-\infty}^{\infty} G(A; B \,|\, \omega)e^{-i\omega(t-t')}d\omega$$

$$= -i \int_{-\infty}^{\infty} \omega\, G(A; B \,|\, \omega)e^{-i\omega(t-t')}d\omega.$$

Using the GF equation of motion as quoted previously in Equation (3.15), along with the integral representation for the delta function in Equation (3.27), we deduce that

$$-i \int_{-\infty}^{\infty} \omega\, G(A; B \,|\, \omega)e^{-i\omega(t-t')}d\omega = -i\,\langle[A, B]_\varepsilon\rangle \frac{1}{2\pi} \int_{-\infty}^{\infty} e^{-i\omega(t-t')}d\omega$$

$$+ i \int_{-\infty}^{\infty} G([\overline{\mathcal{H}}, A]; B \,|\, \omega)e^{-i\omega(t-t')}d\omega.$$

Finally, on equating the terms appearing as the integrands on both sides of the equation (and cancelling out common factors), we arrive at the important result that

$$\omega\, G(A; B \mid \omega) = \frac{1}{2\pi} \langle [A, B]_\varepsilon \rangle - G([\overline{\mathcal{H}}, A]; B \mid \omega). \tag{3.66}$$

This will be taken to represent our standard form of the equation of motion for real-time GFs. In a few special cases, we might find that the operator $[\overline{\mathcal{H}}, A]$ is straightforwardly related to the operator A, and then we can just solve the preceding equation of motion for the original GF $G(A; B \mid \omega)$. More typically, however, the new GF appearing on the right-hand side of Equation (3.66) is more complicated than the original GF. We could next repeat this process by writing down *its* equation of motion, generating yet another GF. Sometimes this process might terminate after a finite number of stages (as we shall see in examples later), or in principle it might just seem to continue indefinitely.

In practice, in the absence of any algebraic termination leading to a closed set of coupled equations, we will usually try to get an approximate solution by imposing a termination to the process after a chosen finite number of steps: this will involve looking for an approximation to simplify the GF that appears on the right-hand side of the last equation. This is usually called a *decoupling approximation*.

We will illustrate this process with examples in the following chapters, starting with cases where the GF equations are exactly solvable (in Chapter 4) and next considering approximation methods (starting in Chapter 5).

Problems

3.1. The Hamiltonian $\mathcal{H}$ for a spinless particle of mass m confined in $3D$ to the interior of a cubical box with sides of length L is $\mathcal{H} = -(\hbar^2/2m)\nabla^2$. It is assumed that the walls of the box are inpenetrable and the potential energy inside the box is zero. Write down the QM energy eigenvalues for the particle and hence an expression for the partition function Q in a canonical ensemble. Then evaluate Q assuming that the summation over states can be replaced by integral(s). Finally show that the thermal average (expectation value) for the energy of the particle is

$$\langle \mathcal{H} \rangle = \frac{3}{2} k_B T$$

in accordance with the equipartition theorem.

3.2. Verify the results quoted in Subsections 3.3.2 for the form taken by the advanced and causal GFs when they are reexpressed in the frequency representation. Do this by analogy with the proofs given for the retarded GF $G_r(\omega)$ in the previous subsection.

3.3. Prove that the relationship

$$g^*(A; B \mid t - t') = \varepsilon g(A^\dagger; B^\dagger \mid t - t')$$

applies for both retarded and advanced GFs. Here ε is the ± 1 factor appearing in the GF definitions, and the $*$ denotes complex conjugate. Deduce also the relationship between the retarded and advanced GFs $G(A; B \mid \omega)$ and $G(B; A \mid \omega)$ in the frequency representation.

3.4. By using the method of contour integration (or otherwise), verify the quoted result in Equation (3.39) for the inverse Kramers–Kronig relation in the case of the retarded GF.

3.5. Suppose that an analytically continued GF $G(A; B \mid \tilde{\omega})$ at the complex frequency label $\tilde{\omega}$ is given by

$$G(A; B \mid \tilde{\omega}) = \frac{2\tilde{\omega}}{(\omega_1 - \tilde{\omega})(\omega_2 - \tilde{\omega})},$$

where ω_1 and ω_2 are real frequencies that satisfy $\omega_2 > \omega_1 > 0$. What is the corresponding *retarded* GF at any real frequency label ω? Deduce the spectral intensity $J(\omega)$ in the preceding case. You might find it helpful to use partial fractions in reexpressing the preceding GF.

3.6. Prove the Fourier transformation result for the imaginary-time GFs as quoted in Equation (3.55), starting from the previous equation in the text.

3.7. The two-time GFs defined in this chapter can be generalized to cases where three or more operators are involved. For the imaginary-time GFs we may take the definition

$$g_M(\tau_1, \tau_2, \tau_3, \ldots, \tau_n) = (-i)^n \left\langle \hat{T}_W \left\{ \check{A}(\tau_1) \check{B}(\tau_2) \check{C}(\tau_3) \cdots \right\} \right\rangle,$$

where $\check{A}$ is given by Equation (3.49) and $\hat{T}_W$ is the Wick time-ordering operator. Show that if $0 < \tau_i \leq \beta$ we have

$$g_M(\tau_1, \tau_2, \ldots, \tau_i = 0, \ldots, \tau_n) = -\varepsilon g_M(\tau_1, \tau_2, \ldots, \tau_i = \beta, \ldots, \tau_n),$$

where $\varepsilon = \pm 1$.

3.8. Provide the detailed steps for the derivation of the Lehmann representation for the retarded GF $G_r(\omega)$ that was broadly outlined in the text. Specifically, you should derive Equation (3.58) and show how this can be used to obtain the final result in Equation (3.59).

3.9. The final expression for the retarded GF at frequency ω in the Lehmann representation is given by Equation (3.59). Use this equation to find the spectral function $J(\omega)$ in terms of the Lehmann representation. Then show that in the case of boson or fermion systems with $A = a$ and $B = a^\dagger$, the spectral function $J(\omega)$ satisfies

$$\int_{-\infty}^{\infty} d\omega J(\omega) = 1,$$

which is an example of a sum rule.

3.10. Starting from Equation (3.11) derive the equation of motion for the causal GF and show that the *final* form of the result in the frequency representation is just the same as that for the retarded GF in Equation (3.66) of Subsection 3.6.2.

3.11. It is known from Maxwell's equation that the magnetic vector and scalar potentials for an EM wave of angular frequency ω satisfy, in general, the inhomogeneous Helmholtz equation. The corresponding classical GF $g(\mathbf{r}, \mathbf{r}')$ in 3D therefore satisfies

$$\nabla^2 g(\mathbf{r}, \mathbf{r}') + k^2 g(\mathbf{r}, \mathbf{r}') = \delta(\mathbf{r} - \mathbf{r}'),$$

where $k = \omega/c$ and c is the vacuum light speed. Assuming there is only an outward wave solution as $|\mathbf{r} - \mathbf{r}'| \to \infty$, show that the GF is

$$g(\mathbf{r}, \mathbf{r}') = -\frac{e^{ik|\mathbf{r}-\mathbf{r}'|}}{4\pi |\mathbf{r} - \mathbf{r}'|}.$$

4

Exact Methods for Green's Function

In this chapter we present some calculations for real-time Green's functions (GFs), as well as related quantities such as the spectral intensities and correlation functions, where the results may be obtained *exactly* for the particular model Hamiltonian under consideration without any further approximation being involved. Such models or systems are necessarily relatively simple, and may have involved physical approximations or the neglect of certain effects at an earlier stage in arriving at the model. Nevertheless, they are instructive to consider as an illustration of the general methods and to act as a springboard for the subsequent extensions to be considered in later chapters.

For the present we shall follow the GF equation-of-motion method that was outlined in Section 3.6. Various calculations will be given where the equations of motion have an exact solution, enabling us to explore the basic properties of the GFs in the complex frequency plane and the correlations in the system. In some cases, there may be several coupled equations of motion (e.g., for "mixed" excitations involving more than one boson or fermion field). The initial examples in this chapter will be covered in some detail in order to establish the methodology.

4.1 Noninteracting Gas of Bosons or Fermions

On setting $v(\mathbf{q}) = 0$ for the pairwise interaction term in the earlier Hamiltonian expression derived for a boson or fermion gas (see Subsection 1.4.2), we are left just with the kinetic energy term in the Hamiltonian and we may write

$$\overline{\mathcal{H}} = \mathcal{H} - \mu \mathcal{N} = \sum_{\mathbf{k}} E_{\mathbf{k}} a_{\mathbf{k}}^{\dagger} a_{\mathbf{k}}. \qquad (4.1)$$

Here $E_{\mathbf{k}} = (k^2/2m) - \mu$ denotes the particle energy expressed relative to the chemical potential μ. We now choose to take $A = a_{\mathbf{k}}$ and $B = a_{\mathbf{k}}^{\dagger}$ in Equation (3.66) to evaluate retarded GFs of the type $G(a_{\mathbf{k}}; a_{\mathbf{k}}^{\dagger} \mid \omega)$ at frequency ω. The behavior in the real-time domain will also be discussed. Afterward, as a simple exercise to demonstrate consistency we show how this may be done for the corresponding imaginary-time GFs, where we employ a direct evaluation from the definitions, instead of the equation of motion as was done for the real-time GFs.

4.1.1 Case of Bosons

For bosons we choose the parameter $\varepsilon = 1$ so that we get commutation relations in the definitions of the GFs in Equations (3.6), (3.7), and (3.11). For the terms appearing on the right-hand side of Equation (3.66) we have $\langle [A, B]_\varepsilon \rangle = \langle [a_{\mathbf{k}}, a_{\mathbf{k}}^\dagger] \rangle = 1$ and

$$[\overline{\mathcal{H}}, A] = \sum_{\mathbf{k}'} E_{\mathbf{k}'} [a_{\mathbf{k}'}^\dagger a_{\mathbf{k}'}, a_{\mathbf{k}}] = E_{\mathbf{k}} [a_{\mathbf{k}}^\dagger a_{\mathbf{k}}, a_{\mathbf{k}}] = -E_{\mathbf{k}} a_{\mathbf{k}}. \tag{4.2}$$

Hence the equation of motion for this GF becomes

$$\omega \, G(a_{\mathbf{k}}; a_{\mathbf{k}}^\dagger \mid \omega) = \frac{1}{2\pi} + E_{\mathbf{k}} \, G(a_{\mathbf{k}}; a_{\mathbf{k}}^\dagger \mid \omega),$$

so we clearly have a situation in which the new GF appearing on the right-hand side is the same as the one with which we started. We have simply

$$(\omega - E_{\mathbf{k}}) \, G(a_{\mathbf{k}}; a_{\mathbf{k}}^\dagger \mid \omega) = \frac{1}{2\pi}. \tag{4.3}$$

If we have $\omega \neq E_{\mathbf{k}}$, we can just divide both sides of the preceding equation by $(\omega - E_{\mathbf{k}})$. This would not, however, yield the general solution, valid for all real frequency ω, which takes the form

$$G(a_{\mathbf{k}}; a_{\mathbf{k}}^\dagger \mid \omega) - \left(\frac{1}{2\pi} \right) \frac{1}{\omega - E_{\mathbf{k}}} + f \delta(\omega - E_{\mathbf{k}}). \tag{4.4}$$

Here f is a constant, representing just the arbitrary constant expected in the general solution of the first-order differential equation for the GF in the time representation. One way to deduce f is to substitute Equation (4.4) back into the defining expression for the real-time GF (whether it be retarded, advanced, or causal). We expect to find a different value of f for each kind of GF. An alternative (and usually much better) way is to write down directly the analytically continued GF, which was introduced in Subsection 3.4.3. Because this GF has a discontinuity on the real axis and it is analytic elsewhere in the complex $\tilde{\omega}$ plane, we have in this case

$$G(a_{\mathbf{k}}; a_{\mathbf{k}}^\dagger \mid \tilde{\omega}) = \left(\frac{1}{2\pi} \right) \frac{1}{\tilde{\omega} - E_{\mathbf{k}}}, \tag{4.5}$$

This has a simple pole corresponding to $\tilde{\omega} = E_{\mathbf{k}}$ on the real axis at the particle energy $E_{\mathbf{k}}$.

Now, for the retarded GF, we previously obtained in Equation (3.44) the relationship that $G(\omega + i\eta) = G_r(\omega)$. Therefore, we have

$$
\begin{aligned}
G_r(a_{\mathbf{k}}; a_{\mathbf{k}}^\dagger \mid \omega) &= \left(\frac{1}{2\pi}\right) \frac{1}{\omega - E_{\mathbf{k}} + i\eta} \\
&= \frac{1}{2\pi} \mathcal{P}\left(\frac{1}{\omega - E_{\mathbf{k}}}\right) - \frac{i}{2}\delta(\omega - E_{\mathbf{k}}).
\end{aligned}
\tag{4.6}
$$

Here η denotes a positive infinitesimal as before, and we have used the identity in Equation (3.31) to obtain the second line of the preceding equation. The result corresponds to $f = -i/2$ in Equation (4.4).

Similarly, for the advanced GF, we previously deduced in Equation (3.45) that $G(\omega - i\eta) = G_a(\omega)$, and so

$$
G_a(a_{\mathbf{k}}; a_{\mathbf{k}}^\dagger \mid \omega) = \left(\frac{1}{2\pi}\right) \frac{1}{\omega - E_{\mathbf{k}} - i\eta}.
\tag{4.7}
$$

We conclude that this corresponds to $f = i/2$ in Equation (4.4). Finally, to get the value of f for the causal GF we would need to go back to the definition of the causal GF. This is much less convenient than for the retarded or advanced GFs, and it is found that the result is

$$
f = -\frac{i}{2} \coth\left(\frac{1}{2}\beta E_{\mathbf{k}}\right).
\tag{4.8}
$$

Note also that the causal GF has an explicit temperature dependence (in the imaginary part), unlike the other types of GFs.

We can now use any of the previously mentioned GF results to deduce the spectral intensity $J(\omega)$ by application of the fluctuation-dissipation theorem as in Equation (3.40). Doing this in terms of the retarded GF, we find

$$
J(\omega) = \frac{-2}{\left(e^{\beta\omega} - 1\right)} \operatorname{Im} G_r(\omega) = \frac{-1}{\pi\left(e^{\beta\omega} - 1\right)} \operatorname{Im}\left(\frac{1}{\omega - E_{\mathbf{k}} + i\eta}\right).
\tag{4.9}
$$

Therefore, we conclude that

$$
J(\omega) = \frac{1}{\left(e^{\beta\omega} - 1\right)}\delta(\omega - E_{\mathbf{k}}),
\tag{4.10}
$$

and so, for a noninteracting boson gas, the spectral intensity consists of a single delta-function spike, i.e., only one Fourier component is present corresponding to the frequency $\omega = E_{\mathbf{k}}$. We notice that the total (or integrated) spectral intensity is finite because

$$
\int_{-\infty}^{\infty} J(\omega)\, d\omega = \int_{-\infty}^{\infty} \frac{d\omega}{e^{\beta\omega} - 1}\delta(\omega - E_{\mathbf{k}}) = \frac{1}{e^{\beta E_{\mathbf{k}}} - 1}.
$$

Next we can deduce the time correlation functions. For example, using Equation (3.17) we have

$$\langle a_{\mathbf{k}}^{\dagger}(t) a_{\mathbf{k}}(t') \rangle = \int_{-\infty}^{\infty} J(\omega) e^{-i\omega(t-t')} d\omega = \frac{\exp\left[-iE_{\mathbf{k}}(t-t')\right]}{\left[\exp(\beta E_{\mathbf{k}}) - 1\right]}. \tag{4.11}$$

As a special case, if we now take the limiting situation of the time labels becoming equal, we obtain the result that the number of bosons with wave vector $\mathbf{k}$ is

$$n_{\mathbf{k}} \equiv n(E_{\mathbf{k}}) = \langle a_{\mathbf{k}}^{\dagger} a_{\mathbf{k}} \rangle = \frac{1}{\left[\exp(\beta E_{\mathbf{k}}) - 1\right]}. \tag{4.12}$$

In other words, we recover the well-known Bose–Einstein distribution function, as expected. As a comment on notation, we will express the full form of the distribution function as $n(E_{\mathbf{k}})$ at energy $E_{\mathbf{k}}$, but when there is no ambiguity we will often use the shorthand form $n_{\mathbf{k}}$.

We note that it is usually preferable to solve for the GFs in the frequency domain, as done in this subsection. Even if our objective is to find a time correlation function (e.g., for comparison with an experiment), the typical procedure is to deduce the spectral intensity $J(\omega)$ from the GF by using the fluctuation-dissipation theorem and then transform to the time-dependent correlation functions.

In the present example, we could alternatively obtain the GF in the time domain by using Equation (3.15), which becomes here

$$\frac{d}{dt} g_r(a_{\mathbf{k}}; a_{\mathbf{k}}^{\dagger} \mid t - t') = -i\delta(t - t') \langle [a_{\mathbf{k}}, a_{\mathbf{k}}^{\dagger}] \rangle + i g_r([\overline{\mathcal{H}}, a_{\mathbf{k}}]; a_{\mathbf{k}}^{\dagger} \mid t - t').$$

The two terms on the right-hand side simplify because $[a_{\mathbf{k}}, a_{\mathbf{k}}^{\dagger}] = 1$ and $[\overline{\mathcal{H}}, a_{\mathbf{k}}] = -E_{\mathbf{k}} a_{\mathbf{k}}$ (see, e.g., Problem 1.8). Therefore, we have after some minor rearrangement

$$\left(i\frac{d}{dt} - E_{\mathbf{k}}\right) g_r(a_{\mathbf{k}}; a_{\mathbf{k}}^{\dagger} \mid t - t') = \delta(t - t').$$

The solution for the time-dependent GF is simply

$$g_r(a_{\mathbf{k}}; a_{\mathbf{k}}^{\dagger} \mid t - t') = -i\theta(t - t') \exp[-iE_{\mathbf{k}}(t - t')], \tag{4.13}$$

as can be verified by substituting back into the preceding differential equation.

4.1.2 Case of Fermions

We make the same choices for the Hamiltonian and for the operators in the GF as in the previous subsection, except that now we choose $\varepsilon = -1$ to have anticommutation relations in the definitions of the GF. It must be kept in mind, however, that the general equation of motion always has a term involving a *commutator* of one of the operators with the Hamiltonian.

It is easy to prove that Equations (4.3)–(4.7) for the GFs apply in the fermion case, just as in the boson case, although some of the intermediate steps are different. Hence the expressions for the retarded and advanced GFs in the energy representation are formally the same as before (with the same values of the constant f). However, the result for the causal function is different, leading to

$$f = -\frac{i}{2} \tanh\left(\frac{1}{2}\beta E_{\mathbf{k}}\right). \tag{4.14}$$

The spectral intensity in the fermion case is found to be

$$J(\omega) = \frac{1}{\left(e^{\beta\omega} + 1\right)}\delta(\omega - E_{\mathbf{k}}). \tag{4.15}$$

Again it is proportional to a delta function at $\omega = E_{\mathbf{k}}$, but it now becomes multiplied by a Fermi–Dirac (FD) factor, rather than the Bose–Einstein (BE) factor as previously.

Similarly for the time correlation function we have

$$\langle a_{\mathbf{k}}^{\dagger}(t)a_{\mathbf{k}}(t')\rangle = \int_{-\infty}^{\infty} J(\omega)e^{-i\omega(t-t')}d\omega = \frac{\exp\left[-iE_{\mathbf{k}}(t-t')\right]}{\left[\exp(\beta E_{\mathbf{k}}) + 1\right]}, \tag{4.16}$$

whereas for the thermal average the FD distribution function is recovered as

$$n_{\mathbf{k}} \equiv n(E_{\mathbf{k}}) = \langle a_{\mathbf{k}}^{\dagger}a_{\mathbf{k}}\rangle = \frac{1}{\left[\exp(\beta E_{\mathbf{k}}) + 1\right]}. \tag{4.17}$$

The same comment regarding notation applies as after Equation (4.12).

4.1.3 The Green's Functions in the Imaginary-Time Formalism

We now reexamine the calculation for noninteracting bosons by using in this case the imaginary-time (or Matsubara) GF formalism. The Hamiltonian is again taken to be as in Equation (4.1), while from Equation (3.50) the imaginary-time GF can be written as

$$\begin{aligned} g_M(\tau) &= -\langle \hat{T}_W \check{a}_{\mathbf{k}}(\tau)\check{a}_{\mathbf{k}}^{\dagger}(0)\rangle \\ &= -\theta(\tau)\langle \check{a}_{\mathbf{k}}(\tau)\check{a}_{\mathbf{k}}^{\dagger}\rangle - \theta(-\tau)\langle \check{a}_{\mathbf{k}}^{\dagger}\check{a}_{\mathbf{k}}(\tau)\rangle. \end{aligned} \tag{4.18}$$

To obtain $\check{a}_{\mathbf{k}}(\tau)$ we employ its definition in Equation (3.49) and we differentiate with respect to the τ label, giving

$$\frac{d}{d\tau}\check{a}_{\mathbf{k}}(\tau) = \frac{d}{d\tau}\left(e^{\overline{\mathcal{H}}\tau}a_{\mathbf{k}}e^{-\overline{\mathcal{H}}\tau}\right) = [\overline{\mathcal{H}}, \check{a}_{\mathbf{k}}(\tau)]$$

$$= \sum_{\mathbf{k}'} E_{\mathbf{k}'}[a_{\mathbf{k}'}^{\dagger}a_{\mathbf{k}'}, \check{a}_{\mathbf{k}}(\tau)]$$

$$= \sum_{\mathbf{k}'} E_{\mathbf{k}'}(-\delta_{\mathbf{k},\mathbf{k}'})\check{a}_{\mathbf{k}'}(\tau) = -E_{\mathbf{k}}\check{a}_{\mathbf{k}}(\tau).$$

Thus, after integrating, we have the simple property that

$$\check{a}_{\mathbf{k}}(\tau) = e^{-E_{\mathbf{k}}\tau}\check{a}_{\mathbf{k}}(0) = e^{-E_{\mathbf{k}}\tau}a_{\mathbf{k}}. \tag{4.19}$$

Next, by substituting (4.19) into (4.18) for the GF, we obtain

$$g_M(\tau) = -\theta(\tau)\langle a_{\mathbf{k}}a_{\mathbf{k}}^{\dagger}\rangle e^{-E_{\mathbf{k}}\tau} - \theta(-\tau)\langle a_{\mathbf{k}}^{\dagger}a_{\mathbf{k}}\rangle e^{-E_{\mathbf{k}}\tau}.$$

For bosons we have $\langle a_{\mathbf{k}}^{\dagger}a_{\mathbf{k}}\rangle = n_{\mathbf{k}} = 1/[\exp(\beta E_{\mathbf{k}}) - 1]$ and $\langle a_{\mathbf{k}}a_{\mathbf{k}}^{\dagger}\rangle = 1 + n_{\mathbf{k}}$. Therefore, the preceding GF result can be expressed more simply as

$$g_M(\tau) = -e^{-E_{\mathbf{k}}\tau}\{\theta(\tau)[1 + n_{\mathbf{k}}] + \theta(-\tau)n_{\mathbf{k}}\}. \tag{4.20}$$

By using Equation (3.55) the frequency Fourier transform of this GF can be found as

$$\mathcal{G}(i\omega_m) = \int_0^{\beta} d\tau\, e^{i\omega_m\tau}g_M(\tau)$$

$$= -[1 + n_{\mathbf{k}}]\int_0^{\beta} d\tau\, e^{i\omega_m\tau}e^{-E_{\mathbf{k}}\tau}$$

$$= -[1 + n_{\mathbf{k}}]\frac{e^{(i\omega_m - E_{\mathbf{k}})\beta} - 1}{i\omega_m - E_{\mathbf{k}}}.$$

Next we use the property of Matsubara boson frequencies that $e^{i\omega_m\beta} = 1$, and following some other straightforward algebra the final GF expression is found to be (see Problem 4.2)

$$\mathcal{G}(i\omega_m) = \frac{1}{i\omega_m - E_{\mathbf{k}}}. \tag{4.21}$$

By inspection, it is clear that if the replacement $i\omega_m \to \omega + i\eta$ is made in the preceding equation and an overall factor of $(1/2\pi)$ is introduced, the result for the retarded GF in (4.6) is regained. This is in accordance with the general property established in Subsection 3.5.2.

For fermions we may follow step-by-step the same approach as used previously for the boson case. The only differences are some changes of sign due to having anticommutation relations between the fermion operators instead of boson commutation relations. It is found, however, that Equation (4.19) for the transformed operators still applies in the fermion case, but Equation (4.20) becomes modified to

$$g_M(\tau) = e^{-E_{\mathbf{k}}\tau} \left\{ -\theta(\tau) [1 - n_{\mathbf{k}}] + \theta(-\tau) n_{\mathbf{k}} \right\}, \tag{4.22}$$

where we now have $n_{\mathbf{k}} = 1/[\exp(\beta E_{\mathbf{k}}) + 1]$ as the Fermi–Dirac (FD) distribution function. For the frequency Fourier transform of the GF we obtain

$$\mathcal{G}(i\omega_m) = \int_0^\beta d\tau \, e^{i\omega_m \tau} g_M(\tau) = [n_{\mathbf{k}} - 1] \int_0^\beta d\tau \, e^{i\omega_m \tau} e^{-E_{\mathbf{k}}\tau}$$

$$= [n_{\mathbf{k}} - 1] \frac{e^{(i\omega_m - E_{\mathbf{k}})\beta} - 1}{i\omega_m - E_{\mathbf{k}}} = \frac{1}{i\omega_m - E_{\mathbf{k}}}. \tag{4.23}$$

We notice that the final result for $\mathcal{G}(i\omega_m)$ is formally the same as in Equation (4.21) for the boson case.

4.2 Green's Functions for a Graphene Sheet

Already in Section 2.7 we calculated the dispersion relation for the electronic excitation energy $E_{\mathbf{k}}$ as a function of the 2D wave vector $\mathbf{k}$ in graphene by using the operator equation-of-motion method. Now we use the GF equations of motion to obtain the retarded GFs with the form $G(a_{\mathbf{k}}; a_{\mathbf{k}}^\dagger \,|\, \omega)$, $G(a_{\mathbf{k}}; b_{\mathbf{k}}^\dagger \,|\, \omega)$, $G(b_{\mathbf{k}}; a_{\mathbf{k}}^\dagger \,|\, \omega)$, and $G(b_{\mathbf{k}}; b_{\mathbf{k}}^\dagger \,|\, \omega)$, where the a and b operators refer to the A and B sublattice sites, respectively, of graphene.

On constructing the GF equations of motion for each of the preceding four GFs using Equation (3.66), we obtain the following expressions:

$$\omega G(a_{\mathbf{k}}; a_{\mathbf{k}}^\dagger \,|\, \omega) = \frac{1}{2\pi} + t F_{\mathbf{k}} G(b_{\mathbf{k}}; a_{\mathbf{k}}^\dagger \,|\, \omega), \tag{4.24}$$

$$\omega G(a_{\mathbf{k}}; b_{\mathbf{k}}^\dagger \,|\, \omega) = t F_{\mathbf{k}} G(b_{\mathbf{k}}; b_{\mathbf{k}}^\dagger \,|\, \omega), \tag{4.25}$$

$$\omega G(b_{\mathbf{k}}; a_{\mathbf{k}}^\dagger \,|\, \omega) = t F_{\mathbf{k}}^* G(a_{\mathbf{k}}; a_{\mathbf{k}}^\dagger \,|\, \omega), \tag{4.26}$$

$$\omega G(b_{\mathbf{k}}; b_{\mathbf{k}}^\dagger \,|\, \omega) = \frac{1}{2\pi} + t F_{\mathbf{k}}^* G(a_{\mathbf{k}}; b_{\mathbf{k}}^\dagger \,|\, \omega), \tag{4.27}$$

where $F_{\mathbf{k}}$ is a complex factor defined as

$$F_{\mathbf{k}} = 2 \exp\left(\frac{1}{2} i k_x a_0\right) \cos\left(\frac{\sqrt{3}}{2} k_y a_0\right) + \exp\left(-i k_x a_0\right) \tag{4.28}$$

and the other notation is the same as before in Section 2.7.

The solutions of the preceding coupled equations for the individual GFs (in the analytically continued form) are then easily found to be given by

$$G(a_{\mathbf{k}};a_{\mathbf{k}}^{\dagger}\,|\,\tilde{\omega}) = G(b_{\mathbf{k}};b_{\mathbf{k}}^{\dagger}\,|\,\tilde{\omega}) = \frac{\tilde{\omega}}{2\pi\,(\tilde{\omega}^2 - E_{\mathbf{k}}^2)}, \tag{4.29}$$

$$G(a_{\mathbf{k}};b_{\mathbf{k}}^{\dagger}\,|\,\tilde{\omega}) = \frac{F_{\mathbf{k}}}{2\pi\,(\tilde{\omega}^2 - E_{\mathbf{k}}^2)}, \tag{4.30}$$

$$G(b_{\mathbf{k}};a_{\mathbf{k}}^{\dagger}\,|\,\tilde{\omega}) = \frac{F_{\mathbf{k}}^{*}}{2\pi\,(\tilde{\omega}^2 - E_{\mathbf{k}}^2)}. \tag{4.31}$$

We have denoted $E_{\mathbf{k}} = t\,|F_{\mathbf{k}}|$, which can be shown after some simple algebra to be identical to the dispersion relation written explicitly in Equation (2.74). As might be expected, all the GFs have energy poles at $\pm E_{\mathbf{k}}$.

The spectral intensities corresponding to each of the preceding GFs may readily be obtained, e.g., by first replacing $\tilde{\omega} \to \omega + i\eta$ to get the retarded GFs and then applying the fluctuation-dissipation theorem given in Equation (3.40). For example, corresponding to $G(a_{\mathbf{k}};a_{\mathbf{k}}^{\dagger}\,|\,\omega)$ we have for the spectral intensity

$$J_{a-a}(\omega) = \frac{1}{4\,(e^{\beta\omega} + 1)}\,\{\delta\,(\omega - E_{\mathbf{k}}) + \delta\,(\omega + E_{\mathbf{k}})\}. \tag{4.32}$$

Hence there are two delta-function spikes in this case and they are weighted by different thermal factors (when evaluated at the two values of ω specified by the delta functions). We note that a useful mathematical identity for delta functions that can be employed when obtaining the preceding result (and elsewhere) is

$$\delta\{f(x)\} = \sum_{i} \frac{1}{|f'(x_i)|}\,\delta(x - x_i) \tag{4.33}$$

for a function $f(x)$ of a real variable x, where $f' = df/dx$ and $\{x_i\}$ denotes the distinct values of x (with $i = 1, 2, \ldots$) at which $f(x) = 0$.

4.3 Interaction of Light with Atoms

As another GF application we consider the interaction of light (photons) with atoms. Here we examine the simplest case when just one two-level atom interacts with a quantized single-mode of an optical cavity, as represented schematically in Figure 4.1(a). This example can be well described in the framework of the Jaynes–Cummings (JC) model [60–64], which is extensively employed in quantum optics [13, 14, 65–67]. This model, as well as its extension to the Dicke model treated later, is mathematically simple.

4.3.1 Derivation of Jaynes–Cummings Model

In Subsection 1.1.2 we showed that the quantized optical (electromagnetic) field can be described by a simple harmonic oscillator. Also, it was established that

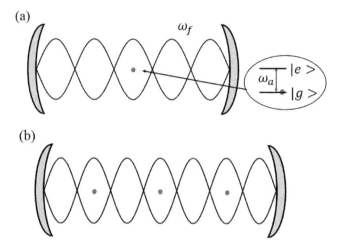

Figure 4.1 Schematic illustration of (a) the Jaynes–Cummings model of a simple two-level atom placed in an optical cavity; (b) the Dicke model in which there are several atoms.

the Hamiltonian for the field may be written as $\mathcal{H}_f = \omega_f a^\dagger a$, where ω_f is the angular frequency of the single-mode quantized field (photon). The boson operator $a^\dagger$ creates a photon and the operator a annihilates a photon. The ground state $|0\rangle$ can be interpreted as the vacuum state with no photon and the eigenstate $|n\rangle$ will correspond to the state of the EM field with n photons. We next consider introducing a two-level atom into the system. This is an atom that consists of only two eigenstates: the ground state denoted by $|g\rangle$ and the excited state denoted by $|e\rangle$. In this Hilbert space of atomic states we can use the Pauli spin matrices for spin half as

$$\sigma^x = \begin{pmatrix} 0 & 1 \\ 1 & 0 \end{pmatrix}, \quad \sigma^y = \begin{pmatrix} 0 & -i \\ i & 0 \end{pmatrix}, \quad \sigma^z = \begin{pmatrix} 1 & 0 \\ 0 & -1 \end{pmatrix}, \tag{4.34}$$

and

$$\sigma^+ = \begin{pmatrix} 0 & 1 \\ 0 & 0 \end{pmatrix}, \quad \sigma^- = \begin{pmatrix} 0 & 0 \\ 1 & 0 \end{pmatrix}. \tag{4.35}$$

The operators σ^+ and σ^- are the raising and lowering operators of the atom with the properties that $\sigma^+ = |e\rangle \langle g|$ and $\sigma^- = |g\rangle \langle e|$, while σ^z is the atomic inversion operator given by

$$\sigma^z = |e\rangle \langle e| - |g\rangle \langle g|. \tag{4.36}$$

The Hamiltonian for the atom can then be specified by $\mathcal{H}_a = \frac{1}{2}\omega_a \sigma^z$, with ω_a representing the energy difference between the ground state $|g\rangle$ and the excited state $|e\rangle$.

There is also another term required in the total Hamiltonian of the system because of the interaction between the atom and the field. A possible interaction process is that one photon is absorbed, and thus the atom becomes excited from its ground state $|g\rangle$ to the state $|e\rangle$. The operator describing this process is $\sigma^+ a$. The other possibility is the reverse process with the emission of a photon while the atom drops back to the ground state. This latter process is described by the operator $\sigma^- a^\dagger$. Note that both of these processes conserve the energy. In general, we can include two other processes given by $\sigma^+ a^\dagger$ and $\sigma^- a$ in the Hamiltonian, which are not energy conserving but are allowed by QM. The interaction part of the Hamiltonian is then simply written as $\mathcal{H}_{af} = g(\sigma^+ + \sigma^-)(a + a^\dagger)$, where g is the coupling constant for the strength of the interaction. The complete JC Hamiltonian is then

$$\mathcal{H}_{JC} = \mathcal{H}_f + \mathcal{H}_a + \mathcal{H}_{af}$$

$$= \omega_f a^\dagger a + \frac{1}{2}\omega_a \sigma^z + g(\sigma^+ + \sigma^-)(a + a^\dagger). \tag{4.37}$$

To study the time dependence of the previously mentioned processes it is convenient to go to the interaction picture. According to the definition in Equation (2.14) we have

$$\mathcal{H}_{af}^{int} = e^{i\mathcal{H}_0 t}\mathcal{H}_{af}e^{-i\mathcal{H}_0 t}, \tag{4.38}$$

where $\mathcal{H}_0 = \mathcal{H}_f + \mathcal{H}_a$ is the unperturbed part. It is found that

$$\mathcal{H}_{af}^{int} = \exp\left\{i\left[\omega_f a^\dagger a + \frac{1}{2}\omega_a \sigma^z\right]t\right\}[g(\sigma^+ + \sigma^-)(a + a^\dagger)]$$

$$\times \exp\left\{-i\left[\omega_f a^\dagger a + \frac{1}{2}\omega_a \sigma^z\right]t\right\}.$$

Because the field operators $a^\dagger$ and a commute with the atom operators $\sigma^{\pm,z}$, we may rearrange the exponential operator terms and utilize the 2×2 matrix representation as

$$\mathcal{H}_{af}^{int} = \exp\left\{i\omega_f a^\dagger a t\right\}\exp\left\{i\frac{1}{2}\omega_a \sigma^z t\right\}g(\sigma^+ + \sigma^-)(a + a^\dagger)$$

$$\times \exp\left\{-i\omega_f a^\dagger a t\right\}\exp\left\{-i\frac{1}{2}\omega_a \sigma^z t\right\}$$

$$= g\exp\left\{i\omega_f a^\dagger a t\right\}(a + a^\dagger)\exp\left\{-i\omega_f a^\dagger a t\right\} \tag{4.39}$$

$$\times \exp\left[it\begin{pmatrix}\frac{1}{2}\omega_a & 0 \\ 0 & -\frac{1}{2}\omega_a\end{pmatrix}\right]\begin{pmatrix}0 & 1 \\ 1 & 0\end{pmatrix}\exp\left[-it\begin{pmatrix}\frac{1}{2}\omega_a & 0 \\ 0 & -\frac{1}{2}\omega_a\end{pmatrix}\right].$$

The result after some further simplification becomes (see Problem 4.5)

$$\mathcal{H}_{af}^{int} = g\left\{\sigma^+ a\exp\left[-i\left(\omega_f - \omega_a\right)t\right] + \sigma^- a^\dagger \exp\left[i\left(\omega_f - \omega_a\right)t\right]\right.$$

$$+ \sigma^- a\exp\left[-i\left(\omega_f + \omega_a\right)t\right] + \sigma^+ a^\dagger \exp\left[i\left(\omega_f + \omega_a\right)t\right]\right\}. \tag{4.40}$$

We may now introduce an approximation to the preceding expression. This is done on the physical basis that two of the terms included (those proportional to $\sigma^- a$ and $\sigma^+ a^\dagger$) are rapidly oscillating terms and thus their effect can be neglected compared with the other two more slowly evolving terms. This is known as the rotating wave approximation (RWA). We may revert to the Schrödinger picture after carrying out the approximation, and we obtain the total JC Hamiltonian under RWA as

$$\mathcal{H}_{JC} = \omega_f a^\dagger a + \frac{1}{2}\omega_a \sigma^z + g\left(\sigma^+ a + \sigma^- a^\dagger\right). \tag{4.41}$$

This completes our derivation of the JC model Hamiltonian.

A generalization of the JC model to systems consisting of N two-level atoms is known as the Dicke model or Tavis–Cummings model [68–70]. A schematic illustration is presented in Figure 4.1(b). It is the basic model for describing the collective interaction of light and matter. The Dicke model reduces to the JC model for the $N = 1$ case. In the Dicke model, N atoms interact cooperatively with a quantized single-mode of the optical field (with frequency ω_f), and the Dicke Hamiltonian in the RWA can be written as

$$\mathcal{H}_D = \omega_f a^\dagger a + \omega_a S^z + g\left(S^+ a + S^- a^\dagger\right). \tag{4.42}$$

Here ω_a is the frequency of any one of the atoms, while $S^{\pm,z}$ are collective atomic operators defined by

$$S^\pm = \sum_{i=1}^{N} \sigma_{(i)}^\pm \quad \text{and} \quad S^z = \frac{1}{2}\sum_{i=1}^{N} \sigma_{(i)}^z. \tag{4.43}$$

The new operators, by analogy with Equation (4.36), are

$$\sigma_{(i)}^+ = |e_i\rangle\langle g_i|, \quad \sigma_{(i)}^- = |g_i\rangle\langle e_i|, \quad \text{and} \quad \sigma_{(i)}^z = |e_i\rangle\langle e_i| - |g_i\rangle\langle g_i|, \tag{4.44}$$

with $|g_i\rangle$ and $|e_i\rangle$ being the ground and excited states, respectively, of the ith atom. The collective atomic operators satisfy the usual angular momentum commutation relations in QM [1, 2].

4.3.2 Green's Functions of the Jaynes–Cummings Model

For simplicity, we now employ the JC model with the Hamiltonian in Equation (4.41) to obtain the GFs. We take $A = \sigma^-$ and $B = \sigma^+$ in Equation (3.66) to evaluate real-time GFs of the type $G(\sigma^-; \sigma^+ \mid \omega)$, and also we choose $\varepsilon = 1$ so that we get commutation relations in the definitions of the GFs. The equation of motion of the preceding GF is found from

$$\omega G(\sigma^-; \sigma^+ \mid \omega) = \frac{1}{2\pi}\left\langle\left[\sigma^-, \sigma^+\right]_\varepsilon\right\rangle - G\left(\left[\mathcal{H}_{JC}, \sigma^-\right]; \sigma^+ \mid \omega\right). \tag{4.45}$$

For the commutators, it is easily verified that $[\sigma^+, \sigma^-] = \sigma^z$ and

$$[\mathcal{H}_{JC}, \sigma^-] = -\omega_a \sigma^- + g\sigma^z a, \tag{4.46}$$

so the GF equation of motion is

$$\omega G(\sigma^-; \sigma^+ \,|\, \omega) = -\frac{1}{2\pi}\langle \sigma^z \rangle + \omega_a G(\sigma^-; \sigma^+ \,|\, \omega)$$
$$- g\langle \sigma^z \rangle G(a; \sigma^+ \,|\, \omega). \tag{4.47}$$

The expectation value of the atomic inversion operator $\langle \sigma^z \rangle$ is -1 at zero temperature and we assume this to be the case throughout the rest of the calculation.

We see that Equation (4.47) contains another GF of the form $G(a; \sigma^+ \,|\, \omega)$. Hence, by making the choices $A = a$ and $B = \sigma^+$, the equation of motion of this other GF is

$$\omega G(a; \sigma^+ \,|\, \omega) = -G([\mathcal{H}_{J-C}, a] ; \sigma^+ \,|\, \omega)$$
$$= \omega_f G(a; \sigma^+ \,|\, \omega) + g G(\sigma^-; \sigma^+ \,|\, \omega).$$

To summarize, we find that the two GFs (in their analytically continued form) are related by

$$G(a; \sigma^+ \,|\, \tilde{\omega}) = \frac{g}{(\tilde{\omega} - \omega_f)} G(\sigma^-; \sigma^+ \,|\, \tilde{\omega}), \tag{4.48}$$

and after inserting this into Equation (4.47) we obtain the required GF as

$$G(\sigma^-; \sigma^+ \,|\, \tilde{\omega}) = \frac{1/2\pi}{\tilde{\omega} - \omega_a - [g^2/(\tilde{\omega} - \omega_f)]}. \tag{4.49}$$

This is consistent with a result obtained in [71].

By identifying the poles of the GF as the solutions for real ω at which the denominator of the GF vanishes, we get the excitation spectrum from the condition that

$$\omega - \omega_a - \frac{g^2}{\omega - \omega_f} = 0. \tag{4.50}$$

This produces two solutions $\omega = \omega^\pm$ corresponding to

$$\omega^\pm = \frac{\omega_f + \omega_a}{2} \pm \sqrt{(1/4)(\omega_f - \omega_a)^2 + g^2}$$
$$= \omega_a - \frac{\delta}{2} \pm \frac{1}{2}\sqrt{\delta^2 + 4g^2}. \tag{4.51}$$

In the second form of the preceding expression for $\omega^\pm$ the notation is that $\omega_R \equiv \sqrt{\delta^2 + 4g^2}$ is called the Rabi splitting, and $\delta \equiv (\omega_a - \omega_f)$ is a detuning parameter. The behavior of the modes when the frequencies are plotted versus the detuning

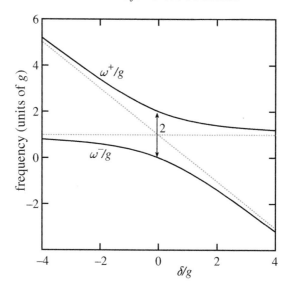

Figure 4.2 Illustration of the energy spectrum plotted versus δ/g for the Jaynes–Cummings (JC) model. When $\delta = 0$ the two branches are separated by an amount $\omega_R = 2g$ corresponding to the Rabi splitting ω_R at that value of δ.

factor is illustrated in Figure 4.2. The modes are seen to be strongly coupled (and split by an amount $\omega_R = 2g$) when $\delta = 0$, but they are relatively decoupled (and unperturbed) otherwise.

We conclude from Equations (4.49) and (4.51) that the corresponding retarded GF can be expressed as

$$G_r(\sigma^-;\sigma^+|\omega) = \frac{1}{2\pi} \frac{\left(\omega - \omega_f + i\eta\right)}{(\omega - \omega^+ + i\eta)(\omega - \omega^- + i\eta)}. \qquad (4.52)$$

To deduce other quantities, such as the spectral intensity $J(\omega)$, we apply the fluctuation-dissipation theorem, as quoted in Equation (3.40), to the preceding retarded GF. This is left to be done as Problem 4.6.

If the GFs and coupled modes are to be generalized to cases in which there are multiple two-level atoms, it is appropriate to use the Dicke model with the Hamiltonian given by Equation (4.42). This calculation is more complicated than in the JC case, but it is partially explored in Problem 4.7.

4.4 Dipole-Exchange Ferromagnet

In this section we consider a ferromagnet with an applied magnetic field along the direction of net magnetization. The Hamiltonian is taken to include both short-range exchange and long-range dipole-dipole interactions as in Section 2.6 (see also

[39, 72]). We recall the result established earlier that, after applying the Holstein–Primakoff (HP) transformation, the expression for the Hamiltonian is given in terms of boson operators by Equation (2.62).

To find the GFs, thereby supplementing our earlier results for the SW frequencies, we now choose $A = a_{\mathbf{k}}$ and $B = a_{\mathbf{k}}^{\dagger}$ in Equation (3.66). This will enable us to evaluate GFs of the type $G(a_{\mathbf{k}}; a_{\mathbf{k}}^{\dagger} \mid \omega)$. Also in this case we choose $\varepsilon = 1$ so that we have commutation relations in the definitions of the GFs. For the terms appearing on the right-hand side of Equation (3.66) we use $\langle [A, B]_{\varepsilon} \rangle = \langle [a_{\mathbf{k}}, a_{\mathbf{k}}^{\dagger}] \rangle = 1$ and

$$[\mathcal{H}, A] = \sum_{\mathbf{k}'} \left\{ P(\mathbf{k}')[a_{\mathbf{k}'}^{\dagger} a_{\mathbf{k}'}, a_{\mathbf{k}}] + Q(\mathbf{k}')[a_{\mathbf{k}'}^{\dagger} a_{-\mathbf{k}'}^{\dagger}, a_{\mathbf{k}}] + Q^{*}(\mathbf{k}')[a_{\mathbf{k}'} a_{-\mathbf{k}'}, a_{\mathbf{k}}] \right\}$$

$$= -\sum_{\mathbf{k}'} \left\{ P(\mathbf{k}') \delta_{\mathbf{k}, \mathbf{k}'} a_{\mathbf{k}'} + Q(\mathbf{k}') a_{\mathbf{k}'}^{\dagger} \delta_{\mathbf{k}, -\mathbf{k}'} + Q(\mathbf{k}') \delta_{\mathbf{k}, \mathbf{k}'} a_{-\mathbf{k}'}^{\dagger} \right\}$$

$$= -P(\mathbf{k}) a_{\mathbf{k}} - [Q(-\mathbf{k}) + Q(\mathbf{k})] a_{-\mathbf{k}}^{\dagger}.$$

Hence the GF equation of motion becomes

$$\omega G(a_{\mathbf{k}}; a_{\mathbf{k}}^{\dagger} \mid \omega) = \frac{1}{2\pi} + P(\mathbf{k}) G(a_{\mathbf{k}}; a_{\mathbf{k}}^{\dagger} \mid \omega)$$
$$+ [Q(-\mathbf{k}) + Q(\mathbf{k})] G(a_{-\mathbf{k}}^{\dagger}; a_{\mathbf{k}}^{\dagger} \mid \omega). \qquad (4.53)$$

There is a new GF on the right-hand side which in turn needs to be calculated. Then we choose $A = a_{-\mathbf{k}}^{\dagger}$ and $B = a_{\mathbf{k}}^{\dagger}$ in Equation (3.66) to evaluate the GF $G(a_{-\mathbf{k}}^{\dagger}; a_{\mathbf{k}}^{\dagger} \mid \omega)$. For the terms appearing on the right-hand side of the equation of motion for the GF we have $\langle [A, B]_{\varepsilon} \rangle = \langle [a_{-\mathbf{k}}^{\dagger}, a_{\mathbf{k}}^{\dagger}] \rangle = 0$ and

$$[\mathcal{H}, A] = \sum_{\mathbf{k}'} \left\{ P(\mathbf{k}')[a_{\mathbf{k}'}^{\dagger} a_{\mathbf{k}'}, a_{-\mathbf{k}}^{\dagger}] + Q(\mathbf{k}')[a_{\mathbf{k}'}^{\dagger} a_{-\mathbf{k}'}^{\dagger}, a_{-\mathbf{k}}^{\dagger}] + Q^{*}(\mathbf{k}')[a_{\mathbf{k}'} a_{-\mathbf{k}'}, a_{-\mathbf{k}}^{\dagger}] \right\}$$

$$= \sum_{\mathbf{k}'} \left\{ P(\mathbf{k}') a_{\mathbf{k}'}^{\dagger} \delta_{-\mathbf{k}, \mathbf{k}'} + Q^{*}(\mathbf{k}') a_{\mathbf{k}'} \delta_{\mathbf{k}, \mathbf{k}'} + Q^{*}(\mathbf{k}') \delta_{-\mathbf{k}, \mathbf{k}'} a_{-\mathbf{k}'} \right\}$$

$$= P(-\mathbf{k}) a_{-\mathbf{k}}^{\dagger} + [Q^{*}(\mathbf{k}) + Q^{*}(-\mathbf{k})] a_{\mathbf{k}}.$$

Therefore, the equation of motion in this case is

$$\omega G(a_{-\mathbf{k}}^{\dagger}; a_{\mathbf{k}}^{\dagger} \mid \omega) = -P(-\mathbf{k}) G(a_{-\mathbf{k}}^{\dagger}; a_{\mathbf{k}}^{\dagger} \mid \omega)$$
$$- [Q^{*}(\mathbf{k}) + Q^{*}(-\mathbf{k})] G(a_{\mathbf{k}}; a_{\mathbf{k}}^{\dagger} \mid \omega), \qquad (4.54)$$

and hence the corresponding analytically continued GFs are related by

$$G(a_{-\mathbf{k}}^{\dagger}; a_{\mathbf{k}}^{\dagger} \mid \tilde{\omega}) = -\frac{[Q^{*}(\mathbf{k}) + Q^{*}(-\mathbf{k})]}{(\tilde{\omega} + P(-\mathbf{k}))} G(a_{\mathbf{k}}; a_{\mathbf{k}}^{\dagger} \mid \tilde{\omega}). \qquad (4.55)$$

Substituting Equation (4.55) into Equation (4.53) leads to

$$G(a_{\mathbf{k}}; a_{\mathbf{k}}^{\dagger} \mid \tilde{\omega}) = \frac{1/2\pi}{\tilde{\omega} - P(\mathbf{k}) + \{|2Q(\mathbf{k})|^2 / [\tilde{\omega} + P(\mathbf{k})]\}}. \tag{4.56}$$

where we assumed $P(\mathbf{k}) = P(-\mathbf{k})$ and also $Q(\mathbf{k}) = Q(-\mathbf{k})$. To find the excitation energy at real ω, we consider the condition for the denominator of the preceding equation to vanish. This occurs when

$$\omega - P(\mathbf{k}) + \frac{|2Q(\mathbf{k})|^2}{\omega + P(\mathbf{k})} = 0,$$

giving the quadratic equation $\omega^2 - (P(\mathbf{k}))^2 + 4|Q(\mathbf{k})|^2 = 0$. This has the solutions $\omega = \pm E_{\mathbf{k}}$, where $E_{\mathbf{k}}$ represents the dispersion relation for the dipole-exchange SW modes, and is the same as obtained in Equation (2.66) using the operator equation-of-motion method. The result in Equation (4.56) can be rewritten in a more convenient form for the corresponding retarded GF as

$$G_r(a_{\mathbf{k}}; a_{\mathbf{k}}^{\dagger} \mid \omega) = \left(\frac{1}{2\pi}\right) \frac{\omega + P(\mathbf{k}) + i\eta}{(\omega + i\eta)^2 - E_{\mathbf{k}}^2}. \tag{4.57}$$

In the limiting case of the Heisenberg ferromagnet (when the dipolar terms are put equal to zero) this GF result simplifies to

$$G_r(a_{\mathbf{k}}; a_{\mathbf{k}}^{\dagger} \mid \omega) = \left(\frac{1}{2\pi}\right) \frac{1}{\omega - E_{\mathbf{k}} + i\eta}, \tag{4.58}$$

where $E_{\mathbf{k}} = P(\mathbf{k})$ now represents the SW energy and it is defined by Equation (1.93). Also, in this limit, it can be seen that $G_r(a_{-\mathbf{k}}^{\dagger}; a_{\mathbf{k}}^{\dagger} \mid \omega) = 0$, which is related to the property that the spin precession has become circular (rather than elliptical).

4.5 Paramagnet with Crystal-Field Anisotropy

In paramagnets the magnetic interactions between different spin sites, which typically are due to exchange and/or magnetic dipole-dipole effects, are negligibly small. Consequently, there is no spontaneous magnetization in zero applied magnetic field, as there is in a ferromagnet below T_C.

The spin Hamiltonian for a paramagnet typically consists of the Zeeman term due to an applied magnetic field together with (in some cases) effects of the magnetic anisotropy at any atomic spin site in the crystal lattice. An example is the single-ion crystal-field anisotropy [26, 73], which arises from a coupling between the atomic spin-orbit interaction and the crystalline electric fields. We assume here that the spin Hamiltonian has the form

$$\mathcal{H} = -g\mu_B B_0 \sum_i S_i^z - D_0 \sum_i (S_i^z)^2, \tag{4.59}$$

where B_0 is the applied magnetic field in the z direction, as in Equation (1.81), and D_0 is a coefficient for the single-ion uniaxial anisotropy. The latter term is typical for many noncubic crystals.

Suppose we now consider evaluating the GF $g(S_n^+; S_m^- | t - t')$, where n and m are site labels, or equivalently the Fourier-transformed GF denoted by $G(S_n^+; S_m^- | \omega)$. The choice of taking $\varepsilon = 1$ is convenient, because the spins satisfy commutation relations. For evaluating the terms in the GF equation of motion (3.66) we require $\langle [S_n^+, S_m^-] \rangle = 2 \langle S_n^z \rangle \delta_{n,m}$ and

$$\left[\mathcal{H}, S_n^+ \right] = -g\mu_B B_0 \sum_i \left[S_i^z, S_n^+ \right] - D_0 \sum_i \left[(S_i^z)^2, S_n^+ \right]$$

$$= -g\mu_B B_0 S_n^+ - D_0 (S_n^z S_n^+ + S_n^+ S_n^z).$$

The second line of the preceding equation is deduced using the spin commutation results stated in Equation (1.84). The required GF equation of motion becomes after rearranging the terms

$$(\omega - g\mu_B B_0) G(S_n^+; S_m^- | \omega) = \frac{1}{\pi} \langle S_n^z \rangle \delta_{n,m} + D_0 G(\{S_n^z S_n^+ + S_n^+ S_n^z\}; S_m^- | \omega). \tag{4.60}$$

We see here that there is a new GF on the right-hand side that is not related in any obvious way to the GF that we seek to evaluate. The general procedure (in the absence of making an approximation) is to form the equation of motion for this new GF. Using the spin commutation relations again, the new GF equation is found to be (see Problem 4.8)

$$(\omega - g\mu_B B_0) G(\{S_n^z S_n^+ + S_n^+ S_n^z\}; S_m^- | \omega)$$

$$= \frac{1}{\pi} \left\{ 3\langle (S_n^z)^2 \rangle - S(S+1) \right\} \delta_{n,m}$$

$$+ D_0 G \left(\{(S_n^z)^2 S_n^+ + 2 S_n^z S_n^+ S_n^z + S_n^+ (S_n^z)^2 \}; S_m^- | \omega \right). \tag{4.61}$$

This has given rise to another (more complicated) GF, so the situation is not encouraging.

It is possible, nevertheless, to proceed further in the calculation if we assume a specific value for the spin quantum number S, as we will illustrate below for the simplest cases of $S = 1/2$ and $S = 1$.

4.5.1 Spin S = 1/2

In this case, we may conveniently employ the representation in terms of the spin-half Pauli matrices, which were quoted in Equation (4.35), taking $S_n^+ \to \frac{1}{2}(\sigma^x + i\sigma^y)$, $S_n^- \to \frac{1}{2}(\sigma^x - i\sigma^y)$, and $S_n^z \to \frac{1}{2}\sigma^z$. Then, on replacing the spin operators by their representation as 2×2 matrices, we have

$$(S_n^z S_n^+ + S_n^+ S_n^z) \to \frac{1}{4}\begin{pmatrix} 1 & 0 \\ 0 & -1 \end{pmatrix}\begin{pmatrix} 0 & 1 \\ 0 & 0 \end{pmatrix} + \frac{1}{4}\begin{pmatrix} 0 & 1 \\ 0 & 0 \end{pmatrix}\begin{pmatrix} 1 & 0 \\ 0 & -1 \end{pmatrix}.$$

When the right-hand side of the preceding equation is simplified using the matrix multiplication properties, it is seen that it reduces to zero. Consequently, the GF $G(\{S_n^z S_n^+ + S_n^+ S_n^z\}; S_m^- \,|\, \omega)$ in Equation (4.60) vanishes, enabling us to obtain the required $G(S_n^+; S_m^- \,|\, \omega)$ directly. The next steps then become similar to the earlier GF calculation for the noninteracting bosons. In particular, for the analytically continued GF (in the complex $\tilde{\omega}$ plane) we have the result

$$G(S_n^+; S_m^- \,|\, \tilde{\omega}) = \frac{\langle S_n^z \rangle}{\pi\,(\tilde{\omega} - g\mu_B B_0)}\delta_{nm}. \tag{4.62}$$

There is a simple pole in the GF equal to $g\mu_B B_0$, which is just the spacing between the two quantized energy levels of the paramagnet (noting that the eigenvalues of S^z are $\pm 1/2$ in units such that $\hbar = 1$). The fact that the term proportional to D_0 in the Hamiltonian has no effect is to be expected because the eigenvalues of $(S^z)^2$ have the same value of $1/4$ for both of the spin-up and spin-down quantized levels.

From Equation (4.62) it is then straightforward to find the spectral intensity and the time correlation functions.

4.5.2 Spin S = 1

In the next case of $S = 1$ the spin operators can be represented in terms of 3×3 matrices (just as for the QM angular momentum operators), where here we need only

$$S_n^+ \to \sqrt{2}\begin{pmatrix} 0 & 1 & 0 \\ 0 & 0 & 1 \\ 0 & 0 & 0 \end{pmatrix}, \quad S_n^z \to \begin{pmatrix} 1 & 0 & 0 \\ 0 & 0 & 0 \\ 0 & 0 & -1 \end{pmatrix}. \tag{4.63}$$

We may use these relationships in the GF Equations (4.60) and (4.61). First we find in this case that the spin combination $\{S_n^z S_n^+ + S_n^+ S_n^z\}$ is non-vanishing, so the term proportional to the anisotropy coefficient D_0 does not disappear from the GF

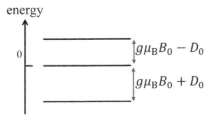

Figure 4.3 The unequal spacing between the energy eigenvalues of a $S = 1$ paramagnet with crystal-field anisotropy as found from Equation (4.59).

equations. Also it follows that there is a simplification for the products of matrices given by

$$((S_n^z)^2 S_n^+ + 2S_n^z S_n^+ S_n^z + S_n^+ (S_n^z)^2) \rightarrow \sqrt{2} \begin{pmatrix} 0 & 1 & 0 \\ 0 & 0 & 1 \\ 0 & 0 & 0 \end{pmatrix} \rightarrow S_n^+. \qquad (4.64)$$

Therefore, the extra GF Equation (4.61) reduces to

$$(\omega - g\mu_B B_0) G(\{S_n^z S_n^+ + S_n^+ S_n^z\}; S_m^- \mid \omega) = \frac{1}{\pi} \left\{ 3\langle (S_n^z)^2 \rangle - 2 \right\} \delta_{n,m}$$

$$+ D_0 G(S_n^+; S_m^- \mid \omega). \qquad (4.65)$$

We now see that Equations (4.62) and (4.65) give us a pair of coupled equations that can be solved by eliminating $G(\{S_n^z S_n^+ + S_n^+ S_n^z\}; S_m^- \mid \omega)$ to find the original GF $G(S_n^+; S_m^- \mid \omega)$. After some straightforward algebra the result for the analytically continued GF is found to be

$$G(S_n^+; S_m^- \mid \tilde{\omega}) = \frac{\langle S_n^z \rangle (\tilde{\omega} - g\mu_B B_0) + \left\{ 3\langle (S_n^z)^2 \rangle - 2 \right\} D_0}{\pi (\tilde{\omega} - g\mu_B B_0 - D_0)(\tilde{\omega} - g\mu_B B_0 + D_0)} \delta_{n,m}. \qquad (4.66)$$

This GF now has two distinct poles, occurring at the values $g\mu_B B_0 \pm D_0$. These just correspond to the spacing between the three energy eigenvalues of the Hamiltonian when $S = 1$ (see Figure 4.3). Again it is a straightforward matter to use the fluctuation-dissipation theorem to deduce the spectral weights and the correlation functions (see Problem 4.9).

The method that we have described here for the two lowest spin quantum numbers can be generalized to higher values of S. This necessitates having an increasing number (in fact, $2S$) of coupled GF equations to achieve a closed set. It follows that the approach is practical only for relatively low spin values.

Problems

4.1. Verify in the case of causal GFs, that the values quoted for the constant f are as given in Equations (4.8) and (4.14) for noninteracting boson and fermion systems, respectively.

4.2. Work through the intermediate steps required for the derivation of the Matsubara GF $\mathcal{G}(i\omega_m)$ in Equation (4.21) for noninteracting bosons and fermions, starting from the result for $g_M(\tau)$ in Equation (4.20).

4.3. Consider N weakly interacting boson particles with most of the particles condensed into the $\mathbf{k} = 0$ state. Using the approximate form of the Hamiltonian $\mathcal{H}_R$ as obtained in Equation (1.70), derive the equation of motion for the GF $G(a_{\mathbf{k}}; a_{\mathbf{k}}^{\dagger} \mid \omega)$ in the frequency representation (choose $\varepsilon = 1$). Show that it is necessary to write down *another* GF equation of motion to obtain a closed set of equations. Do this and hence solve for the previously mentioned GF in the frequency representation. Now use the fluctuation-dissipation theorem to obtain the corresponding spectral function. Also deduce an expression for the thermal average $\langle a_{\mathbf{k}}^{\dagger} a_{\mathbf{k}} \rangle$.

4.4. According to Equation (2.50), the quantized longitudinal vibrations of a monatomic lattice of atoms can be described in terms of non-interacting phonons with the Hamiltonian

$$\mathcal{H} = \sum_k \omega_k (a_k^{\dagger} a_k + 1/2),$$

where $a_k^{\dagger}$ and a_k are the phonon creation and annihilation operators at wavenumber k. The frequency ω_k is given by Equation (2.46). Derive the retarded GF $G_r(a_k; a_k^{\dagger} \mid \omega)$, and then use the result to deduce another retarded GF corresponding to $G_r(\{a_k - a_{-k}^{\dagger}\}; \{a_k^{\dagger} - a_{-k}\} \mid \omega)$. This combination commonly occurs in electron–phonon problems.

4.5. Verify the form taken by the interaction Hamiltonian for the Jaynes–Cummings (JC) model, as quoted in Equation (4.40), starting from Equation (4.39).

4.6. Use the JC Hamiltonian in Equation (4.41) and the fluctuation-dissipation theorem to obtain the corresponding spectral function for the GF in Equation (4.49).

4.7. The interaction of N two-level atoms with a single quantized mode of the optical field can be described by the Dicke Hamiltonian in Equation (4.42). Follow the same approach as in Subsection 4.3.2 to obtain the equation of motion for the GF $G(S^-; S^+ \mid \omega)$ using the Dicke Hamiltonian. Show that the right-hand side of the equation of motion contains a new GF (but do not solve for it).

4.8. Assuming a paramagnetic system described by the spin Hamiltonian in Equation (4.59), the equation of motion for the GF $G(S_n^+; S_m^- | \omega)$ is given by Equation (4.60). As discussed, it involves *another* GF that can be written in the form

$$G(\{S_n^z S_n^+ + S_n^+ S_n^z\}; S_m^- | \omega).$$

Obtain the equation of motion for the preceding GF, verifying that it can be expressed as in Equation (4.61) for general value of the spin quantum number S.

4.9. The analytically continued GF for a paramagnet with spin quantum number $S = 1$ is given by Equation (4.66). Use this result to deduce the corresponding spectral function. Obtain results for the behavior of the poles and their contributions to the spectral function when the limit is taken of $D_0 \to 0$. Next discuss the behavior of these quantities when $D_0 \neq 0$ and the limit is taken of zero temperature (assuming $\langle S_n^z \rangle$ and $\langle (S_n^z)^2 \rangle$ both tend to 1 in this case).

4.10. A model Hamiltonian was quoted in Problem 1.13 for the coupling between SWs and phonons. Now, instead of using a diagonalizing transformation as before, derive the GF equations of motion for $G(a_{\mathbf{k}}; a_{\mathbf{k}}^\dagger | \omega)$ and $G(b_{\mathbf{k}}; b_{\mathbf{k}}^\dagger | \omega)$. Show that these equations are coupled to other GFs and solve for the preceding two GFs in the retarded case. Next write down the explicit form taken by these results for the case of an acoustic phonon (with $\Omega_{\mathbf{k}} = v|\mathbf{k}|$ where v is the sound velocity) and a magnon (approximated by $\omega_{\mathbf{k}} = \omega_0$, independent of $\mathbf{k}$). For the interaction term you may assume that $c_{\mathbf{k}} = c_0$, independent of $\mathbf{k}$, and also that $c_0 \ll \omega_0$.

4.11. For the previous Problem 4.10 in the absence of an interaction term in the Hamiltonian, there would be a crossover of the dispersion curves for $\omega_{\mathbf{k}}$ and $\Omega_{\mathbf{k}}$ (plotted versus $|\mathbf{k}|$) at $|\mathbf{k}| = \omega_0/v$. What are the solutions for the quasiparticle energies in the vicinity of the crossover when $c_0 \neq 0$? Make rough sketches of the previously mentioned quantities (versus $|\mathbf{k}|$) going from $|\mathbf{k}| = 0$ up to large $|\mathbf{k}|$. Show that with the previous assumptions there is a stability condition requiring $|\mathbf{k}| > c_0^2/\omega_0 v$ for the lower branch.

4.12. When Cr^{3+} ions in a low concentration are embedded in a suitable host crystalline lattice, they may behave as a paramagnet with the Hamiltonian given by Equation (4.59) and spin quantum number $S = 3/2$. Now extend the GF calculation for $G_1 \equiv G(S_n^+; S_m^- | \tilde{\omega})$ given in Subsection 4.5.2 for $S = 1$ to apply to the Cr^{3+} case with $S = 3/2$. You should expect that *three* coupled GF equations will be required. These equations are for G_1, and also for

$$G_2 = G(\{S_n^z S_n^+ + S_n^+ S_n^z\}; S_m^- | \omega) \quad \text{and}$$
$$G_3 = G(\{(S^z)^2 S^+ + S^z S^+ S^z + S^+ (S^z)^2\}; S^- | \omega).$$

For $S = 3/2$ you will need to introduce a 4×4 matrix representation for the spin operators (as in standard QM text books for the angular momentum states). Use these results to simplify the new GF generated in the third equation of motion, showing that it reduces to a multiple of G_2.

4.13. The particle-hole operators $\rho_q^\dagger(k)$ and $\rho_q^\dagger$ were defined in Section 2.8 for an electron gas. It was shown in Equation (2.79) that the spectrum for density fluctuations for a *noninteracting* gas corresponds simply to $\omega_0(k, q) = E_{k+q} - E_k$, where $E_k = (k^2/2m) - \mu$ in the notation of Section 4.1. Now start by considering the frequency-dependent GF $G(\rho_q^\dagger(k); \rho_q | \omega)$. Obtain its equation of motion for the noninteracting case using Equation (3.66) with $\varepsilon = 1$, and show (by summing over the wave vector k) that the result for the retarded GF $G(\rho_q^\dagger; \rho_q | \omega)$ is

$$G(\rho_q^\dagger; \rho_q | \omega) = \frac{1}{2\pi} \sum_k \frac{n_{k+q}^0 - n_k^0}{E_{k+q} - E_k + \omega + i\eta},$$

where the notation follows Section 2.8. The summation term on right-hand side of the preceding equation is known as the Lindhard function.

5

Green's Functions Using Decoupling Methods

In this chapter we consider some further examples that make use of the real-time Green's functions (GFs) and their equations of motion. These cases will generally involve applications to theoretical models with a greater complexity or with more significant interaction schemes between the particles or spins than was the case in Chapter 4, so that usually an exact calculation is no longer possible. Typically, this is because the coupled GF equations do not form a closed set, and new GFs are generated in each successive equation of motion. Therefore, some form of approximation becomes necessary to obtain a solution.

In many of the examples the approximations may involve introducing a so-called decoupling to simplify products between operators, leaving a finite set of GF equations. These approximations may be physically reasonable, but difficult to justify in advance. The approach will contrast with the rigorous perturbation expansion techniques, which will be considered in Chapters 8 and 9 using the imaginary-time GFs.

5.1 Hartree–Fock Theory for an Interacting Fermion Gas

Here we seek to generalize the GF calculation given in Section 4.1, which included the case of a noninteracting gas of fermions or bosons. Now we will include the effects of the quartic interaction terms, which are assumed to be sufficiently weak. This calculation will be done in the first instance for fermion systems, for which the approximations to be made are more appropriate and are analogous to Hartree–Fock theory for the electron states in atoms. Afterward we will briefly consider what happens if the same approach is applied to the boson case.

We now employ the full form of the second-quantized Hamiltonian as derived earlier in Subsection 1.4.2, where

$$\overline{\mathcal{H}} = \sum_{\mathbf{k}} E_{\mathbf{k}} a_{\mathbf{k}}^{\dagger} a_{\mathbf{k}} + \frac{1}{2} \sum_{\mathbf{k}_1, \mathbf{k}_2, \mathbf{q}} v(\mathbf{q}) a_{\mathbf{k}_1}^{\dagger} a_{\mathbf{k}_2}^{\dagger} a_{\mathbf{k}_2 + \mathbf{q}} a_{\mathbf{k}_1 - \mathbf{q}} \tag{5.1}$$

with $E_{\mathbf{k}} = (k^2/2m) - \mu$. At this stage, we ignore for simplicity the effects due to spin labels. As in the noninteracting case for fermions in Subsection 4.1.2, we evaluate real-time GFs of the type $G(a_{\mathbf{k}}; a_{\mathbf{k}}^\dagger \,|\, \omega)$ in the frequency representation. We employ the general GF equation of motion as in Equation (3.66), and we choose $\varepsilon = -1$ so that we get anticommutation relations in the definitions of the GFs.

In evaluating the terms on the right-hand side of Equation (3.66) we see that $\langle\{a_{\mathbf{k}}, a_{\mathbf{k}}^\dagger\}\rangle = 1$ and also

$$[\overline{\mathcal{H}}, a_{\mathbf{k}}] = \sum_{\mathbf{k}'} E_{\mathbf{k}'}[a_{\mathbf{k}'}^\dagger a_{\mathbf{k}'}, a_{\mathbf{k}}] + \frac{1}{2}\sum_{\mathbf{k}_1, \mathbf{k}_2, \mathbf{q}} v(\mathbf{q})[a_{\mathbf{k}_1}^\dagger a_{\mathbf{k}_2}^\dagger a_{\mathbf{k}_2+\mathbf{q}} a_{\mathbf{k}_1-\mathbf{q}}, a_{\mathbf{k}}]$$

$$= -E_{\mathbf{k}} a_{\mathbf{k}} - \sum_{\mathbf{k}_1, \mathbf{q}} v(\mathbf{q}) a_{\mathbf{k}_1}^\dagger a_{\mathbf{k}_1+\mathbf{q}} a_{\mathbf{k}-\mathbf{q}}. \tag{5.2}$$

Hence the GF equation of motion becomes

$$\omega G(a_{\mathbf{k}}; a_{\mathbf{k}}^\dagger \,|\, \omega) = \frac{1}{2\pi} + E_{\mathbf{k}} G(a_{\mathbf{k}}; a_{\mathbf{k}}^\dagger \,|\, \omega)$$

$$+ \sum_{\mathbf{k}_1, \mathbf{q}} v(\mathbf{q}) G(a_{\mathbf{k}_1}^\dagger a_{\mathbf{k}_1+\mathbf{q}} a_{\mathbf{k}-\mathbf{q}}; a_{\mathbf{k}}^\dagger \,|\, \omega), \tag{5.3}$$

which generalizes the result in Equation (4.3) where it was assumed that $v(\mathbf{q}) = 0$. Next, we could, in turn, proceed by writing down the equation of motion for the new GF on the right-hand side of Equation (5.3). We would find, however, that this gives us a more complicated GF, and so on in a seemingly unending chain.

To find an approximate solution our strategy is to make a decoupling approximation in the series of equations. The simplest such procedure is to look for a suitable decoupling for the GF on the right-hand side of Equation (5.3), i.e., we seek a way to approximate the GF $G(a_{\mathbf{k}_1}^\dagger a_{\mathbf{k}_1+\mathbf{q}} a_{\mathbf{k}-\mathbf{q}}; a_{\mathbf{k}}^\dagger \,|\, \omega)$. With this in mind, we focus on the product of the three operators appearing on the left of the GF, specifically $a_{\mathbf{k}_1}^\dagger a_{\mathbf{k}_1+\mathbf{q}} a_{\mathbf{k}-\mathbf{q}}$.

Before proposing a specific decoupling for this case, it is worthwhile to digress briefly regarding decoupling schemes in general. As a simple example, suppose we are considering a decoupling of a product of just two operators in a product AB on the left of the GF $G(AB; C \,|\, \omega)$. We note that the product can be formally reexpressed as

$$AB = A\langle B\rangle + \langle A\rangle B + (A - \langle A\rangle)(B - \langle B\rangle) - \langle A\rangle\langle B\rangle.$$

Now, provided that at least one of the thermal averages $\langle A\rangle$ or $\langle B\rangle$ is nonzero, it seems reasonable in the spirit of mean-field theory [74] to neglect the fluctuation term $(A - \langle A\rangle)(B - \langle B\rangle)$ in the preceding expression because it is of second order in small quantities. The last term in the preceding equation is just a constant, which will not give rise to a GF term. Hence we arrive at a GF decoupling approximation of the form

$$G(AB; C \mid \omega) \rightarrow \langle A \rangle \, G(B; C \mid \omega) + \langle B \rangle \, G(A; C \mid \omega). \tag{5.4}$$

The preceding mean-field argument for systems where the fluctuations are relatively small will generalize in an obvious manner to products of three or more operators.

Returning now to the problem at hand for the electron gas, a physically appealing approximation is to write initially

$$a_{\mathbf{k}_1}^{\dagger} a_{\mathbf{k}_1+\mathbf{q}} a_{\mathbf{k}-\mathbf{q}} \approx \langle a_{\mathbf{k}_1}^{\dagger} a_{\mathbf{k}_1+\mathbf{q}} \rangle a_{\mathbf{k}-\mathbf{q}} - \langle a_{\mathbf{k}_1}^{\dagger} a_{\mathbf{k}-\mathbf{q}} \rangle a_{\mathbf{k}_1+\mathbf{q}}$$
$$+ a_{\mathbf{k}_1}^{\dagger} \langle a_{\mathbf{k}_1+\mathbf{q}} a_{\mathbf{k}-\mathbf{q}} \rangle. \tag{5.5}$$

Here we are considering each operator in turn and replacing the product of the other two operators by its average value. There are three terms in the preceding expression because there are three ways of doing this. This is following the same arguments as in the previous example leading to Equation (5.4). Also, we remark that the negative sign in one term is due to the anticommutation property of the fermion operators (i.e., we have changed the order of the operators being averaged). A rough justification for the decoupling in this case is that we neglect the fluctuation of $a_{\mathbf{k}_1}^{\dagger} a_{\mathbf{k}_1+\mathbf{q}}$ about its average value $\langle a_{\mathbf{k}_1}^{\dagger} a_{\mathbf{k}_1+\mathbf{q}} \rangle$ for the first term, along with a similar argument for the other terms.

Some further simplification of Equation (5.5) is now possible due to the symmetries occuring in the Hamiltonian $\mathcal{H}$ of the system and the fact that the definition of the equilibrium thermal average depends on $\overline{\mathcal{H}}$ (see Chapter 3). The symmetry properties inherent in each term of $\overline{\mathcal{H}}$ are conservation of the wave vector and conservation of the number of particles, from which it follows that these must be conserved quantities in any equilibrium thermal average. Hence, without any more approximation for the terms in Equation (5.5), these considerations lead us to write

$$\langle a_{\mathbf{k}_1}^{\dagger} a_{\mathbf{k}_1+\mathbf{q}} \rangle = \langle a_{\mathbf{k}_1}^{\dagger} a_{\mathbf{k}_1} \rangle \delta_{\mathbf{q},0} \quad \text{and} \quad \langle a_{\mathbf{k}_1}^{\dagger} a_{\mathbf{k}-\mathbf{q}} \rangle = \langle a_{\mathbf{k}_1}^{\dagger} a_{\mathbf{k}_1} \rangle \delta_{\mathbf{k}_1,\mathbf{k}-\mathbf{q}}$$

because of conservation of wave vector, and

$$\langle a_{\mathbf{k}_1+\mathbf{q}} a_{\mathbf{k}-\mathbf{q}} \rangle = 0$$

because of conservation of the number of particles. The decoupling approximation for the product of the three operators then becomes

$$a_{\mathbf{k}_1}^{\dagger} a_{\mathbf{k}_1+\mathbf{q}} a_{\mathbf{k}-\mathbf{q}} \approx \delta_{\mathbf{q},0} \langle a_{\mathbf{k}_1}^{\dagger} a_{\mathbf{k}_1} \rangle a_{\mathbf{k}} - \delta_{\mathbf{k}_1,\mathbf{k}-\mathbf{q}} \langle a_{\mathbf{k}_1}^{\dagger} a_{\mathbf{k}_1} \rangle a_{\mathbf{k}}, \tag{5.6}$$

or in terms of the GF it gives the simplification

$$G(a_{\mathbf{k}_1}^{\dagger} a_{\mathbf{k}_1+\mathbf{q}} a_{\mathbf{k}-\mathbf{q}}; a_{\mathbf{k}}^{\dagger} \mid \omega) \approx (\delta_{\mathbf{q},0} - \delta_{\mathbf{k}_1,\mathbf{k}-\mathbf{q}}) \langle a_{\mathbf{k}_1}^{\dagger} a_{\mathbf{k}_1} \rangle \, G(a_{\mathbf{k}}; a_{\mathbf{k}}^{\dagger} \mid \omega). \tag{5.7}$$

When this decoupling approximation is substituted back into Equation (5.3) we have an implicit expression for the required GF as

$$
\omega G(a_{\mathbf{k}}; a_{\mathbf{k}}^{\dagger} \,|\, \omega) = \frac{1}{2\pi} + E_{\mathbf{k}}\, G(a_{\mathbf{k}}; a_{\mathbf{k}}^{\dagger} \,|\, \omega)
$$
$$
+ \sum_{\mathbf{k}_1} \{v(0) - v(\mathbf{k} - \mathbf{k}_1)\} \langle a_{\mathbf{k}_1}^{\dagger} a_{\mathbf{k}_1} \rangle\, G(a_{\mathbf{k}}; a_{\mathbf{k}}^{\dagger} \,|\, \omega). \qquad (5.8)
$$

This can be rearranged as

$$
(\omega - W_{\mathbf{k}})\, G(a_{\mathbf{k}}; a_{\mathbf{k}}^{\dagger} \,|\, \omega) = \frac{1}{2\pi}, \qquad (5.9)
$$

where

$$
W_{\mathbf{k}} = E_{\mathbf{k}} + \sum_{\mathbf{k}_1} \{v(0) - v(\mathbf{k} - \mathbf{k}_1)\} \langle a_{\mathbf{k}_1}^{\dagger} a_{\mathbf{k}_1} \rangle. \qquad (5.10)
$$

Equation (5.9) is formally the same result as in the noninteracting case, except that $E_{\mathbf{k}}$ has been replaced by $W_{\mathbf{k}}$. Therefore, all the subsequent results that were obtained previously in the noninteracting case (such as for the different real-time GFs, the spectral function, and the correlation functions) still apply, but with the replacement throughout that $E_{\mathbf{k}} \to W_{\mathbf{k}}$. In summary, instead of having particles with energy $E_{\mathbf{k}}$ as in the noninteracting case, we now have quasiparticles with a modified energy $W_{\mathbf{k}}$.

The thermal average appearing on the right-hand side of Equation (5.10) represents the number of quasiparticles with wave vector $\mathbf{k}_1$ and is given by

$$
\langle a_{\mathbf{k}_1}^{\dagger} a_{\mathbf{k}_1} \rangle = \frac{1}{\exp(\beta W_{\mathbf{k}_1}) + 1} \qquad (5.11)
$$

by analogy with Equation (4.17). From Equations (5.10) and (5.11) it is clear that we have two relationships from which, in principle, a self-consistent solution for $W_{\mathbf{k}}$ can be obtained. They provide us with the *self-consistent* form of the *Hartree–Fock equations*.

In some cases, a further approximation may be appropriate. For example, if the interaction term v in the Hamiltonian is sufficiently small, we may argue that $E_{\mathbf{k}}$ and $W_{\mathbf{k}}$ are not too different, allowing us to replace $W_{\mathbf{k}}$ with $E_{\mathbf{k}}$ in Equation (5.11), i.e., we will just use the same distribution function as for the noninteracting particles. Then from Equation (5.10) we have approximately

$$
W_{\mathbf{k}} = E_{\mathbf{k}} + \sum_{\mathbf{k}_1} \frac{\{v(0) - v(\mathbf{k} - \mathbf{k}_1)\}}{\exp(\beta E_{\mathbf{k}_1}) + 1}. \qquad (5.12)
$$

The GF decoupling approximation made earlier in Equation (5.7) is difficult to justify a priori by rigorous means, although we have presented general reasons for it. Typically, the suitability of any decoupling approximation is supported by general physical arguments, and the "justification" for the approximation comes afterward in terms of the success for the predictions made. Broadly, the preceding Hartree–Fock (or HF) theory is fairly good as an approximation provided the relative correction to the particle energy is small, meaning

$$\left| \sum_{\mathbf{k}_1} \frac{\{v(0) - v(\mathbf{k} - \mathbf{k}_1)\}}{\exp(\beta E_{\mathbf{k}_1}) + 1} \right| \ll \frac{k^2}{2m}. \tag{5.13}$$

In conclusion, it is worth mentioning that if we want to improve on the HF theory, we could in principle defer the decoupling approximation to a later stage. Specifically we could write down an equation of motion for the new GF $G(a_{\mathbf{k}_1}^{\dagger} a_{\mathbf{k}_1+\mathbf{q}} a_{\mathbf{k}-\mathbf{q}}; a_{\mathbf{k}}^{\dagger} | \omega)$ in Equation (5.3) and then employ a decoupling approximation in *its* equation of motion.

Application of the Theory to Interacting Bosons

Suppose now we try applying the HF-type theory to a weakly interacting boson system. We may start with the same form of Hamiltonian and again evaluate GFs of the type $G(a_{\mathbf{k}}; a_{\mathbf{k}}^{\dagger} | \omega)$, but we choose $\varepsilon = 1$ consistent with the commutation properties of bosons. After going through the various steps as described in the preceding text, we find that the GF equation of motion is identical to Equation (5.3). However, the analogous decoupling approximation in this case becomes

$$G(a_{\mathbf{k}_1}^{\dagger} a_{\mathbf{k}_1+\mathbf{q}} a_{\mathbf{k}-\mathbf{q}}; a_{\mathbf{k}}^{\dagger} | \omega) \approx (\delta_{\mathbf{q},0} + \delta_{\mathbf{k}_1,\mathbf{k}-\mathbf{q}}) \langle a_{\mathbf{k}_1}^{\dagger} a_{\mathbf{k}_1} \rangle \, G(a_{\mathbf{k}}; a_{\mathbf{k}}^{\dagger} | \omega).$$

In particular, there is a difference in sign on the right-hand side compared with Equation (5.7) because bosons satisfy commutation relations. Hence the expression for the quasiparticle energy becomes modified to

$$W_{\mathbf{k}} = E_{\mathbf{k}} + \sum_{\mathbf{k}_1} \{v(0) + v(\mathbf{k} - \mathbf{k}_1)\} \langle a_{\mathbf{k}_1}^{\dagger} a_{\mathbf{k}_1} \rangle, \tag{5.14}$$

where the number of boson quasiparticles is now given self-consistently by

$$\langle a_{\mathbf{k}_1}^{\dagger} a_{\mathbf{k}_1} \rangle = \frac{1}{\exp(\beta W_{\mathbf{k}_1}) - 1}.$$

We conclude that this form of decoupling theory for bosons would be useful only under much more restrictive conditions than for a gas of fermions. It may be valid only provided the temperature is sufficiently high that there are very few bosons in the zero-wave-vector (zero-energy) ground state. In other words, we must have $T \gg T_0$ where T_0 is the Bose–Einstein condensation temperature. If $T < T_0$ the occupation number of the $k = 0$ state becomes macroscopically large ($\sim N$), and the term

$$\sum_{\mathbf{k}_1} \{v(0) + v(\mathbf{k} - \mathbf{k}_1)\} \langle a_{\mathbf{k}_1}^\dagger a_{\mathbf{k}_1} \rangle$$

"blows up" because of the zero momentum term. However, if $T \ll T_0$ most of the particles will be in the zero-energy ground state, and we have the situation already analyzed in Subsection 1.5.1 using diagonalization of the reduced Hamiltonian in second quantization.

5.2 Random Phase Approximation for Ferromagnets

We recall that in Section 1.5.2 the Heisenberg Hamiltonian was introduced as a model for an exchange-dominated ferromagnet with the dipole-dipole interactions being neglected. The spin Hamiltonian was given in Equation (1.81), where the parameters included the exchange interaction $J_{i,j}$ (> 0) between neighboring spin sites i and j and an applied field B_0 acting in the z direction of magnetization.

Previously, we solved for the spin waves (SWs) as the quasiparticles at low temperatures $T \ll T_C$ by using the Holstein–Primakoff (HP) transformation from spin operators to boson operators. The approximate Hamiltonian when expressed in terms of the boson operators was (apart from a constant)

$$\mathcal{H} = \sum_{\mathbf{k}} E_{\mathbf{k}} a_{\mathbf{k}}^\dagger a_{\mathbf{k}}$$

as in Equation (1.91), where the quasiparticle (SW) energy at wave vector $\mathbf{k}$ was found to be

$$E_{\mathbf{k}} = g\mu_R B_0 + S\{J(0) - J(\mathbf{k})\}.$$

Using this bosonic representation the GFs could be written down, as we described for ferromagnets in Section 4.4 with the magnetic dipole-dipole interactions also being taken into account.

An alternative approach, which we want to pursue here, is to calculate the GFs by working directly in terms of the spin operators. This will be done without having to transform to boson operators and without assuming low temperatures, so the GF calculation is potentially of wider applicability.

Specifically, we seek to evaluate the GF Fourier components $G(S_n^+; S_m^- \mid \omega)$ where n and m are site labels, making the choice $\varepsilon = 1$ giving commutation relations, by analogy with the calculation for paramagnets in Section 4.5. To construct the GF equation of motion we again use the commutator average that $\langle [S_n^+, S_m^-] \rangle = 2\langle S_n^z \rangle \delta_{nm}$. Then $[\mathcal{H}, S_n^+]$ must be found using the full spin Hamiltonian $\mathcal{H}$, which can be rewritten in component form as in the second line of Equation (1.87). It follows that we need to evaluate four types of commutators corresponding to the spin terms in this equation. This may be done making use of the operator identity $[XY, Z] = X[Y, Z] + [X, Z]Y$, yielding, eventually (see Problem 5.1),

$$[\mathcal{H}, S_n^+] = \sum_i J_{in}(S_n^z S_i^+ - S_i^z S_n^+) - g\mu_B B_0 S_n^+. \tag{5.15}$$

The GF equation of motion, which is obtained using Equations (3.66) and (5.15) becomes

$$\omega\, G(S_n^+; S_m^- \mid \omega) = \left(\frac{1}{2\pi}\right) 2\langle S_n^z\rangle \delta_{n,m} + g\mu_B B_0\, G(S_n^+; S_m^- \mid \omega)$$
$$- \sum_i J_{in} \left\{ G(S_n^z S_i^+; S_m^- \mid \omega) - G(S_i^z S_n^+; S_m^- \mid \omega) \right\}. \tag{5.16}$$

It is seen that two new GFs of a similar type have appeared on the right-hand side in the preceding equation. If we were to write down their GF equations of motion, we would find that an even more complicated product of operators arises, so we eventually have to resort to an approximation. The simplest way to proceed is to solve Equation (5.16) approximately by decoupling the new GFs on the right-hand side. In the same spirit as in Equation (5.4), along with its application to HF theory earlier in this chapter, we now make the approximation

$$G(S_i^z S_n^+; S_m^- \mid \omega) \approx \langle S_i^z\rangle G(S_n^+; S_m^- \mid \omega). \tag{5.17}$$

In effect, we are ignoring the fluctuations in S_i^z and replacing the operator by its average value $\langle S_i^z\rangle$. We notice that there could be another decoupled term having the form $\langle S_i^+\rangle G(S_n^z; S_m^- \mid \omega)$, but this vanishes because $\langle S_i^+\rangle = \langle S_i^x\rangle + i\langle S_i^y\rangle = 0$ for the system magnetized along the z direction. Similarly, for the other GF in Equation (5.16) we have

$$G(S_n^z S_i^+; S_m^- \mid \omega) \approx \langle S_n^z\rangle G(S_i^+; S_m^- \mid \omega). \tag{5.18}$$

The preceding decoupling is called the *random phase approximation* (or RPA). After applying the decoupling approximation, the equation of motion for the GF is simply

$$\left\{\omega - g\mu_B B_0 - \sum_i \langle S_i^z\rangle J_{in}\right\} G(S_n^+; S_m^- \mid \omega) = \frac{1}{\pi}\langle S_n^z\rangle \delta_{n,m}$$
$$- \sum_i J_{in}\langle S_n^z\rangle\, G(S_i^+; S_m^-, \mid \omega).$$

Because all the sites in an infinite lattice are equivalent to one another, it follows that $\langle S_i^z\rangle = \langle S_n^z\rangle \equiv \langle S^z\rangle$, independent of the site label. Also at this stage we make a further transformation of the GF from the site labels to a wave-vector representation by

$$G(S_n^+; S_m^- \mid \omega) = \frac{1}{N} \sum_{\mathbf{k}} G_{\mathbf{k}}(\omega) \exp[i\mathbf{k} \cdot (\mathbf{r}_n - \mathbf{r}_m)],$$

where N is the total number of spins. In the decoupled GF equation of motion we will denote the Fourier transform of the exchange interaction by $J(\mathbf{k})$, as defined in Equation (1.90). Also for the Kronecker delta we will use

$$\delta_{n,m} = \frac{1}{N} \sum_{\mathbf{k}} \exp[i\mathbf{k} \cdot (\mathbf{r}_n - \mathbf{r}_m)].$$

Then the equation of motion becomes simply

$$\{\omega - g\mu_B B_0 - \langle S^z \rangle J(0)\} G_{\mathbf{k}}(\omega) = \frac{1}{\pi} \langle S^z \rangle - \langle S^z \rangle J(\mathbf{k}) G_{\mathbf{k}}(\omega),$$

which may be rewritten as

$$\{\omega - E_{\mathbf{k}}\} G_{\mathbf{k}}(\omega) = \frac{1}{\pi} \langle S^z \rangle, \tag{5.19}$$

with the generalized definition that now

$$E_{\mathbf{k}} = g\mu_B B_0 + \langle S^z \rangle \{J(0) - J(\mathbf{k})\}. \tag{5.20}$$

We can solve Equation (5.19) for the GFs by following the same procedure as described in Chapter 4. Thus, the analytically continued GF in the complex $\tilde{\omega}$-plane is

$$G_{\mathbf{k}}(\tilde{\omega}) = \frac{\langle S^z \rangle}{\pi\{\tilde{\omega} - E_{\mathbf{k}}\}}. \tag{5.21}$$

As before, if $\tilde{\omega} = \omega + i\eta$ we obtain the retarded GF, and if $\tilde{\omega} = \omega - i\eta$ we obtain the advanced GF.

We see that the GFs have a simple pole at $E_{\mathbf{k}}$, which is given by Equation (5.20), and this is, therefore, the quasiparticle energy or frequency. At low temperatures $(T \ll T_C)$ we have $\langle S^z \rangle \to S$, and $E_{\mathbf{k}}$ becomes the same expression as quoted earlier for the SW energy at low T. Equation (5.20), therefore, represents a simple description of how the SWs become generalized at higher T. So far $\langle S^z \rangle$ is undetermined, but it is an important quantity to evaluate, because it is proportional to the magnetization in the ferromagnet and it appears as a parameter in both Equations (5.20) and (5.21). We can, in fact, obtain it self-consistently using the correlation functions related to the GF, as we will show next.

From the fluctuation-dissipation theorem in Equation (3.40) the spectral intensity $J(\omega)$ is related to the retarded GF, denoted here as $G_{\mathbf{k}}^r(\omega)$, by

$$J(\omega) = \frac{-2}{(e^{\beta\omega} - 1)} \mathrm{Im}\, G_{\mathbf{k}}^r(\omega) = \frac{-2}{(e^{\beta\omega} - 1)} \mathrm{Im}\, \frac{\langle S^z \rangle}{\pi\{\omega - E_{\mathbf{k}} + i\eta\}},$$

where we used Equation (5.21). It follows, therefore, that

$$J(\omega) = \frac{2\langle S^z \rangle}{\left(e^{\beta \omega} - 1\right)} \delta(\omega - E_{\mathbf{k}}),$$

which consists of a single delta-function contribution. Transforming to the time correlation function using Equation (3.17), we find

$$\langle S_{\mathbf{k}}^-(t') S_{\mathbf{k}}^+(t) \rangle = \int_{-\infty}^{\infty} J_{\mathbf{k}}(\omega) e^{-i\omega(t-t')} d\omega = \frac{2\langle S^z \rangle}{\{e^{\beta E_{\mathbf{k}}} - 1\}} e^{-i E_{\mathbf{k}}(t-t')}.$$

If we take the case of equal time labels, this simplifies to give for the static correlation function

$$\langle S_{\mathbf{k}}^- S_{\mathbf{k}}^+ \rangle = \frac{2\langle S^z \rangle}{\{e^{\beta E_{\mathbf{k}}} - 1\}}. \tag{5.22}$$

The Fourier transform back to the site labels is defined by

$$\langle S_n^- S_m^+ \rangle = \frac{1}{N} \sum_{\mathbf{k}} \exp[i\mathbf{k} \cdot (\mathbf{r}_n - \mathbf{r}_m)] \langle S_{\mathbf{k}}^- S_{\mathbf{k}}^+ \rangle,$$

and in the special case of equal sites ($n = m$) this becomes a sum rule

$$\langle S_n^- S_n^+ \rangle = \frac{1}{N} \sum_{\mathbf{k}} \langle S_{\mathbf{k}}^- S_{\mathbf{k}}^+ \rangle. \tag{5.23}$$

Now, the result of combining Equations (5.22) and (5.23) leads to a general *equation of state* for the ferromagnet as

$$\langle S_n^- S_n^+ \rangle = 2\langle S^z \rangle \frac{1}{N} \sum_{\mathbf{k}} \frac{1}{\{e^{\beta E_{\mathbf{k}}} - 1\}}, \tag{5.24}$$

for any site n. We have referred to this result as an equation of state because it can be used to deduce the thermal average $\langle S^z \rangle$ self-consistently.

For simplicity, we outline in the following text how this would be done when the spin quantum number $S = \frac{1}{2}$, but the calculation is capable of generalization to higher spin values (see, e.g., [75]). We make use of the following two identities for spins at the same site:

$$S^+ S^- - S^- S^+ = 2S^z \quad \text{and} \quad S^+ S^- + S^- S^+ = 1. \tag{5.25}$$

The first of these identities is the commutator and holds for any spin S, while the second is specifically for $S = \frac{1}{2}$ (as may be verified using the Pauli spin matrices). Therefore, we have $S^- S^+ = 1/2 - S^z$, which implies on taking averages that $\langle S^- S^+ \rangle = 1/2 - \langle S^z \rangle$. Together with Equation (5.24) this last result leads to

$$\frac{1}{2} - \langle S^z \rangle = 2\langle S^z \rangle \frac{1}{N} \sum_{\mathbf{k}} \frac{1}{\{e^{\beta E_{\mathbf{k}}} - 1\}}, \tag{5.26}$$

which is the specific form of the equation of state when $S = \frac{1}{2}$. We now see that the pair of Equations (5.20) and (5.26) connect the two unknowns $\langle S^z \rangle$ and $E_{\mathbf{k}}$. Thus we can, in principle, solve for these two quantities as a function of the variables such as temperature or applied field B_0. We shall give a broad account below of how to do this for three special cases.

5.2.1 Low Temperatures $(T \ll T_C)$

When $T \ll T_C$ the spins are all well aligned in the z direction, and so we have $\langle S^z \rangle \approx \frac{1}{2}$. Replacing $\langle S^z \rangle$ by $\frac{1}{2}$ on the right-hand side of Equation (5.26) gives the approximate result for the spin deviation as

$$\frac{1}{2} - \langle S^z \rangle = \frac{1}{N} \sum_{\mathbf{k}} n_{\mathbf{k}},$$

where $n_{\mathbf{k}}$ is the Bose–Einstein thermal factor for the number of SWs at temperature T (recalling that we defined $\beta = 1/k_B T$):

$$n_{\mathbf{k}} = \left[\exp \left\{ \beta [g \mu_B B_0 + (1/2)\{J(0) - J(\mathbf{k})\}] \right\} - 1 \right]^{-1}. \tag{5.27}$$

If $B_0 = 0$ for the applied field the sum over $\mathbf{k}$ may be estimated analytically, as described in many of the standard solid-state physics books (see, e.g., [19, 20] for details). Briefly, the argument depends on noting that $n_{\mathbf{k}}$ becomes very large in the region where $k = |\mathbf{k}|$ is small (compared with a Brillouin zone boundary wave vector) because it behaves then as

$$n_{\mathbf{k}} = [\exp(\alpha k^2 / k_B T) - 1]^{-1}, \tag{5.28}$$

where α is a geometric constant that depends on the lattice structure. For example, using the expression for $J(\mathbf{k})$ given in Equation (1.94) for a b.c.c. ferromagnet, we conclude that $\alpha = J a^2 / 2$ in this case. Also, for this (or any other cubic) structure, the sum over $\mathbf{k}$ can be replaced by a 3D integral as

$$\frac{1}{N} \sum_{\mathbf{k}} \frac{1}{\exp(\alpha k^2 / k_B T) - 1} \rightarrow \lambda a^3 \int \frac{d^3 k}{\exp(\alpha k^2 / k_B T) - 1}$$

$$\rightarrow 4\pi \lambda a^3 \int_0^{k_c} \frac{k^2 dk}{\exp(\alpha k^2 / k_B T) - 1}.$$

Here λ is a numerical factor depending on the lattice structure, and the upper limit of integration (the cutoff value k_c) is chosen by approximating the Brillouin zone in $\mathbf{k}$ space by a sphere of radius k_c. In fact, at low enough temperatures we are justified in replacing the integration limit by infinity, and it is then straightforward to conclude that the integral is proportional to $T^{3/2}$. Hence we have the result to leading order that

$$\langle S^z \rangle \approx \frac{1}{2} - c\,T^{3/2}, \tag{5.29}$$

where c is a constant. This is referred to as *Bloch's law*, giving the decrease in the magnetization due to the thermally excited SWs as varying with a 3/2 power of temperature.

5.2.2 High Temperatures $(T \gg T_C)$

In this case $\langle S^z \rangle$ will become vanishingly small, but it will still be nonzero provided we assume there is an applied field B_0. Here we introduce a scaled magnetic susceptibility χ defined by

$$\chi = \lim_{b \to 0} \frac{\langle S^z \rangle}{b}, \tag{5.30}$$

where for convenience we denote $b = g\mu_B B_0$. Because this is a zero-field susceptibility, the preceding definition of χ is equivalent to taking a derivative with respect to b. Equation (5.26) then becomes

$$\frac{1}{2} - \chi b = \frac{1}{N} \sum_{\mathbf{k}} \frac{2\chi b}{\exp[\beta b + \beta \chi b\{J(0) - J(\mathbf{k})\}] - 1}.$$

For small enough b the denominator can be simplified by employing a power-series expansion of the exponential to yield

$$\frac{1}{2} = \frac{1}{N} \sum_{\mathbf{k}} \frac{2\chi b}{\beta b + \beta \chi b\{J(0) - J(\mathbf{k})\} + \mathcal{O}(b^2)},$$

which simplifies in the limit of $b \to 0$ to become

$$\frac{1}{2} = \frac{1}{N} \sum_{\mathbf{k}} \frac{2\chi}{\beta[1 + \chi\{J(0) - J(\mathbf{k})\}]}.$$

Note that this has become independent of b and it provides us with an implicit solution for χ at high temperature (or small β). An explicit solution can be developed as a series expansion in increasing powers of β, giving

$$\chi = \frac{1}{4}\beta + \frac{1}{16}J(0)\beta^2 + \frac{1}{64}\left\{\frac{1}{N}\sum_{\mathbf{k}}[J^2(0) - J^2(\mathbf{k})]\right\}\beta^3 + \mathcal{O}(\beta^4). \tag{5.31}$$

The first two terms on the right-hand side are essentially exact (as compared with the results of rigorous high-T series expansion methods), but the β^3 term deviates from the exact result (see [75]). Nevertheless, the approximate theory is found to be relatively successful.

5.2.3 Estimation of the Critical Temperature ($T \approx T_C$)

We now take $B_0 = 0$ for the applied field in Equation (5.26) and we want to find the value of the temperature T (or the corresponding value of β) at which $\langle S^z \rangle \to 0$ as we approach the critical temperature from below. We start with the zero-field result expressed as

$$\frac{1}{2} - \langle S^z \rangle = 2\langle S^z \rangle \frac{1}{N} \sum_{\mathbf{k}} \frac{1}{\exp\left[\beta \langle S^z \rangle \{J(0) - J(\mathbf{k})\}\right] - 1}.$$

When T is close to but less than T_C, we know that $\langle S^z \rangle$ is very small, and so the term in the exponential is also very small. Expanding the exponential as a power series gives

$$\frac{1}{2} - \langle S^z \rangle = 2\langle S^z \rangle \frac{1}{N} \sum_{\mathbf{k}} \frac{1}{1 + \beta \langle S^z \rangle \{J(0) - J(\mathbf{k})\} + \mathcal{O}(\langle S^z \rangle^2) - 1}$$

$$= \frac{2}{N} \sum_{\mathbf{k}} \frac{1}{\beta\{J(0) - J(\mathbf{k})\} + \mathcal{O}(\langle S^z \rangle)}.$$

Taking the limit that $\langle S^z \rangle \to 0$ as $T \to T_C$ from below, we find

$$\frac{1}{2} = \frac{2}{N} \sum_{\mathbf{k}} \frac{1}{\beta_C \{J(0) - J(\mathbf{k})\}}, \tag{5.32}$$

where $\beta_C = 1/k_B T_C$. Hence the final result is

$$T_C = \frac{1}{4k_B} \left[\frac{1}{N} \sum_{\mathbf{k}} \frac{1}{\{J(0) - J(\mathbf{k})\}} \right]^{-1}. \tag{5.33}$$

This is normally a better estimate for T_C for ferromagnets than that given by mean-field theory (see, e.g., [19]), which is just $J(0)/4k_B$ for $S = 1/2$.

It can also be shown that we arrive at the same estimate of T_C as in Equation (5.33) by letting T approach T_C from above and examining the condition for the susceptibility χ to diverge (again for $B_0 \to 0$). This is considered in Problem 5.2.

5.3 Random Phase Approximation for Antiferromagnets

If the nearest-neighbor exchange interaction in a crystal lattice has the opposite sign from that in the ferromagnetic case, then the preference at low temperatures will be for an antiparallel (rather than parallel) alignment between neighboring spins. Often we can think of simple antiferromagnets in terms of there being two interpenetrating sublattices, which we label arbitrarily as A (for spin "up") and B (for spin "down"), as shown schematically in Figure 5.1. Some examples of two-sublattice antiferromagnets where the dominant exchange is between nearest-neighbor spin sites on

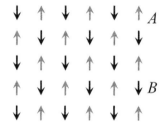

Figure 5.1 Schematic illustration of a simple two-sublattice antiferromagnet (with spins arranged on a 2D square lattice), where the nearest neighbors of spins on sublattice A (spin-up) lie on sublattice B (spin down), and vice versa.

opposite sublattices are FeF_2 (with a tetragonal rutile structure) and $RbMnF_3$ (with a cubic perovskite structure) [76].

For simplicity we shall ignore here the long-range magnetic dipole-dipole interactions. The Heisenberg exchange Hamiltonian can then be written in the form

$$\mathcal{H} = \sum_{i,j} J_{i,j} \mathbf{S}_i \cdot \mathbf{S}_j - g\mu_B B_{an} \sum_i S_i^z + g\mu_B B_{an} \sum_j S_j^z, \qquad (5.34)$$

where $J_{i,j}$ is again positive (but note that we have different factors in front of the summation), with i and j referring to spin sites on sublattices A and B, respectively. We have included Zeeman-type terms to describe an effective anisotropy field of magnitude B_{an}, which acts in the z direction and $-z$ direction for sublattices A and B, respectively. This anisotropy field helps to stabilize the antiferromagnetic ordering, and typically it arises due to the single-ion anisotropy mentioned in Section 4.5. For simplicity, we assume there is no external applied field ($B_0 = 0$), and so by symmetry we may write either $\langle S_n^z \rangle = \langle S^z \rangle$ if site label n corresponds to sublattice A or $\langle S_n^z \rangle = -\langle S^z \rangle$ if n is on sublattice B.

We will now evaluate GFs of the form $G(S^+; S^- | \omega)$ for the operators at various sites. The same method as in the previous section for the ferromagnetic case will be followed, but there will be some differences of sign because the exchange now has a different sign and because the previous applied magnetic field term is replaced by Zeeman terms with $\pm B_{an}$ due to the anisotropy. We will see that once again new GFs of the type $G(S^z S^+; S^- | \omega)$ are formed on the right-hand side of the GF equations of motion, and we may employ the RPA decoupling to simplify them.

For example, starting with $G(S_{n(A)}^+; S_m^- | \omega)$, where $n(A)$ means that the site n is considered to be on sublattice A, we can form the GF equation of motion from Equation (3.66). This can be decoupled using RPA, just as in the ferromagnetic case, and after some straightforward algebra we find that

$$\left\{ \omega - g\mu_B B_{an} + \sum_j \langle S_j^z \rangle J_{j,n(A)} \right\} G(S_{n(A)}^+; S_m^- \mid \omega)$$

$$= \frac{1}{\pi} \langle S_{n(A)}^z \rangle \delta_{n(A),m} + \sum_j J_{j,n(A)} \langle S_{n(A)}^z \rangle G(S_j^+; S_m^- \mid \omega). \qquad (5.35)$$

This simplifies slightly when the thermal averages are replaced by the value appropriate to their sublattice (noting also that j will be on sublattice B) when we assume that the exchange couples only nearest neighbors. Thus, taking $m(A)$ and $m(B)$ to refer to other sites on sublattices A and B, respectively, we have the two cases

$$\left\{ \omega - g\mu_B B_{an} - \langle S^z \rangle \sum_j J_{j,n(A)} \right\} G(S_{n(A)}^+; S_{m(A)}^- \mid \omega)$$

$$= \frac{1}{\pi} \langle S^z \rangle \delta_{n(A),m(A)} + \langle S^z \rangle \sum_j J_{j,n(A)} G(S_j^+; S_{m(A)}^- \mid \omega) \qquad (5.36)$$

and

$$\left\{ \omega - g\mu_B B_{an} - \langle S^z \rangle \sum_j J_{j,n(A)} \right\} G(S_{n(A)}^+; S_{m(B)}^- \mid \omega)$$

$$= \langle S^z \rangle \sum_j J_{j,n(A)} G(S_j^+; S_{m(B)}^- \mid \omega). \qquad (5.37)$$

Another similar pair of coupled GF equations is obtained when we make the other sublattice choice to consider $G(S_{n(B)}^+; S_m^- \mid \omega)$, namely

$$\left\{ \omega + g\mu_B B_{an} + \langle S^z \rangle \sum_i J_{i,n(B)} \right\} G(S_{n(B)}^+; S_{m(B)}^- \mid \omega)$$

$$= -\frac{1}{\pi} \langle S^z \rangle \delta_{n(B),m(B)} - \langle S^z \rangle \sum_i J_{i,n(B)} G(S_i^+; S_{m(B)}^- \mid \omega) \qquad (5.38)$$

and

$$\left\{ \omega + g\mu_B B_{an} + \langle S^z \rangle \sum_i J_{i,n(B)} \right\} G(S_{n(B)}^+; S_{m(A)}^- \mid \omega)$$

$$= -\langle S^z \rangle \sum_i J_{i,n(B)} G(S_i^+; S_{m(A)}^- \mid \omega), \qquad (5.39)$$

where i is on sublattice A.

We can solve the set of Equations (5.36)–(5.39), just as in the ferromagnetic case, by transforming from the site labels to a wave-vector representation. The wave-vector Fourier components are defined as before, except that the previous $G_{\mathbf{k}}(\omega)$ now has sublattice labels for each spin involved in the GF, so there are four related quantities conveniently denoted by $G_{\mathbf{k}}^{AA}(\omega)$, $G_{\mathbf{k}}^{AB}(\omega)$, $G_{\mathbf{k}}^{BA}(\omega)$, and $G_{\mathbf{k}}^{BB}(\omega)$. They form a set of coupled equations that can be reexpressed as

$$\{\omega - g\mu_B B_{an} - \langle S^z \rangle J(0)\} G_{\mathbf{k}}^{AA}(\omega) - \langle S^z \rangle J(\mathbf{k}) G_{\mathbf{k}}^{BA}(\omega) = \frac{\langle S^z \rangle}{\pi},$$

$$\{\omega - g\mu_B B_{an} - \langle S^z \rangle J(0)\} G_{\mathbf{k}}^{AB}(\omega) - \langle S^z \rangle J(\mathbf{k}) G_{\mathbf{k}}^{BB}(\omega) = 0,$$

$$\{\omega + g\mu_B B_{an} + \langle S^z \rangle J(0)\} G_{\mathbf{k}}^{BB}(\omega) + \langle S^z \rangle J(\mathbf{k}) G_{\mathbf{k}}^{AB}(\omega) = -\frac{\langle S^z \rangle}{\pi},$$

$$\{\omega + g\mu_B B_{an} + \langle S^z \rangle J(0)\} G_{\mathbf{k}}^{BA}(\omega) + \langle S^z \rangle J(\mathbf{k}) G_{\mathbf{k}}^{AA}(\omega) = 0. \tag{5.40}$$

These linear equations are easily solved to give the results for the analytically continued GFs at complex frequency $\tilde{\omega}$ as

$$G_{\mathbf{k}}^{AA}(\tilde{\omega}) = G_{\mathbf{k}}^{BB}(-\tilde{\omega}) = \frac{\langle S^z \rangle [g\mu_B B_{an} + \langle S^z \rangle J(0) + \tilde{\omega}]}{\pi [\tilde{\omega}^2 - E_{\mathbf{k}}^2]},$$

$$G_{\mathbf{k}}^{AB}(\tilde{\omega}) = G_{\mathbf{k}}^{BA}(\tilde{\omega}) = \frac{-\langle S^z \rangle^2 J(\mathbf{k})}{\pi [\tilde{\omega}^2 - E_{\mathbf{k}}^2]}, \tag{5.41}$$

where

$$E_{\mathbf{k}} = \left[\{g\mu_B B_{an} + \langle S^z \rangle J(0)\}^2 - \{\langle S^z \rangle J(\mathbf{k})\}^2 \right]^{1/2}. \tag{5.42}$$

Equation (5.42) is the SW dispersion relation for a two-sublattice antiferromagnet at any finite temperature T below the critical temperature, or Néel temperature, T_N. The temperature dependence of $E_{\mathbf{k}}$ arises from the $\langle S^z \rangle$ factors and from the B_{an} term. Both quantities decrease with increasing temperature. Typically, it is found for the anisotropy field that $B_{an} \propto \langle S^z \rangle^s$ where the index s lies between 1 and 2. The dispersion relation is seen to be quite different from the ferromagnetic dispersion relation given by Equation (5.20). At wave vector $\mathbf{k} = 0$ the antiferromagnetic SW energy becomes

$$E_0 = \left[g\mu_B B_{an} \{g\mu_B B_{an} + 2\langle S^z \rangle J(0)\} \right]^{1/2}. \tag{5.43}$$

All four GFs in Equation (5.41) have poles in the complex $\tilde{\omega}$-plane at $\pm E_{\mathbf{k}}$, so there are two branches to the spectrum, compared to just one branch in the ferromagnetic case. They are, however, degenerate in magnitude in this case. In Figure 5.2 we illustrate the form of the SW dispersion curve at low temperatures ($T \ll T_N$) for two values of the anisotropy field assuming an antiferromagnet with a s.c. lattice structure for the spin sites. The two signs for the poles of the GF are a consequence of the fact that the spin precession on one sublattice is in the opposite sense from that on the other sublattice. The expression on the right-hand side of Equation (5.43) corresponds to the antiferromagnetic resonance (AFMR) frequency. Typically (unless B_{an} is very small), its value lies in the infrared region of the electromagnetic spectrum. AFMR and Raman scattering (of light) are both useful

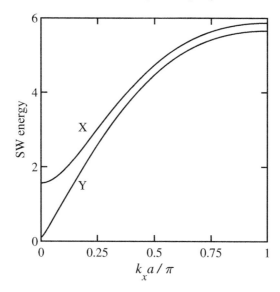

Figure 5.2 Example of the SW dispersion relation $E_{\mathbf{k}}$ in Equation (5.42) in dimensionless units (where $E_{\mathbf{k}}/SJ$ is plotted versus $k_x a/\pi$ and we take $k_y = k_z = 0$) for a s.c. antiferromagnet at $T \ll T_N$. Curves for two values of the effective anisotropy field are shown, corresponding to $g\mu_B B_{an}/SJ = 0.2$ (curve X) and 0.001 (curve Y).

experimental techniques [19] to probe the antiferromagnetic SWs. The degeneracy in the magnitudes of the two SW branches that we have in the present calculation is removed when there is a nonzero applied magnetic field $B_0 \neq 0$ to produce a splitting.

Finally, from the GFs in Equation (5.41) we can straightforwardly work out the spectral intensities and time correlation functions, just as in the ferromagnetic case.

5.4 Electron Correlations and the Hubbard Model

The Hubbard model, which was briefly introduced in Subsection 1.4.3, involves an interplay between the site-to-site hopping in a many-electron system and the Coulomb repulsion effects involving two electrons at the same atomic site (one electron with spin up and the other with spin down). The Hamiltonian was quoted in Equation (1.68) and is applicable when inter-site screening effects are sufficiently large that the Coulomb effects between electrons at different sites can be ignored.

Despite the relative simplicity of the model, it provides useful insights into the physics of strongly correlated electronic systems and the criteria for metal-insulator transitions, which are sometimes referred to as Mott transitions. The model is particularly relevant for narrow-band transition-metal compounds where the electron correlations are large. It embodies many of the same physical concepts as the earlier Stoner model in magnetism, which was thoroughly investigated for materials with wider bands. In his original work Hubbard [21] employed the model that bears his name to investigate how the strong correlations can lead to a splitting of an electronic band into nonoverlapping subbands. Some thorough accounts of this topic can be found in the books by Madelung [77], Mott [78], and Yosida [79], as well as in Hubbard's original papers [21, 80, 81].

Here we will briefly discuss some limiting cases of the model and provide GF approximation schemes that are applicable in the wide and narrow electron band cases, depending on the ratio t/U where t is the nearest-neighbor hopping parameter and U is the Coulomb energy defined in Subsection 1.4.2. We start by considering the effect of the individual terms in Equation (1.68).

First we note that a Hamiltonian consisting only of the hopping term $\mathcal{H}_h$ (proportional to t) is exactly solvable using either the operator equation of motion or the GF equation of motion, by analogy with the calculations carried out for graphene in Chapters 2 and 4, respectively. It is a simple exercise to repeat those calculations for the case of a 3D cubic lattice instead of the 2D bipartite honeycomb lattice applicable for a graphene sheet. The expression for the electronic band energy $\epsilon_{\mathbf{k}}$ at wave vector $\mathbf{k}$ in 3D is

$$\epsilon_{\mathbf{k}} = -2t[\cos(k_x a) + \cos(k_y a) + \cos(k_z a)] \tag{5.44}$$

for a s.c. material with the lattice constant a (see Problem 5.5). Therefore, the total band width is $W = 6t$. Also it can be shown that the single-particle electronic GF will have a simple pole at this same quantity, giving

$$G(a_\sigma; a_\sigma^\dagger \mid \omega) = \left(\frac{1}{2\pi}\right) \frac{1}{\omega - \epsilon_{\mathbf{k}} + i\eta} \tag{5.45}$$

for the retarded GF at angular frequency ω, where $a_\sigma^\dagger$ and a_σ are the fermion creation and annihilation operators correspond to the electron spin state σ (equal to ↑ or ↓).

However, for a Hamiltonian that consists only of the Coulomb repulsion term $\mathcal{H}_U$ in Equation (1.68), the physical outcome depends on the filling factor for the band. We have labeled the different atomic sites by i and we suppose there are N of these. Because only two electrons (one with spin up and the other with spin down) can occupy each site i as a consequence of the Pauli exclusion principle, a completely full band would have $2N$ electrons. The case of special interest, which we will pursue further, is that of a half-filled band, i.e., N electrons distributed between the N sites. Clearly, the state with the lowest overall energy will be when every site is singly occupied. Further, the motion of any charge would lead to an unoccupied site and a doubly occupied site, consequently increasing the energy by an amount U. For this reason the system would behave as an insulator at low temperature.

Next we will examine two physical situations where both terms of the Hubbard Hamiltonian contribute, taking initially the case in which hopping has only a small effect compared with the dominant Coulomb term.

5.4.1 Case in Which the Coulomb Term Is Dominant

This is the case when $t/U \ll 1$ (or $W/U \ll 1$) and we may use QM perturbation theory to include the hopping effects. The unperturbed part of the Hamiltonian is taken as $\mathcal{H}_U$, and the perturbation is $\mathcal{H}_h$. The correction to the electron energy at any given site i comes in second order of perturbation through the following virtual process. An electron at site i can transfer (hop) to one of the nearest-neighbor sites i', which becomes doubly occupied. Then one of the electrons at i' does a return hop to site i, resulting in a net exchange of electrons. The process is only possible, however, if the electron spins at the two sites were originally *antiparallel* because the parallel case is excluded by the Pauli exclusion principle. The energy correction is just $-2t^2/U$ for the pair of sites, and so the effective Hamiltonian in second order of perturbation is

$$\mathcal{H}^{(2)} = -\frac{2t^2}{U} \sum_{\langle i,i' \rangle} \sum_{\sigma,\sigma'} a^\dagger_{i,\sigma'} a_{i',\sigma'} a^\dagger_{i',\sigma} a_{i,\sigma} , \tag{5.46}$$

where $\langle i,i' \rangle$ indicates that the summations are taken over *distinct* nearest-neighbor sites.

It is interesting that the preceding approximate Hamiltonian can be related to the Heisenberg exchange model [79, 82] because it can be reexpressed (as we show next) in terms of spin operators as

$$\mathcal{H}^{(2)} = \frac{4t^2}{U} \sum_{\langle i,i' \rangle} \left\{ \mathbf{S}_i \cdot \mathbf{S}_{i'} - \frac{1}{4} \right\}. \tag{5.47}$$

One way to establish the equivalence between Equations (5.46) and (5.47) is to make use of the following relationship between the spin components S_i^+, S_i^-, and S_i^z at site i and the fermion creation and annihilation operators for the electrons:

$$S_i^+ = a^\dagger_{i,\uparrow} a_{i,\downarrow}, \quad S_i^- = a^\dagger_{i,\downarrow} a_{i,\uparrow},$$

$$S_i^z = \frac{1}{2} \left(a^\dagger_{i,\uparrow} a_{i,\uparrow} - a^\dagger_{i,\downarrow} a_{i,\downarrow} \right). \tag{5.48}$$

We notice that the defining expressions for S_i^+ and S_i^- involve spin-flip combinations, as expected for the spin raising and lowering operators, while the expression for S_i^z gives the net spin polarization. This result in Equation (5.48) is sometimes referred to as the *coupled-fermion representation* (see, e.g., [27]). It may easily be verified, using the standard properties of fermion operators, that the spin operators defined in Equation (5.48) correctly reproduce the spin commutation properties as quoted in Equation (1.84). This is left for Problem 5.6, together with the subsequent steps that are needed to prove the equivalence of Equations (5.46) and (5.47).

To summarize the preceding result, we have demonstrated that the Hubbard model with one electron per site (the half-filling case) and strong Coulomb interactions ($U \gg W$) reduces to the antiferromagnetic Heisenberg Hamiltonian typical of an insulator. Also we see from Equation (5.47) that the antiferromagnetic exchange constant J_{AF} is given by

$$J_{AF} = \frac{4t^2}{U} = \frac{W^2}{9U} > 0. \tag{5.49}$$

The assumed half-filling is a necessary requirement for this result to hold.

5.4.2 Case of Hartree–Fock-Type Decoupling

Both terms of the Hubbard Hamiltonian in second quantization are included here, and we place no restriction on the electronic filling factor. Instead we assume that the Coulomb term will be approximated by making a decoupling in the same spirit as the HF theory of Section 5.1. However, instead of proceeding as before it is advantageous to utilize the property that the Coulombic term $\mathcal{H}_U$ in Equation (1.68) is already factorized as a product of the number operators due to correlation effects. This suggests a decoupling approximation of the form (in terms of site labels i and j)

$$n_{i,\uparrow} n_{i,\downarrow} \rightarrow \langle n_{i,\uparrow} \rangle n_{i,\downarrow} + n_{i,\uparrow} \langle n_{i,\downarrow} \rangle + \langle n_{i,\uparrow} \rangle \langle n_{i,\downarrow} \rangle. \tag{5.50}$$

The last term is just a constant and can be henceforth ignored for the excitation spectrum. When the preceding result is combined with the hopping term in the Hubbard Hamiltonian we find (apart from the constant term) that

$$\mathcal{H} = \sum_{i,j,\sigma} \left\{ t_{i,j} + U \langle n_{i,\bar{\sigma}} \rangle \right\} a_{i,\sigma}^{\dagger} a_{j,\sigma}, \tag{5.51}$$

where $\bar{\sigma}$ denotes the spin projection that is opposite to that for the state with the label σ.

Despite the simplification achieved by the preceding result, it is still very complicated to continue further because of the different possibilities for $\langle n_{i,\uparrow} \rangle$ and $\langle n_{i,\downarrow} \rangle$, which in turn determine the magnetic phase. For example, if the numbers of electrons with spin up and spin down are equal, without there being any long-range order, we have a *paramagnetic* phase. Other possible states are *ferromagnetic* (when the numbers of electrons with spin up and spin down are different, giving a net magnetization) and *antiferromagnetic* (when the numbers of electrons with spin up and spin down are equal, but there is short-range order as in a two-sublattice model).

Thorough review accounts of the complexities involved in obtaining self-consistent solutions from Equation (5.51) or similar results, including descriptions of the various conflicting results in the published literature, have been given by Herring [83] and Mahan [47]. Qualitatively, the spin-up and spin-down electronic bands (each with band width W) are split in energy by an amount of order U by the Coulomb term. We can conclude that if $U \ll W$ the bands will still overlap and a metallic behavior is expected. However, an insulating behavior would be associated with $U \gg W$ when the bands are well separated. At some intermediate stage, when U and W are comparable with one another, the possibility of a metal-insulator transition arises.

As a simplified calculation we outline in the following text how the theory develops further in the particular case when it is assumed that $\langle n_{i,\sigma} \rangle \equiv \langle n_\sigma \rangle = \langle a_\sigma^\dagger a_\sigma \rangle$ independent of the site label i. It then follows that Equation (5.51) can be simplified slightly and rewritten in terms of effective spin-up and spin-down band energies as

$$\mathcal{H} = \sum_{\mathbf{k}} \left[\{\epsilon_{\mathbf{k}} + U\langle n_\downarrow \rangle\} \, a_{\mathbf{k},\uparrow}^\dagger a_{\mathbf{k},\uparrow} + \{\epsilon_{\mathbf{k}} + U\langle n_\uparrow \rangle\} \, a_{\mathbf{k},\downarrow}^\dagger a_{\mathbf{k},\downarrow} \right]. \tag{5.52}$$

We have transformed to the wave-vector $\mathbf{k}$ representation for the operators.

The $\langle n_\uparrow \rangle$ and $\langle n_\downarrow \rangle$ factors can now be determined self-consistently by using, for example,

$$\langle n_\uparrow \rangle = \frac{1}{N} \sum_{\mathbf{k}} \langle a_{\mathbf{k},\uparrow}^\dagger a_{\mathbf{k},\uparrow} \rangle = -\frac{1}{N} \sum_{\mathbf{k}} \frac{n(\omega)}{\pi} \operatorname{Im} \frac{1}{\omega - \epsilon_{\mathbf{k}} - U\langle n_\downarrow \rangle + i\eta}$$

$$= \frac{1}{N} \sum_{\mathbf{k}} n(\epsilon_{\mathbf{k}} + U\langle n_\downarrow \rangle) \delta(\omega - \epsilon_{\mathbf{k}} - U\langle n_\downarrow \rangle), \tag{5.53}$$

where we have applied the fluctuation-dissipation theorem from Equation (3.40) to relate the correlation function to the imaginary part of the relevant GF. Here η denotes a positive infinitesimal, as before, and $n(\omega)$ denotes the Fermi–Dirac distribution factor (which includes the chemical potential μ). In the last line of the preceding equation the imaginary part was taken using the identity in Equation (3.31).

As a final step, if we introduce a density-of-states function $Z(\epsilon)$ for an electron with energy ϵ in the unperturbed band, we obtain for $\langle n_\uparrow \rangle$ and similarly for $\langle n_\downarrow \rangle$

$$\langle n_\uparrow \rangle = \int Z(\epsilon)\, n\big(\epsilon_k + U \langle n_\downarrow \rangle\big)\, d\epsilon, \qquad (5.54)$$

$$\langle n_\downarrow \rangle = \int Z(\epsilon)\, n\big(\epsilon_k + U \langle n_\uparrow \rangle\big)\, d\epsilon. \qquad (5.55)$$

The preceding two equations, when taken together with

$$n_0 \equiv \langle n_\uparrow \rangle + \langle n_\downarrow \rangle \qquad (5.56)$$

as specifying the total band filling factor, are sufficient in principle to enable a solution for $\langle n_\uparrow \rangle$, $\langle n_\downarrow \rangle$, and n_0 to be obtained (assuming a model form for the density-of-states function is adopted). Various examples are given in the references cited earlier.

5.5 The Anderson Model for Localized States in Metals

In this section, we continue to discuss other properties of the electrons in metals, along with their associated magnetic moments. A topic that has been of ongoing interest concerns the electron correlations and magnetic states when metallic impurities are present in another host metal or when a dilute alloy is formed. Thus, when a low concentration of a transition metal element is dissolved in a "nonmagnetic" metal (one that is not ferromagnetic, antiferromagnetic, or ferrimagnetic), the resulting material either may or may not exhibit a localized magnetic moment. For example, the introduction of Cr or Fe into a host of Au or Cu results in localized moments, but the introduction of the same materials into Al does not (see, e.g., [85]).

A simplified, but nevertheless useful, model for the behavior of localized spins (typically d-state electrons) interacting with the conduction (or s-state) electrons in a host metal was introduced by Anderson [84]. Clear critical reviews of this model can also be found, for example, in [9, 26, 85–87].

In the basic form of the Anderson model, the Hamiltonian contains a term $\mathcal{H}_0$ for the kinetic energy of the conduction (s) electron band in the metal, with states filled up to the Fermi energy ϵ_F, as mentioned in Subsection 2.8.1. Similarly, there is a term $\mathcal{H}_d$ that represents the contribution of a localized state (such as a d electron), treated for simplicity as nondegenerate. In addition, two extra terms are required. One of these gives the interaction between the conduction electrons and the localized state, and will be denoted by $\mathcal{H}_{sd}$. It will produce a mixing, or hybridization, of the individual s and d states. The other term is a Coulomb repulsion term $\mathcal{H}_U$ acting between two electrons on an impurity site, by analogy with a similar term in the Hubbard model. The total Hamiltonian can therefore be written as $\mathcal{H} = \mathcal{H}_0 + \mathcal{H}_d + \mathcal{H}_{sd} + \mathcal{H}_U$, where the component terms are

$$\mathcal{H}_0 = \sum_{\mathbf{k},\sigma} \epsilon_{\mathbf{k}} a_{\mathbf{k},\sigma}^\dagger a_{\mathbf{k},\sigma},$$

$$\mathcal{H}_d = \sum_\sigma \epsilon_d d_\sigma^\dagger d_\sigma,$$

$$\mathcal{H}_{sd} = \sum_{\mathbf{k},\sigma} \left(V_{\mathbf{k}} a_{\mathbf{k},\sigma}^\dagger d_\sigma + V_{\mathbf{k}}^* a_{\mathbf{k},\sigma} d_\sigma^\dagger \right),$$

$$\mathcal{H}_U = U n_{d,\uparrow} n_{d,\downarrow}. \tag{5.57}$$

Here $a_{\mathbf{k},\sigma}^\dagger$ and $a_{\mathbf{k},\sigma}$ are the creation and annihilation operators for the conduction electrons of the host material with energy $\epsilon_{\mathbf{k}}$, wave vector $\mathbf{k}$, and spin projection σ. The operators $d_\sigma^\dagger$ and d_σ describe an impurity electron with energy level ϵ_d, which can couple with the conduction electrons through the sd interaction term $V_{\mathbf{k}}$, which is taken to be positive. Finally, U is the onsite Coulomb repulsion and $n_{d,\sigma} = d_\sigma^\dagger d_\sigma$ is the number operator for the d-electrons with spin projection σ.

Before starting the GF calculation it is helpful if we try to anticipate what might occur in this physical situation. We suppose, for simplicity, that there is just *one* localized electron, which of course will have two possible spin orientations. If we assume for the moment that it has spin up, then another electron with spin down will "feel" the Coulomb repulsion of the spin-up electron, and so its unperturbed energy (which necessarily lies below the Fermi energy) will be increased by an amount U. This may possibly take it to an energy state *above* the Fermi energy. Furthermore, the mixing of s and d states (through the term $\mathcal{H}_{sd}$) may act to raise the energy of a spin-up state and lower the energy of a spin-down state, so it is plausible that a cooperative effect may occur with a persisting localized moment. This situation is illustrated schematically in Figure 5.3, which we will also refer to again later in this section. The preceding scenario will now be explored, and we model it mathematically using the GF method, starting with the special case of $U = 0$.

5.5.1 Zero Onsite Coulomb Energy

We first look at the case when the effects of U are absent because it turns out that we may then obtain exact expressions for the required GFs. We start with the GF $G(d_\sigma; d_\sigma^\dagger \mid \omega)$ in terms of the operators for the impurity electrons and use the equation of motion (3.66). Also we choose $\varepsilon = -1$ so that we get anticommutation relations in the definitions of the GFs. The equation of motion is found from

$$\omega G(d_\sigma; d_\sigma^\dagger \mid \omega) = \frac{1}{2\pi} \langle \{d_\sigma, d_\sigma^\dagger\} \rangle - G([\mathcal{H}, d_\sigma]; d_\sigma^\dagger \mid \omega). \tag{5.58}$$

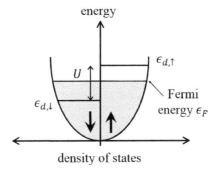

energy

density of states

Figure 5.3 Illustration of some basic features of the Anderson model for metals. The density of states for the spin-up and spin-down s bands are shown, together with the split levels for the d-states. See the text in Section 5.5 for further discussion.

For the commutator on the right-hand side it is easily shown that

$$[\mathcal{H}, d_\sigma] = \sum_{\sigma'} \epsilon_d [d_{\sigma'}^\dagger d_{\sigma'}, d_\sigma] + \sum_{k,\sigma'} [(V_k a_{k,\sigma'}^\dagger d_{\sigma'} + V_k^* a_{k,\sigma'} d_{\sigma'}^\dagger), d_\sigma]$$

$$= -\left(\epsilon_d d_\sigma + \sum_k V_k^* a_{k,\sigma} \right).$$

On inserting this result into Equation (5.58) we find

$$\omega\, G(d_\sigma; d_\sigma^\dagger \mid \omega) = \frac{1}{2\pi} + \epsilon_d\, G(d_\sigma; d_\sigma^\dagger \mid \omega) + \sum_k V_k^*\, G(a_{k,\sigma}; d_\sigma^\dagger \mid \omega). \qquad (5.59)$$

We see that the right-hand side contains another GF corresponding to a mixing between the conduction and impurity electrons. Next, this new GF is found from its equation of motion as

$$\omega\, G(a_{k,\sigma}; d_\sigma^\dagger \mid \omega) = 0 - G([\mathcal{H}, a_{k,\sigma}]; d_\sigma^\dagger \mid \omega).$$

Evaluating the commutator for the term on the right-hand side gives

$$[\mathcal{H}, a_{k,\sigma}] = -(\epsilon_k a_{k,\sigma} + V_k d_\sigma),$$

and so we obtain the new GF equation of motion as

$$\omega\, G(a_{k,\sigma}; d_\sigma^\dagger \mid \omega) = \epsilon_k\, G(a_{k,\sigma}; d_\sigma^\dagger \mid \omega) + V_k\, G(d_\sigma; d_\sigma^\dagger \mid \omega).$$

This can be rearranged (for the analytically continued GFs) as

$$G(a_{k,\sigma}; d_\sigma^\dagger \mid \tilde{\omega}) = \frac{V_k}{\tilde{\omega} - \epsilon_k}\, G(d_\sigma; d_\sigma^\dagger \mid \tilde{\omega}). \qquad (5.60)$$

Then it is seen that Equations (5.59) and (5.60) lead to

$$\tilde{\omega}\, G(d_\sigma; d_\sigma^\dagger \mid \tilde{\omega}) = \frac{1}{2\pi} + \epsilon_d\, G(d_\sigma; d_\sigma^\dagger \mid \tilde{\omega}) + \sum_{\mathbf{k}} \frac{|V_\mathbf{k}|^2}{\tilde{\omega} - \epsilon_\mathbf{k}} G(d_\sigma; d_\sigma^\dagger \mid \tilde{\omega}).$$

After some rearrangement we obtain the result for the analytically continued GF at complex $\tilde{\omega}$ as

$$G(d_\sigma; d_\sigma^\dagger \mid \tilde{\omega}) = \left(\frac{1}{2\pi}\right) \frac{1}{\tilde{\omega} - \epsilon_d - \Sigma(\tilde{\omega})}. \tag{5.61}$$

with

$$\Sigma(\tilde{\omega}) = \sum_{\mathbf{k}} \frac{|V_\mathbf{k}|^2}{\tilde{\omega} - \epsilon_\mathbf{k}}. \tag{5.62}$$

At first sight we seem to have a GF that is formally rather similar to that quoted in Equation (4.49) for the Jaynes–Cummings (JC) model. There is, however, an important difference because the quantity $\Sigma(\tilde{\omega})$ in the present case involves a summation over a wave vector $\mathbf{k}$, so we cannot just multiply by $(\tilde{\omega} - \epsilon_\mathbf{k})$ to simplify the denominator of the GF, as we did previously. Instead, we need to examine the role of $\Sigma(\tilde{\omega})$, which is sometimes known as a *self-energy* for reasons that will become apparent in Chapter 8. Here Σ appears as an extra term in the denominator of Equation (5.61), so it looks like a "correction" term to the impurity energy ϵ_d. This requires more interpretation at this stage because Σ is, in fact, a complex quantity with a real and imaginary part.

The procedure to follow is that explained in Chapter 3, i.e., we obtain the retarded GF by everywhere making the replacement that $\tilde{\omega} \to \omega + i\eta$. For the self-energy term this leads to

$$\Sigma(\omega + i\eta) = \sum_{\mathbf{k}} \frac{|V_\mathbf{k}|^2}{\omega - \epsilon_\mathbf{k} + i\eta}$$

$$= \mathcal{P} \sum_{\mathbf{k}} \frac{|V_\mathbf{k}|^2}{\omega - \epsilon_\mathbf{k}} - i\pi \sum_{\mathbf{k}} |V_\mathbf{k}|^2 \delta(\omega - \epsilon_\mathbf{k}), \tag{5.63}$$

where we have used the identity in Equation (3.31). The real (principal value) part is often relatively small or can be interpreted as a modification (a slight shift) to ϵ_d for the pole in Equation (5.61). The effect of the imaginary part, however, is of greater significance [90]. In the case of host materials with wide structureless conduction bands, it is often a good empirical approximation to write the preceding imaginary term as a constant denoted by $-i\Gamma$ with any ω dependence of Γ being ignored. This value can be related to the density of states in the conduction band because

$$\Gamma = \pi \sum_{\mathbf{k}} |V_{\mathbf{k}}|^2 \delta(\omega - \epsilon_{\mathbf{k}}) \approx \pi |\overline{V}|^2 \sum_{\mathbf{k}} \delta(\omega - \epsilon_{\mathbf{k}}) = \pi |\overline{V}|^2 \, Z. \tag{5.64}$$

Here $|\overline{V}|$ denotes an average value of $|V_{\mathbf{k}}|$ for the wide conduction band, and in the last step of the preceding equation we have used the result that the density of states Z for an excitation with dispersion relation $\epsilon_{\mathbf{k}}$ is equal to the wave-vector summation of the delta function (see, e.g., [20]).

An effect of the self-energy function (approximated as mentioned in the preceding text) is to give rise to a broadening of the spectral function, provided the density of states Z is nonzero. The spectral intensity $J_\sigma(\omega)$ corresponding to the GF in Equation (5.61) now can be deduced by application of the fluctuation-dissipation theorem. First, Equation (5.61) can be rewritten as

$$G(d_\sigma; d_\sigma^\dagger \,|\, \omega) = \left(\frac{1}{2\pi}\right) \frac{1}{\omega - \epsilon_d + i\Gamma} = \left(\frac{1}{2\pi}\right) \frac{\omega - \epsilon_d - i\Gamma}{(\omega - \epsilon_d)^2 + \Gamma^2}.$$

This gives us the result at effectively zero temperature (meaning $k_B T \ll \epsilon_F$, which is typically a good approximation for a metal) that

$$J_\sigma(\omega) = \frac{\Gamma/\pi}{(\omega - \epsilon_d)^2 + \Gamma^2}. \tag{5.65}$$

This represents a Lorentzian-like intensity function, centered at $\omega = \epsilon_d$ and with width 2Γ. At this stage there is no dependence of $J_\sigma(\omega)$ on σ.

5.5.2 Inclusion of the Onsite Coulomb Energy

Now we consider the general case in which $U \neq 0$ bringing in an additional term in the Hamiltonian. We will find that exact solutions for the GFs are no longer possible from the equations of motion. The starting point is again $G(d_\sigma; d_\sigma^\dagger \,|\, \omega)$. When forming its equation of motion we find that the commutator of the Hamiltonian with d_σ has an extra term, becoming

$$[\mathcal{H}, d_\sigma] = -\left(\epsilon_d d_\sigma + \sum_{\mathbf{k}} V_{\mathbf{k}}^* a_{\mathbf{k},\sigma} + U d_\sigma n_{d,\overline{\sigma}}\right),$$

where the notation $\overline{\sigma}$ again indicates the spin projection that is opposite to σ. Hence the modified GF equation of motion, which replaces Equation (5.59), is

$$\omega \, G(d_\sigma; d_\sigma^\dagger \,|\, \omega) = \frac{1}{2\pi} + \epsilon_d \, G(d_\sigma; d_\sigma^\dagger \,|\, \omega) + \sum_{\mathbf{k}} V_{\mathbf{k}}^* G(a_{\mathbf{k},\sigma}; d_\sigma^\dagger \,|\, \omega)$$

$$+ U \, G(d_\sigma n_{d,\overline{\sigma}}; d_\sigma^\dagger \,|\, \omega). \tag{5.66}$$

The other GF equation obtained previously for $G(a_{\mathbf{k},\sigma}; d_\sigma^\dagger \,|\, \omega)$ is found to be unchanged, so Equation (5.60) still holds.

It is evident that Equation (5.66) contains another new GF on the right-hand side, and we now seek a solution by using a decoupling approximation for it that is similar to those made in the previous sections of this chapter (e.g., the RPA decoupling for ferromagnets). Hence we write

$$G(d_\sigma n_{d,\bar{\sigma}}; d_\sigma^\dagger \mid \omega) \approx \langle n_{d,\bar{\sigma}} \rangle \, G(d_\sigma; d_\sigma^\dagger \mid \omega),$$

where we are ignoring fluctuations in the number operator for the d electrons and replacing the operator by its average value. This is again a mean-field type of approximation. The other term arising in the decoupling of the product $d_\sigma n_{d,\bar{\sigma}}$ would give a contribution proportional to $\langle d_\sigma \rangle$, which is equal to zero. The decoupled equation of motion for the GF in Equation (5.66) is therefore

$$\omega \, G(d_\sigma; d_\sigma^\dagger \mid \omega) = \frac{1}{2\pi} + [\epsilon_d + U \langle n_{d,\bar{\sigma}} \rangle] \, G(d_\sigma; d_\sigma^\dagger \mid \omega)$$

$$+ \sum_{\mathbf{k}} V_{\mathbf{k}}^* \, G(a_{\mathbf{k},\sigma}; d_\sigma^\dagger \mid \omega). \tag{5.67}$$

The use of Equations (5.66) and (5.67) then leads us to conclude that Equation (5.61) for the analytically continued GF is generalized to

$$G(d_\sigma; d_\sigma^\dagger \mid \tilde{\omega}) = \left(\frac{1}{2\pi} \right) \frac{1}{\tilde{\omega} - \epsilon_d - U \langle n_{d,\bar{\sigma}} \rangle - \Sigma(\tilde{\omega})}, \tag{5.68}$$

where the expression for the self-energy Σ in Equation (5.62) still applies.

Next we may again make an analytic continuation to obtain the retarded GF as described in the previous subsection. By analogy with the analysis given there, we now have the results that

$$G(d_\sigma; d_\sigma^\dagger \mid \omega) = \left(\frac{1}{2\pi} \right) \frac{\omega - \epsilon_d' - U \langle n_{d,\bar{\sigma}} \rangle - i\Gamma}{[\omega - \epsilon_d' - U \langle n_{d,\bar{\sigma}} \rangle]^2 + \Gamma^2} \tag{5.69}$$

and

$$J_\sigma(\omega) = \frac{\Gamma/\pi}{[\omega - \epsilon_d' - U \langle n_{d,\bar{\sigma}} \rangle]^2 + \Gamma^2}. \tag{5.70}$$

In the preceding two equations we have denoted $\epsilon_d' = \epsilon_d + \mathrm{Re}(\Sigma)$, where the real part of the self-energy, $\mathrm{Re}(\Sigma)$, can be found from the right-hand side of Equation (5.63) and provides a small correction (or shift) to the energy ϵ_d. The infinitesimal quantity η does not appear in the preceding equations because it can be assumed that $\eta \ll \Gamma$.

We now see that, in general, there are distinct Lorentzian peaks predicted at energies $\epsilon_d' + U \langle n_{d,\downarrow} \rangle$ when $\sigma = \uparrow$ and at $\epsilon_d' + U \langle n_{d,\uparrow} \rangle$ when $\sigma = \downarrow$. Hence there may be a splitting, depending on the $\langle n_{d,\sigma} \rangle$ values. For example, if $\langle n_{d,\downarrow} \rangle = 1$ and $\langle n_{d,\uparrow} \rangle = 0$, we have the situation depicted in Figure 5.3 (but without the peak broadening effects due to Γ). It can easily be deduced that in the limit when $V_\mathbf{k} \to 0$ (but $U \neq 0$), the spectral function $J_\sigma(\omega)$ has two delta-function spikes at the values $\omega = \epsilon_d$ and $\omega = \epsilon_d + U$ (see Problem 5.7).

A useful result following from Equation (5.70) is that it can be used to deduce the magnetization due to the electron-spin polarizations. The magnetization is proportional to the difference in the average numbers of up and down spins, so we denote

$$m = \langle n_{d,\uparrow} \rangle - \langle n_{d,\downarrow} \rangle,$$

where

$$\langle n_{d,\uparrow} \rangle = \int d\omega \, n(\omega) J_\uparrow(\omega) = \int \frac{d\omega \, n(\omega)(\Gamma/\pi)}{[\omega - \epsilon_d' - U \langle n_{d,\downarrow} \rangle]^2 + \Gamma^2}. \tag{5.71}$$

At very low (effectively zero) temperature the Fermi–Dirac thermal occupation function $n(\omega)$ can be replaced by a step function that limits the range of integration to $\omega < 0$, and so we obtain

$$\langle n_{d,\uparrow} \rangle = \int_{-\infty}^{0} \frac{d\omega(\Gamma/\pi)}{[\omega - \epsilon_d' - U \langle n_{d,\downarrow} \rangle]^2 + \Gamma^2}$$

$$= \frac{1}{\pi} \cot^{-1} \left(\frac{\epsilon_d' + U \langle n_{d,\downarrow} \rangle}{\Gamma} \right). \tag{5.72}$$

From this, together with a similar result for the spin-down occupation number $\langle n_{d,\downarrow} \rangle$, we end up with a pair of coupled equations given by

$$\cot(\pi n_{d,\uparrow}) = \left(n_{d,\downarrow} + \frac{\epsilon_d'}{U} \right) \frac{U}{\Gamma},$$

$$\cot(\pi n_{d,\downarrow}) = \left(n_{d,\uparrow} + \frac{\epsilon_d'}{U} \right) \frac{U}{\Gamma}. \tag{5.73}$$

From an analysis of the condition for the preceding pair of equations to have a self-consistent solution with $m \neq 0$ (or $n_{d,\downarrow} \neq n_{d,\uparrow}$), we have a model for studying the conditions for the occurrence of a net magnetization in this system (see Problem 5.8).

5.6 Microscopic Theory of Superconductivity

The microscopic theory of superconductivity was developed in the 1950s, culminating in the BCS theory due to Bardeen, Cooper, and Schrieffer [88], who were awarded the 1972 Nobel Prize in physics. Various alternative formulations of the BCS theory were made soon after its publication [89–91] and here we will outline some calculations using the equation-of-motion method for the operators and real-time GFs. There are two stages in the development of the theory: one is to establish a model Hamiltonian in second quantization to account for super-conductivity, and the other is to obtain the solutions of this Hamiltonian for the superconducting state.

The starting point for obtaining a Hamiltonian to describe the interacting electrons in a superconductor has some broad similarities to the material in Subsection 1.4.2. In second quantization there could be single-particle terms (including the kinetic energy and the potential energy of the lattice) and a two-particle interaction that is quartic in the fermion operators and involves an interaction (denoted here by V). However, instead of the interaction depending on a single wave-vector transfer $\mathbf{q}$ (as depicted in Figure 1.3), a more complicated wave-vector dependence might be expected as well as a spin dependence. It was suggested by Cooper [92], in work leading up to the BCS theory, that a "pairing effect" occurs between an electron with wave vector and spin projection specified by $(\mathbf{k}, \uparrow)$ and another electron with $(-\mathbf{k}, \downarrow)$, or vice versa.

The Cooper pairs are electrons interacting through a net *attractive* term that involves virtual phonons in the surrounding lattice. The basic idea is that, due to the electron–phonon interaction, a phonon can be emitted by one electron and reabsorbed by the other member of the pair to produce this effect. We will defer the discussion of interaction processes of this type until Chapter 9 using diagrammatic perturbation methods for GFs. The BCS Hamiltonian corresponds to making two main assumptions: one is to ignore any background lattice potential, and the second is to introduce the pairing effect just mentioned as the dominant interaction. We will write

$$\mathcal{H} = \sum_{\mathbf{k},\sigma} \epsilon_{\mathbf{k}} a_{\mathbf{k},\sigma}^{\dagger} a_{\mathbf{k},\sigma} + \sum_{\mathbf{k},\mathbf{k}'} V(\mathbf{k},\mathbf{k}') a_{\mathbf{k}\uparrow}^{\dagger} a_{-\mathbf{k}\downarrow}^{\dagger} a_{-\mathbf{k}'\downarrow} a_{\mathbf{k}'\uparrow}. \tag{5.74}$$

From the dynamics of the electron–phonon interaction, it may be deduced (see Chapter 9) that $V(\mathbf{k},\mathbf{k}')$ is effectively nonzero only within a small range of values for $\mathbf{k}$ and $\mathbf{k}'$ close to the Fermi surface [92]. Often it is assumed, for simplicity, that

$$V(\mathbf{k},\mathbf{k}') = \begin{cases} -V_0 & \text{if } |\epsilon_{\mathbf{k}}| < \omega_D \text{ and } |\epsilon_{\mathbf{k}'}| < \omega_D, \\ 0 & \text{otherwise.} \end{cases} \tag{5.75}$$

Here V_0 is a positive constant, the particle energy $\epsilon_{\mathbf{k}}$ is measured relative to the Fermi energy, and ω_D denotes the Debye energy (see, e.g., [19, 20]) for the phonons participating in the process. Henceforth, in this section, we shall follow a convenient shorthand notation used in the superconductivity field by writing $\mathbf{k}$ for the state specified by $(\mathbf{k}, \uparrow)$ and $-\mathbf{k}$ for the state $(-\mathbf{k}, \downarrow)$. Then the BCS Hamiltonian in Equation (5.74) can be reexpressed as

$$\mathcal{H} = \sum_{\mathbf{k}} \epsilon_{\mathbf{k}}(a_{\mathbf{k}}^{\dagger} a_{\mathbf{k}} + a_{-\mathbf{k}}^{\dagger} a_{-\mathbf{k}}) - V_0 \sum_{\mathbf{k},\mathbf{k}'} a_{\mathbf{k}}^{\dagger} a_{-\mathbf{k}}^{\dagger} a_{-\mathbf{k}'} a_{\mathbf{k}'} \tag{5.76}$$

within the subspace of the paired electrons.

By using Equation (2.5) we will obtain the equations of motion for the operators $a_{\mathbf{k}}$ and $a_{-\mathbf{k}}^{\dagger}$, which are found to represent a coupled pair. On replacing the time derivatives in the equations according to $d/dt \to -i\omega$ for the excitations, we find

$$(\omega - \epsilon_{\mathbf{k}})a_{\mathbf{k}} = a_{-\mathbf{k}}^{\dagger} V_0 \sum_{\mathbf{k}'} a_{-\mathbf{k}'} a_{\mathbf{k}'}, \tag{5.77}$$

$$(\omega + \epsilon_{\mathbf{k}})a_{-\mathbf{k}}^{\dagger} = a_{\mathbf{k}} V_0 \sum_{\mathbf{k}'} a_{\mathbf{k}'}^{\dagger} a_{-\mathbf{k}'}^{\dagger}. \tag{5.78}$$

Next, provided the condition $|\epsilon_{\mathbf{k}}| < \omega_D$ required in Equation (5.75) is satisfied, we may replace the products of the pair of operators within the $\mathbf{k}'$ summations in Equations (5.77) and (5.78) by their averages. This is a mean-field type of simplification made in the same spirit as the decoupling approximation for the HF theory in Section 5.1, where there is also an average taken for products of two operators. Here we define the complex quantities

$$\Delta = V_0 \sum_{\mathbf{k}'} \langle a_{-\mathbf{k}'} a_{\mathbf{k}'} \rangle \quad \text{and} \quad \Delta^* = V_0 \sum_{\mathbf{k}'} \langle a_{\mathbf{k}'}^{\dagger} a_{-\mathbf{k}'}^{\dagger} \rangle, \tag{5.79}$$

and we ignore any weak $\mathbf{k}$-dependence that may be associated with Δ. After the decoupling approximation has been made, Equations (5.77) and (5.78) may be written in a matrix form as

$$\begin{pmatrix} \epsilon_{\mathbf{k}} & \Delta \\ \Delta^* & -\epsilon_{\mathbf{k}} \end{pmatrix} \begin{pmatrix} a_{\mathbf{k}} \\ a_{-\mathbf{k}}^{\dagger} \end{pmatrix} = \omega \begin{pmatrix} a_{\mathbf{k}} \\ a_{-\mathbf{k}}^{\dagger} \end{pmatrix}. \tag{5.80}$$

The nontrivial solutions of the preceding equation are simply $\omega = \pm E_{\mathbf{k}}$ corresponding to the eigenvalues of the 2×2 matrix, where

$$E_{\mathbf{k}} = \sqrt{\epsilon_{\mathbf{k}}^2 + \Delta_0^2} \tag{5.81}$$

and we have defined $\Delta_0 = |\Delta| > 0$. We now see that that Δ_0 represents an energy gap in the spectrum of electronic excitations. The prediction of a gap above the ground state is among the major achievements of the BCS theory and a necessary condition for superconductivity.

The eigenvectors of the preceding 2×2 matrix, which are the quasiparticle operators denoted as $\alpha_{\mathbf{k}}$ and $\alpha_{-\mathbf{k}}^{\dagger}$, are related to the original operators through a linear transformation (a Bogoliubov type of transformation similar to that in Subsection 1.5.1). We may express the transformation in the present case as

$$a_{\mathbf{k}} = u_{\mathbf{k}}\alpha_{\mathbf{k}} + v_{\mathbf{k}}\alpha_{-\mathbf{k}}^{\dagger}, \quad a_{-\mathbf{k}} = u_{\mathbf{k}}\alpha_{-\mathbf{k}} - v_{\mathbf{k}}\alpha_{\mathbf{k}}^{\dagger}. \tag{5.82}$$

It is easily shown that the new operators are also fermion operators satisfying the usual anticommutation relations, provided the condition $u_{\mathbf{k}}^2 + v_{\mathbf{k}}^2 = 1$ is satisfied. It is assumed that $u_{\mathbf{k}}$ and $v_{\mathbf{k}}$ are, respectively, even and odd functions of the wave vector: they satisfy $u_{\mathbf{k}} = u_{-\mathbf{k}}$ and $v_{\mathbf{k}} = -v_{-\mathbf{k}}$. It follows from Equations (5.80) and (5.82) that

$$\epsilon_{\mathbf{k}} u_{\mathbf{k}} + \Delta v_{\mathbf{k}} = E_{\mathbf{k}} u_{\mathbf{k}}$$

for the coefficients defining the quasiparticle operators. After taking the square modulus of each side and rearranging the terms, we find

$$\frac{2 u_{\mathbf{k}} v_{\mathbf{k}}}{u_{\mathbf{k}}^2 - v_{\mathbf{k}}^2} = \frac{\Delta_0}{\epsilon_{\mathbf{k}}}.$$

If we now write $u_{\mathbf{k}} = \cos\theta_{\mathbf{k}}$ and $v_{\mathbf{k}} = \sin\theta_{\mathbf{k}}$, the preceding result implies that the angle $\theta_{\mathbf{k}}$ is related to the energy gap by

$$\tan(2\theta_{\mathbf{k}}) = \Delta_0 / \epsilon_{\mathbf{k}}. \tag{5.83}$$

Several other properties follow from the preceding analysis, and we refer to [9] for a more detailed account. Thus, for example, it may be shown that the BCS ground state (denoted by $|\text{BCS}\rangle$), is related to the vacuum state $|0\rangle$ by

$$|\text{BCS}\rangle = \prod_{\mathbf{k}} (-1/v_{\mathbf{k}}) \alpha_{-\mathbf{k}} \alpha_{\mathbf{k}} |0\rangle = \prod_{\mathbf{k}} (u_{\mathbf{k}} + v_{\mathbf{k}} a_{\mathbf{k}}^\dagger a_{-\mathbf{k}}^\dagger) |0\rangle. \tag{5.84}$$

Here the factor of $(-1/v_{\mathbf{k}})$ in the first expression is included just to ensure the normalization condition that $\langle \text{BCS}|\text{BCS}\rangle = 1$. Also the second step has been made by using the inverse transformation to Equation (5.82), as well as the property that $a_{\mathbf{k}}|0\rangle = a_{-\mathbf{k}}|0\rangle = 0$ for the vacuum state (see Problem 5.9). Next we can use Equation (5.84) to evaluate $\alpha_{\mathbf{k}}|\text{BCS}\rangle$, which leads to

$$\alpha_{\mathbf{k}} |\text{BCS}\rangle = \alpha_{\mathbf{k}} \prod_{\mathbf{k}'} (-1/v_{\mathbf{k}'}) \alpha_{-\mathbf{k}'} \alpha_{\mathbf{k}'} |0\rangle$$

$$= \left(\prod_{\mathbf{k}' \neq \mathbf{k}} (-1/v_{\mathbf{k}'}) \alpha_{-\mathbf{k}'} \alpha_{\mathbf{k}'} \right) \times (1/v_{\mathbf{k}}) \alpha_{-\mathbf{k}} \alpha_{\mathbf{k}} \alpha_{\mathbf{k}} |0\rangle = 0.$$

The last part vanishes because it contains the fermion operator product $\alpha_{\mathbf{k}} \alpha_{\mathbf{k}}$ with the same wave vector. The preceding result proves that $|\text{BCS}\rangle$ acts as the "vacuum" state for the quasiparticle operator $\alpha_{\mathbf{k}}$, as might be anticipated.

Another result that arises from this version of the BCS theory is the self-consistent calculation of the gap Δ_0 by using Equation (5.79). We now see that

$$\langle a_{-\mathbf{k}} a_{\mathbf{k}} \rangle = \langle \text{BCS}|(u_{\mathbf{k}} \alpha_{-\mathbf{k}} - v_{\mathbf{k}} \alpha_{\mathbf{k}}^\dagger)(u_{\mathbf{k}} \alpha_{\mathbf{k}} + v_{\mathbf{k}} \alpha_{-\mathbf{k}}^\dagger)|\text{BCS}\rangle, \tag{5.85}$$

where the preceding average is taken with respect to the BCS ground state. When the products of operators are considered, the only surviving term is

$$\langle \text{BCS} | (u_\mathbf{k} \alpha_{-\mathbf{k}})(v_\mathbf{k} \alpha_{-\mathbf{k}}^\dagger) | \text{BCS} \rangle.$$

This can be manipulated into the form $u_\mathbf{k} v_\mathbf{k} \langle \text{BCS} | (1 - \alpha_{-\mathbf{k}}^\dagger \alpha_{-\mathbf{k}}) | \text{BCS} \rangle$, which simply reduces to $u_\mathbf{k} v_\mathbf{k}$. Thus, we have the result from Equation (5.79) that

$$\Delta = V_0 \sum_{\mathbf{k}'} u_{\mathbf{k}'} v_{\mathbf{k}'} = \frac{1}{2} V_0 \sum_{\mathbf{k}'} \sin(2\theta_{\mathbf{k}'}). \tag{5.86}$$

On noting that the right-hand side of the preceding equation is real and using Equation (5.83), we find that the energy gap in the zero-temperature limit satisfies the condition

$$1 = \frac{1}{2} V_0 \sum_{\mathbf{k}'} \left(\frac{1}{E_{\mathbf{k}'}} \right) = \frac{1}{2} V_0 \sum_{\mathbf{k}'} \left(\frac{1}{\sqrt{\epsilon_{\mathbf{k}'}^2 + \Delta_0^2}} \right). \tag{5.87}$$

An estimate of Δ_0 can be obtained from the preceding result if the summation over $\mathbf{k}'$ is replaced by an integral. It can be shown (see, e.g., [9, 23]) that an expression of the approximate form

$$\Delta_0 \simeq \omega_D \left\{ \sinh \left(\frac{1}{Z_F V_0} \right) \right\}^{-1} \tag{5.88}$$

is obtained, where Z_F is the density of energy states near the Fermi surface.

It should be pointed out that, although we have been following an equation-of-motion method for the operators in this section, a generalization to evaluate the analogous GF equations of motion can be made. For example, we can form the equation of motion for $G(a_\mathbf{k}; a_\mathbf{k}^\dagger | \omega)$. In doing so, we would find after a similar decoupling approximation to the one described in the preceding text that this GF is coupled to $G(a_{-\mathbf{k}}^\dagger; a_\mathbf{k}^\dagger | \omega)$. In turn, the equation of motion for this latter GF couples to $G(a_\mathbf{k}; a_\mathbf{k}^\dagger | \omega)$, and it is an easy matter to calculate the solution for each GF (see Problem 5.10). As might be expected, the GFs have poles at $\omega = \pm E_\mathbf{k}$.

Finally, we comment that the outstanding development in superconductivity since the advent of the BCS theory was the experimental discovery of the so-called high-temperature superconductors (HTSCs) by Bednorz and Müller [93] in 1986. The "conventional" superconductors prior to this discovery had critical temperatures T_c below about 30 K (a limitation related to the Cooper pairing mechanism involving phonons), but some of the new superconductors have T_c values well in excess of 100 K. The development of theories for HTSCs is still a matter of ongoing research.

Problems

5.1. Consider an exchange-dominated ferromagnet represented by the spin Hamiltonian in Equation (1.87). Verify that the commutation relation $[\mathcal{H}, S_n^+]$ can be written in the form quoted in Equation (5.15).

5.2. Starting from the equation-of-state result in Equation (5.26) for a Heisenberg ferromagnet, prove that the same expression for the Curie temperature T_C as specified in Equation (5.33) is obtained by applying the condition for the magnetic susceptibility χ to diverge as the temperature T approaches T_C from above (assuming a nonzero applied field B_0, which is eventually allowed to tend to zero).

5.3. Consider a Heisenberg ferromagnet in which there is an additional hyperfine coupling between each electronic spin $\mathbf{S}_i$ and the nuclear spin denoted by $\mathbf{I}_i$ at the same site. The Hamiltonian is

$$\mathcal{H} = -\frac{1}{2}\sum_{i,j} J_{i,j}\mathbf{S}_i \cdot \mathbf{S}_j + \sum_i A\mathbf{I}_i \cdot \mathbf{S}_i - g\mu_B B_0 \sum_i S_i^z - g_N \mu_N B_0 \sum_i I_i^z,$$

where $J_{i,j}$ is the exchange interaction, A is the hyperfine coupling constant, B_0 is an applied magnetic field in the z direction, g and μ_B are the g-factor and Bohr magneton for an electron, and g_N and μ_N are the corresponding quantities for a nuclear spin. The electronic and nuclear spins each satisfy the usual commutation relations among themselves, and they are assumed to commute with each other. Derive the equation of motion satisfied by the GF $G(S_m^+; S_n^- \mid \omega)$, and make a decoupling approximation (as in RPA) to decouple products of any spins of different types and/or at different sites. Show that the resulting equation contains another GF of the form $G(I_m^+; S_n^- \mid \omega)$ and write down its equation of motion. Next, prove that the pair of equations form a closed set and solve them by transforming to a wave-vector representation to obtain the original GF. Discuss the form of the SW solution(s) corresponding to the poles of this GF.

5.4. The four coupled equations of motion for a Heisenberg antiferromagnet in zero applied magnetic field ($B_0 = 0$) after RPA decoupling are quoted in Equation (5.40). Use these results to verify Equation (5.41) as the solutions for the analytically continued GFs and Equation (5.42) for the SW dispersion relation $E_\mathbf{k}$.

5.5. By following the analogous steps as used for graphene in Sections 2.7 and 4.2, prove the dispersion relation quoted in Equation (5.44) for the electronic bands in a 3D s.c. solid and the GF result in Equation (5.45). Note in this case that there is just a single lattice of equivalent sites.

5.6. Verify the result stated in Subsection 5.4.1 that Equation (5.48) is a valid representation of the spin-half operators (i.e., check that the usual spin commutation relations are recovered). Next, prove that Equations (5.46) and (5.47) are equivalent, as stated. The property that $(a_{i,\uparrow}^{\dagger} a_{i,\uparrow} + a_{i,\downarrow}^{\dagger} a_{i,\downarrow}) = 1$ may be useful.

5.7. Consider the Anderson model in Section 5.5 in the absence of the hybridization term ($V_{\mathbf{k}} = 0$) but including the Coulomb repulsion ($U \neq 0$). Extend the results given there to obtain the spectral intensity function in this special case.

5.8. By solving the coupled Equations (5.73) in the Anderson model of Section 5.5, deduce the regions of magnetic ($m \neq 0$) and nonmagnetic ($m = 0$) behavior in terms of the parameters of the model.

5.9. Show that the relationship between the BCS ground state, denoted as $|\text{BCS}\rangle$, and the vacuum state $|0\rangle$ is as quoted in Equation (5.84).

5.10. Considering the BCS Hamiltonian as in Equation (5.76), form the equation of motion for $G(a_{\mathbf{k}}; a_{\mathbf{k}}^{\dagger} | \omega)$. Show that it involves a new GF of the form $G(a_{-\mathbf{k}}^{\dagger}; a_{\mathbf{k}}^{\dagger} | \omega)$, and use the RPA decoupling to simplify the result. Next obtain the equation of motion for the new GF and show that it couples to GFs of the form $G(a_{\mathbf{k}}; a_{\mathbf{k}}^{\dagger} | \omega)$. Solve the set of coupled GF equations to find the solutions for the individual GFs. Verify that each GF has poles at $\omega = \pm E_{\mathbf{k}}$, with $E_{\mathbf{k}}$ given by Equation (5.81).

5.11. In Problem 4.13 the GFs $G(\rho_{\mathbf{q}}^{\dagger}(\mathbf{k}); \rho_{\mathbf{q}} | \omega)$ and $G(\rho_{\mathbf{q}}^{\dagger}; \rho_{\mathbf{q}} | \omega)$ were calculated for a fermion gas in the absence of interactions. Now, by using the GF equation-of-motion method (or otherwise), calculate these GFs with the effects of the interaction $v(\mathbf{q})$ included through a decoupling approximation of the same form as in the text following Equation (2.83). In particular, show that the expression for $G(\rho_{\mathbf{q}}^{\dagger}; \rho_{\mathbf{q}} | \omega)$ has an additional pole (compared to the noninteracting case), giving rise to the plasmon branch.

6

Linear Response Theory and Green's Functions

We now consider physical situations in which there is a time-dependent perturbation applied to a system that initially is described by a time-independent Hamiltonian and is in thermal equilibrium. In doing so, we will be extending various results for the time evolution of quantum many-body systems that were established in Chapter 2 in an operator formalism. Here we will be particularly concerned with studying the role of Green's functions (GFs) in determining the time dependences.

This topic will lead us directly into linear response theory, in which we consider the time evolution of a system in response to a sufficiently *weak* time-dependent perturbation. One of the main objectives will be to establish a formal connection between linear response functions and GFs. On the one hand, this may provide us with another method to calculate real-time GFs, as an alternative to the equation-of-motion method used so far. On the other hand, if we independently know the relevant GFs, it leads us to useful results for the overall time evolution of the system.

An important mathematical tool in carrying out the calculations will be the density matrix (or density operator). This was introduced into quantum mechanics by von Neumann in the 1920s (see [94]), and later found applications in statistical physics, e.g., see the excellent review by ter Haar [95]. The density matrix is useful for studying the ensemble averages and other properties of operators in nonequilibrium situations. After employing the density matrix here to develop linear response theory, we establish the useful connections with GFs and we also discuss the related concept of generalized susceptibility functions. We will provide examples of calculating GFs by the linear response method. This topic has applications to transport properties through deriving the Kubo formula, e.g., an electric field might be switched on causing a flow of electrons, and we would want to calculate the electrical conductivity. It will also be shown how some scattering problems can be treated within a similar formalism.

6.1 The Density Matrix

The density matrix is introduced following the approach in statistical physics textbooks (see, e.g., [4, 56, 96, 97]). We envisage an ensemble consisting of many ($\mathcal{R} \gg 1$) identically prepared systems, all of which are characterized by a Hamiltonian $\mathcal{H}$ that may be time dependent in general. We denote by $|\psi_k\rangle$ the time-dependent normalized wave function corresponding to the kth system in this ensemble. The time-dependent Schrödinger equation can be written as

$$i\frac{d\,|\psi_k\rangle}{dt} = \mathcal{H}\,|\psi_k\rangle\,, \qquad (k = 1, 2, \ldots, \mathcal{R}). \tag{6.1}$$

We also introduce a complete set of orthonormal functions $|n\rangle$, so that the wave functions at any time t can be expanded as

$$|\psi_k\rangle = \sum_n c_n^k(t)\,|n\rangle\,. \tag{6.2}$$

Here the time-dependent coefficients satisfy $c_n^k(t) = \langle n\,|\,\psi_k\rangle$. The state of the kth system in the ensemble can now be described in terms of the set of coefficients $\{c_n^k(t)\}$ for different n, and it follows that

$$i\frac{d}{dt}c_n^k(t) = i\frac{d}{dt}\langle n\,|\,\psi_k\rangle = i\langle n\,|\,\frac{d}{dt}\psi_k\rangle = \langle n\,|\,\mathcal{H}\,|\psi_k\rangle$$
$$= \langle n\,|\mathcal{H}\sum_m c_m^k(t)\,|m\rangle = \sum_m \langle n\,|\mathcal{H}\,|m\,\rangle c_m^k(t). \tag{6.3}$$

As a further property we note that

$$\sum_n |c_n^k(t)|^2 = 1$$

for the complete and orthonormal set. Therefore, each $|c_n^k(t)|^2$ factor can be interpreted as the probability of finding the kth system in the state $|n\rangle$ at time t.

We now define the *density matrix*, denoted by ρ, as the operator with matrix elements given by

$$\rho_{mn} = \frac{1}{\mathcal{R}}\sum_{k=1}^{\mathcal{R}} c_m^k(t)\big\{c_n^k(t)\big\}^*. \tag{6.4}$$

In other words, the (m, n) matrix element of ρ is the ensemble average of $c_m^k(t)\{c_n^k(t)\}^*$. We note, in particular, that a diagonal element ρ_{nn} of this matrix is just the ensemble average of the probability $|c_n^k(t)|^2$. Because the probabilities add to give unity, it follows that ρ has the property that

$$\text{Tr}(\rho) = 1. \tag{6.5}$$

Another property, which we will employ extensively in this chapter, is that the mean value (over the ensemble) for any operator A is given by

$$\langle A \rangle = \text{Tr}(\rho A).$$ (6.6)

This follows by starting from the definition of the mean value from QM that

$$\langle A \rangle = \frac{1}{\mathcal{R}} \sum_{k=1}^{\mathcal{R}} \langle \psi_k | A | \psi_k \rangle.$$

Then, from the expansion in Equation (6.2) and the definition of ρ, we have

$$\langle A \rangle = \frac{1}{\mathcal{R}} \sum_{k=1}^{\mathcal{R}} \sum_{m,n} c_m^k (c_n^k)^* \langle n | A | m \rangle$$

$$= \sum_{m,n} \rho_{mn} A_{nm} = \sum_{m} (\rho A)_{mm}.$$

The last term is just $\text{Tr}(\rho A)$ by definition, proving the stated result. In the special case of thermal equilibrium we know from statistical mechanics that the mean value is given by Equation (3.4), and so it follows in the equilibrium case that

$$\rho = \frac{e^{-\beta(\mathcal{H}-\mu\mathcal{N})}}{\text{Tr}\left\{e^{-\beta(\mathcal{H}-\mu\mathcal{N})}\right\}} = \frac{1}{Q} e^{-\beta(\mathcal{H}-\mu\mathcal{N})},$$ (6.7)

where we recall that $\beta = 1/k_B T$ is related to the temperature T. Of course, Equation (6.7) will *not* apply in a nonequilibrium situation.

An important result that we require for later use is that the density matrix ρ satisfies the following equation of motion:

$$i\frac{d\rho}{dt} = [\mathcal{H}, \rho].$$ (6.8)

This can be proved using the definition of ρ in Equation (6.4), which gives

$$i\frac{d\rho_{mn}}{dt} = \frac{1}{\mathcal{R}} \sum_{k=1}^{\mathcal{R}} i\frac{d}{dt} \left(c_m^k(t)\{c_n^k(t)\}^*\right)$$

$$= \frac{1}{\mathcal{R}} \sum_{k=1}^{\mathcal{R}} i \left(\frac{dc_m^k(t)}{dt}\{c_n^k(t)\}^* + c_m^k(t)\frac{d\{c_n^k(t)\}^*}{dt}\right).$$

Then, using Equation (6.3) and the definition of ρ one more time, this becomes

$$i\frac{d\rho_{mn}}{dt} = \sum_{p} (\langle m|\mathcal{H}|p\rangle \rho_{pn} - \rho_{mp}\langle p|\mathcal{H}|n\rangle) = (\mathcal{H}\rho)_{mn} - (\rho\mathcal{H})_{mn},$$

which proves the stated property. It is important to notice the difference in sign for the preceding result in Equation (6.8) compared with Equation (2.5) for the equation of motion of an operator in the Heisenberg picture.

6.2 Linear Response Theory

Here we will consider calculating the response of a system (that is otherwise unperturbed and in equilibrium) to a small time-varying external perturbation. We write for the total Hamiltonian

$$\mathcal{H} = \mathcal{H}_0 + \mathcal{H}_1,$$

where $\mathcal{H}_0$ describes the unperturbed system and $\mathcal{H}_1$ is the time-dependent perturbation. As a boundary condition, we assume that at $t = -\infty$ the system is in equilibrium (with a density matrix equal to ρ_0) and it is unperturbed. Because $d\rho_0/dt = 0$, it follows from Equation (6.8) that $[\mathcal{H}_0, \rho_0] = 0$. The perturbation is then switched on, starting at $t = -\infty$, and at any later time t we denote $\rho = \rho_0 + \rho_1$. From Equation (6.8) we have

$$i\frac{d\rho}{dt} = [\mathcal{H}_0 + \mathcal{H}_1, \rho_0 + \rho_1],$$

and so it follows that

$$i\frac{d\rho_1}{dt} = [\mathcal{H}_0, \rho_0] + [\mathcal{H}_0, \rho_1] + [\mathcal{H}_1, \rho_0] + [\mathcal{H}_1, \rho_1].$$

The first term on the right-hand side is zero, as just noted, while the last term will be neglected because it is of second order in the small quantities $\mathcal{H}_1$ and ρ_1. Therefore, making a *linear approximation*, we have

$$i\frac{d\rho_1}{dt} = [\mathcal{H}_0, \rho_1] + [\mathcal{H}_1, \rho_0]. \tag{6.9}$$

We now seek to solve this equation for ρ_1 without further approximation, using the boundary condition that $\mathcal{H}_1 = 0$ and $\rho_1 = 0$ at $t = -\infty$. To accomplish this, we introduce a new operator defined by

$$\tilde{\rho}_1 = e^{i\mathcal{H}_0 t}\, \rho_1\, e^{-i\mathcal{H}_0 t}. \tag{6.10}$$

On differentiating this with respect to time t, we find

$$\frac{d}{dt}\tilde{\rho}_1 = i\mathcal{H}_0 e^{i\mathcal{H}_0 t}\rho_1 e^{-i\mathcal{H}_0 t} - i e^{i\mathcal{H}_0 t}\rho_1\mathcal{H}_0 e^{-i\mathcal{H}_0 t} + e^{i\mathcal{H}_0 t}\frac{d\rho_1}{dt}e^{-i\mathcal{H}_0 t}.$$

Then, by using Equation (6.9) and rearranging the terms, it is easy to show that

$$\frac{d}{dt}\tilde{\rho}_1 = -i e^{i\mathcal{H}_0 t}[\mathcal{H}_1, \rho_0]e^{-i\mathcal{H}_0 t}.$$

Integrating both sides of this equation from time $-\infty$ to time t gives

$$\tilde{\rho}_1 = -i \int_{-\infty}^{t} e^{i\mathcal{H}_0 t'} [\mathcal{H}_1, \rho_0] e^{-i\mathcal{H}_0 t'} dt',$$

and on transforming back to ρ_1 we arrive at the formal solution

$$\rho_1 = -i \int_{-\infty}^{t} e^{i\mathcal{H}_0 (t'-t)} [\mathcal{H}_1, \rho_0] e^{-i\mathcal{H}_0 (t'-t)} dt'. \tag{6.11}$$

It follows, therefore, for the time-dependent average corresponding to any operator A, that Equation (6.6) gives

$$\langle A \rangle = \mathrm{Tr}(\rho A) = \mathrm{Tr}(\rho_0 A) + \mathrm{Tr}(\rho_1 A) = \langle A \rangle_0 + \mathrm{Tr}(\rho_1 A), \tag{6.12}$$

where $\langle A \rangle_0$ denotes an average for the unperturbed system (i.e., using the Hamiltonian $\mathcal{H}_0$). Then, from the preceding expression for ρ_1, we obtain

$$\langle A \rangle = \langle A \rangle_0 - i \int_{-\infty}^{t} \mathrm{Tr} \left\{ e^{i\mathcal{H}_0 (t'-t)} [\mathcal{H}_1, \rho_0] e^{-i\mathcal{H}_0 (t'-t)} A \right\} dt'. \tag{6.13}$$

This is an important result in its own right, and we will employ it later in applications to transport theory. For other applications, however, including making a connection with GFs, it is useful to rewrite the Tr term in Equation (6.13). Denoting this by I we have

$$I = \mathrm{Tr} \left\{ e^{i\mathcal{H}_0 (t'-t)} \mathcal{H}_1 \rho_0 e^{-i\mathcal{H}_0 (t'-t)} A - e^{i\mathcal{H}_0 (t'-t)} \rho_0 \mathcal{H}_1 e^{-i\mathcal{H}_0 (t'-t)} A \right\}$$

$$= \mathrm{Tr} \left\{ e^{i\mathcal{H}_0 (t'-t)} \mathcal{H}_1 e^{-i\mathcal{H}_0 (t'-t)} \rho_0 A - \rho_0 e^{i\mathcal{H}_0 (t'-t)} \mathcal{H}_1 e^{-i\mathcal{H}_0 (t'-t)} A \right\},$$

where the second line in the preceding equation follows because ρ_0 and $\mathcal{H}_0$ commute. Next we define

$$\mathcal{H}_1(t) = e^{i\mathcal{H}_0 t} \mathcal{H}_1 e^{-i\mathcal{H}_0 t}, \tag{6.14}$$

which is like transforming to the interaction picture in QM (see Chapter 2), so that we can write

$$I = \mathrm{Tr} \left\{ \mathcal{H}_1(t' - t) \rho_0 A - \rho_0 \mathcal{H}_1(t' - t) A \right\}$$

$$= \mathrm{Tr} \left\{ \rho_0 A \mathcal{H}_1(t' - t) - \rho_0 \mathcal{H}_1(t' - t) A \right\}.$$

Here we used the property that products of operators under the Tr are invariant under cyclic permutation. Finally we obtain I in the form

$$I = \mathrm{Tr} \left\{ \rho_0 [A, \mathcal{H}_1(t' - t)] \right\} = \langle [A, \mathcal{H}_1(t' - t)] \rangle_0.$$

As a result of all these manipulations, we conclude that Equation (6.13) for the time-dependent average can be expressed alternatively as

$$\langle A \rangle = \langle A \rangle_0 - i \int_{-\infty}^{t} \langle [A, \mathcal{H}_1(t' - t)] \rangle_0 \, dt'. \tag{6.15}$$

In summary, this last result tells us that to calculate an average $\langle A \rangle$ for a nonequilibrium system at time t we may instead calculate a different average, namely the correlation function $\langle [A, \mathcal{H}_1(t' - t)] \rangle_0$, which is for the *unperturbed* system. Then this average may be used in Equation (6.15). The unperturbed correlation function could be evaluated, if necessary, by the methods discussed earlier for equilibrium systems, e.g., by calculating the corresponding GF from its equation of motion. Equation (6.15) leads us more directly, however, to the introduction of linear response functions.

6.3 Response Functions and Green's Functions

We now suppose, in our previous development of linear response theory, that the perturbation $\mathcal{H}_1$ has the specific form

$$\mathcal{H}_1 = -Bf(t), \tag{6.16}$$

where B is a QM operator and $f(t)$ is a scalar function that describes the time dependence of the interaction. For example, if we consider a particle moving along the x axis in 1D and subject to a position-independent, time-varying force $f(t)$, then the potential energy corresponding to $\mathcal{H}_1$ is $-xf(t)$. Alternatively, if we have a magnetic system and a time-varying magnetic field $B^x(t)$ acting in the x direction is switched on, such that $b(t) = g\mu_B B^x(t)$, then the Zeeman energy of interaction is

$$\mathcal{H}_1 = -\sum_i g\mu_B B^x(t) S_i^x = -b(t) \sum_i S_i^x,$$

where the sum is over all magnetic sites. This is the physical situation in experiments for ferromagnetic resonance (FMR), which we will return to later, as well as other examples.

Substituting Equation (6.16) into (6.15) for the average $\langle A \rangle$, and assuming for simplicity that $\langle A \rangle_0 = 0$ for the unperturbed system, we have

$$\langle A \rangle = -i \int_{-\infty}^{t} \langle [A, B(t' - t)] \rangle_0 \, f(t') dt'.$$

From Equation (3.6) we recall the definition of a retarded GF for an "unperturbed" system (with Hamiltonian $\mathcal{H}_0$) in the commutator case as being

$$g_r(A; B \,|\, t - t') = -i\theta(t - t') \langle [A(t), B(t')] \rangle$$

when ε is chosen as 1. The right-hand side of the GF is just the same as $-i\theta(t - t')$ $\langle [A, B(t' - t)] \rangle$ because the dependence is only on the difference in time labels (see Subsection 3.1.2). Therefore, the preceding linear response result can be reexpressed as

$$\langle A \rangle = -\int_{-\infty}^{\infty} g_r(A; B \,|\, t - t') f(t') dt'. \tag{6.17}$$

Essentially, this expression tells us that the "response" produced in a local observable A (as measured by $\langle A \rangle$) is related linearly through an integral to the perturbing "force term" $f(t')$, which couples to another operator B. Consequently, the GF $g_r(A; B \,|\, t - t')$ can be thought of as a "response function." This interpretation is reminiscent of the role of the classical GFs in mathematics (see Subsection 3.6.1). For example, there is a formal analogy between the preceding Equation (6.17) and Equation (3.62).

The integral on the right-hand side of Equation (6.17) can be thought of as a convolution in the time domain between two functions, one of which depends on t' and the other (the GF) depends on the difference $t - t'$. This suggests that there may be a simplification if we take Fourier transforms from the time to the frequency domain. The Fourier transform of the GF was defined previously in Equation (3.65), and now we transform the time-dependent quantities $\langle A \rangle$ and $f(t)$ in a similar way by writing

$$\langle A \rangle = \int_{-\infty}^{\infty} \langle A \rangle_\omega e^{-i\omega t} d\omega, \tag{6.18}$$

$$f(t) = \int_{-\infty}^{\infty} F(\omega) e^{-i\omega t} d\omega. \tag{6.19}$$

The inverse transformation to Equation (6.18) gives

$$\langle A \rangle_\omega = \frac{1}{2\pi} \int_{-\infty}^{\infty} \langle A \rangle e^{i\omega t} dt$$
$$= -\frac{1}{2\pi} \int_{-\infty}^{\infty} \int_{-\infty}^{\infty} g_r(A; B \,|\, t - t') f(t') e^{i\omega t} dt' dt, \tag{6.20}$$

where we employed Equation (6.17) to obtain the second line in the preceding equation. After substituting Equations (3.65) and (6.19) into Equation (6.20) we eventually conclude (as in Problem 6.1) that

$$\langle A \rangle_\omega = -2\pi G(A; B \,|\, \omega) F(\omega). \tag{6.21}$$

This result can be useful in either of two ways, depending on the context. On the one hand, if we can calculate the GF $G(A; B \,|\, \omega)$, then we can find the time-dependent average $\langle A \rangle$, initially in terms of its Fourier components $\langle A \rangle_\omega$, for any given choice of time dependence in the perturbation term that transforms to $F(\omega)$. On the other hand, if we can find $\langle A \rangle_\omega$ by some other method, then the GF can be found from

$$G(A; B \mid \omega) = \left(-\frac{1}{2\pi}\right) \frac{\langle A \rangle_\omega}{F(\omega)}. \tag{6.22}$$

The preceding expression illustrates the role of the GF as being proportional to a linear response function (or more simply a response function): a perturbation specified by $F(\omega)$ is applied to the system and the response in terms of $\langle A \rangle_\omega$ is found, which gives (apart from a proportionality factor) the required GF. In the following text, we first mention some properties of the response function, and next we define a generalized susceptibility function. Then we examine some examples to illustrate the general principles.

• *Causality Relation*

Causality is a statement of the temporal relation between cause and effect. The function that describes the response is necessarily zero before the source input has been applied. It means that the response function $g(A; B \mid t - t')$, or more simply $g(t - t')$, is zero for $t < t'$, which is evident from the definition adopted in Section 3.1 for the *retarded* GF. Therefore, from the expression for the Fourier transform of $g(t - t')$ in Equation (3.22), we must have

$$\int_{-\infty}^{\infty} G(\omega) e^{-i\omega(t-t')} d\omega = 0 \quad \text{for } t < t'. \tag{6.23}$$

If we extend this to become a contour integral in the complex frequency plane, so that part of the contour is along the real axis and the contour is completed around a semicircle at infinity, it is evident when $t < t'$ that we should choose the contour to be in the *upper* half-plane for convergence (like the contour C_2 in Figure 3.1). Then we conclude for the contour integral to vanish there cannot be any poles of the integrand in the upper half-plane. This means $G(\omega)$ is analytic in the upper half-plane, and this is a direct consequence of the causality.

• *Time Interval Invariance*

We have already shown in Chapter 3 that the time correlation functions (and hence the GFs) depend on the labels t and t' only through the time interval $(t - t')$. The choice for the zero (or origin) for measuring the time is arbitrary. This is already evident from the form of Equation (6.17).

• *Impulse Response Functions*

If the perturbing force is an impulsive spike in time, applied at say time $t' = 0$, and represented as a unit delta function $f(t') = \delta(t')$, we obtain $\langle A \rangle = -g(t)$ from Equation (6.17). Therefore, $-g(t)$ describes the behavior of the system in the presence of an abrupt perturbation. In this case, $-g(t)$ would often be called the impulse response function. Again there is an analogy with the classical GFs in Subsection 3.6.1, in particular with Equation (3.64).

A different assumption for the time dependence of $f(t')$ may also be adopted. For example, a harmonic form as $f(t') \propto e^{-i\omega t'}$ is often convenient when studying the excitation frequencies in a system (as for the operator equations of motion in Chapter 2).

6.4 Response Functions and Applications

To illustrate further some of the basic concepts of linear response theory, we next introduce the concept of the generalized susceptibility as an application of the method, and then we consider a 1D damped harmonic oscillator as a classical example.

6.4.1 The Generalized Susceptibility

The response of the system to the perturbation is typically related to an external field through a response function or equivalently a generalized susceptibility. To obtain an expression for the susceptibility we consider the perturbation in the form given by Equation (6.16). The perturbation involves an operator B and a scalar time-dependent function $f(t)$, for which we seek a response in terms of $\langle A(t) \rangle$ for operator A. In the absence of the perturbation we will assume that the QM average of A is zero, but in the presence of the perturbation $\langle A(t) \rangle$ becomes nonzero with a value that can be found following Equation (6.17) as

$$\langle A(t) \rangle = -\int_{-\infty}^{\infty} g(t - t') f(t') dt', \tag{6.24}$$

where g is the appropriate GF providing the response. We now know, however, that causality requires that $g(t - t') = 0$ for $t < t'$. To satisfy this property Equation (6.24) becomes expressible as

$$\langle A(t) \rangle = -\int_{-\infty}^{t} g(t - t') f(t') dt', \tag{6.25}$$

which is just the same as

$$\langle A(t) \rangle = -\int_{0}^{\infty} g(\xi) f(t - \xi) d\xi. \tag{6.26}$$

Here we have changed variables by defining $\xi = t - t'$ (and thus we have $d\xi = -dt'$).

By means of a Fourier transformation, any time-dependent function can be thought of as being built up of many single-frequency components, each of which has a time-dependent factor like $e^{-i\omega t}$. For such a monochromatic driving function, on putting $f(t) = F(\omega)e^{-i\omega t}$, Equation (6.26) becomes

$$\langle A(t) \rangle = -\int_0^\infty g(\xi)F(\omega)e^{-i\omega(t-\xi)}d\xi$$

$$= -F(\omega)e^{-i\omega t}\int_0^\infty g(\xi)e^{i\omega\xi}d\xi. \tag{6.27}$$

We note that the only time dependence of $\langle A(t) \rangle$ is through the factor of $e^{-i\omega t}$, because the integration over ξ is independent of t. Thus a harmonic driving force $f(t) = F(\omega)e^{-i\omega t}$ produces a harmonic response that can be written as $\langle A(t) \rangle = \langle A \rangle_\omega e^{-i\omega t}$, where

$$\langle A \rangle_\omega = -F(\omega)\int_0^\infty g(\xi)e^{i\omega\xi}d\xi.$$

We may now follow convention [56] by defining a *generalized susceptibility* function $\chi(\omega)$ as

$$\chi(\omega) = \frac{\langle A \rangle_\omega}{F(\omega)}, \tag{6.28}$$

which in this simple case is just a response function related to the GF by

$$\chi(\omega) = -\int_0^\infty g(\xi)e^{i\omega\xi}d\xi. \tag{6.29}$$

The concept of a generalized susceptibility is more typically employed in cases where the driving function $f(t)$ is *real* (unlike in the example just given), and so the response produced in the system is real. An effect of the force is to change of the state of the system, along with there being an absorption of energy that is dissipated within the system. We can develop the preceding arguments to show the connection between the generalized susceptibility or response function $\chi(\omega)$ and the absorption (or dissipation) of energy due to the force f (see, e.g., [56]). To illustrate this point, we will consider a case in which the function $f(t)$ is real and takes the simple form

$$f(t) = \frac{1}{2}(f_0 e^{-i\omega t} + f_0^* e^{i\omega t}). \tag{6.30}$$

Using the definition of $\chi(\omega)$ we may then prove that

$$\langle A(t) \rangle = \frac{1}{2}\left[\chi(\omega)f_0 e^{-i\omega t} + \chi^*(\omega)f_0^* e^{i\omega t}\right]. \tag{6.31}$$

We see from the preceding example, and from Equation (6.26) in general, that the frequency-dependent generalized susceptibility $\chi(\omega)$ may be complex, but it satisfies

$$\chi(\omega) = \chi^*(-\omega). \tag{6.32}$$

The real and imaginary parts of $\chi(\omega)$ must therefore obey

$$\mathrm{Re}[\chi(\omega)] = \mathrm{Re}[\chi(-\omega)] \quad \text{and} \quad \mathrm{Im}[\chi(\omega)] = -\mathrm{Im}[\chi(-\omega)],$$

so the real part is a symmetric (or even) function of the frequency, and the imaginary part is an antisymmetric (or odd) function.

It follows that the average energy dissipation per unit time, or dW/dt, due to the force in Equation (6.30) can be found from the average value of the time derivative of the Hamiltonian in Equation (6.16):

$$\frac{dW}{dt} = \frac{d\mathcal{H}_1}{dt} = -\frac{df(t)}{dt}\langle A(t)\rangle.$$

By inserting Equation (6.31) for $\langle A(t)\rangle$ into the preceding expression and averaging over the period $2\pi/\omega$ of the force, we find (see Problem 6.2)

$$\frac{dW}{dt} = \frac{i\omega}{4}\left[\chi^*(\omega) - \chi(\omega)\right]|f_0|^2 = \frac{\omega}{2}\mathrm{Im}\left[\chi(\omega)\right]|f_0|^2. \tag{6.33}$$

We see that the imaginary part of $\chi(\omega)$ is directly related to the energy dissipation, which is an expression of the result obtained more generally as the fluctuation-dissipation theorem for the GFs (see Subsection 3.4.2).

6.4.2 Dissipation and Response for an Oscillator

We consider the damped harmonic oscillator as an example to illustrate the dissipative process. The classical equation of motion in 1D for an oscillator in the presence of a driving force $f(t)$ has the form

$$\frac{d^2u}{dt^2} + \Gamma\frac{du}{dt} + \omega_0^2 u = f(t). \tag{6.34}$$

Here u specifies the coordinate being driven, Γ is the friction (or damping) constant, and ω_0 denotes angular frequency for the natural resonance. The average displacement $\langle u(t)\rangle$ is related to the force $f(t)$ following Equation (6.17), and so in this case

$$\langle u(t)\rangle = -\int_{-\infty}^{\infty} g_r(t - t')f(t')dt', \tag{6.35}$$

where $g_r(t - t')$ is the retarded GF for the system. To find $g_r(t)$ for this system we first consider the Fourier transform relationship

$$g_r(t) = \int_{-\infty}^{\infty} d\omega \, G_r(\omega) e^{-i\omega t}. \tag{6.36}$$

By substituting Equations (6.35) and (6.36) into (6.34), we deduce that

$$G_r(\omega) = \left(-\frac{1}{2\pi}\right) \frac{1}{\omega_0^2 - \omega^2 - i\Gamma\omega}. \tag{6.37}$$

The corresponding generalized susceptibility function is, therefore

$$\chi(\omega) = \frac{1}{\omega_0^2 - \omega^2 - i\Gamma\omega}, \tag{6.38}$$

It is clear by inspection that the symmetry property stated in Equation (6.32) is satisfied.

The real and imaginary parts of $G_r(\omega)$ are easily written down from Equation (6.37), giving

$$\mathrm{Re}[G_r(\omega)] = \left(-\frac{1}{2\pi}\right) \frac{\omega_0^2 - \omega^2}{\left(\omega_0^2 - \omega^2\right)^2 + \Gamma^2\omega^2},$$

$$\mathrm{Im}[G_r(\omega)] = \left(-\frac{1}{2\pi}\right) \frac{\Gamma\omega}{\left(\omega_0^2 - \omega^2\right)^2 + \Gamma^2\omega^2}. \tag{6.39}$$

These quantities are shown plotted as functions of ω in Figure 6.1. In accordance with general principles, $\mathrm{Re}[G_r(\omega)]$ and $\mathrm{Im}[G_r(\omega)]$ must obey the Kramers–Kronig relations given in Equation (3.36).

Finally, we consider some consequences of the functional form in the complex $\tilde{\omega}$-plane of the generalized susceptibility in Equation (6.38). The complex poles of this function correspond to

$$\tilde{\omega} = -\frac{i\Gamma}{2} \pm \sqrt{\omega_0^2 - \frac{\Gamma^2}{4}}, \tag{6.40}$$

so the two main cases that arise are

- *The underdamped regime*: $\omega_0^2 > \Gamma^2/4$. Both of the poles have a negative imaginary part, and so they lie in the lower half plane (in agreement with causality).
- *The overdamped regime*: $\omega_0^2 < \Gamma^2/4$. Both of the poles are pure imaginary and lie on the negative imaginary axis (again in agreement with causality).

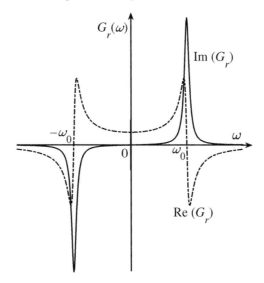

Figure 6.1 The real part (dashed line) and imaginary, or dissipative, part (solid line) of the GF in Equation (6.38) for the damped harmonic oscillator, plotted as a function of ω. For illustration we have taken $\Gamma/\omega_0 = 0.1$.

6.5 Phonons in an Infinite Elastic Medium

As an example of response functions in condensed matter systems, we consider an infinite, isotropic elastic medium with a longitudinal vibrational wave (longitudinal acoustic phonon) propagating in one direction taken as the z direction. This calculation will introduce properties relating to the spatial dependence of the response function or GF. The relevant acoustic wave equation in elasticity is well known [98] as

$$\rho \frac{\partial^2 u}{\partial t^2} - \rho v_L^2 \frac{\partial^2 u}{\partial z^2} = 0, \tag{6.41}$$

where u is the displacement in the z direction, ρ is the density of the medium, and v_L is the acoustic velocity. Suppose we now add a driving term (acting as a perturbation), which we choose to represent a harmonic point force of angular frequency ω applied at position z'. In other words, we choose

$$f(z,t) = \frac{f_0}{\overline{A}} e^{-i\omega t} \, \delta(z - z').$$

Here f_0 is an amplitude and $\overline{A}$ is the (macroscopically large) area of the medium in the xy-plane perpendicular to the wave propagation. This force term corresponds to an interaction energy

$$\mathcal{H}_1 = -\int\int\int u(z)f(z,t)dxdydz$$

$$= -f_0 e^{-i\omega t}\int u(z)\delta(z-z')dz = -f_0 e^{-i\omega t}u(z').$$

We now add the preceding force term to the wave equation of motion and we solve for $u(z)$. Because $\mathcal{H}_1$ couples to $u(z')$ and we are using a harmonic time representation, we expect to deduce the GF $G(u(z);u(z')\,|\,\omega)$ from the response function. So, with the driving force included, the wave equation in Equation (6.41) becomes generalized to

$$\rho\frac{\partial^2 u}{\partial t^2} + \rho\Gamma\frac{\partial u}{\partial t} - \rho v_L^2\frac{\partial^2 u}{\partial z^2} = \frac{f_0}{A}e^{-i\omega t}\delta(z-z'), \qquad (6.42)$$

where we have also added a small damping term (with coefficient $\Gamma > 0$). Assuming a time dependence as $\exp(-i\omega t)$, this simplifies to become

$$\frac{d^2 u}{dz^2} + q^2 u = -\frac{f_0}{\rho v_L^2 A}\delta(z-z'), \qquad (6.43)$$

where we have defined a complex quantity q by

$$q^2 = \frac{\omega(\omega + i\Gamma)}{v_L^2}. \qquad (6.44)$$

There are two solutions for q, and without loss of generality we may choose the root with $\mathrm{Re}(q) > 0$. It is easy to show that this also has $\mathrm{Im}(q) > 0$.

The solution of the preceding differential equation for the z dependence of u is found, as usual, from a complementary function plus a particular integral. The complementary function is

$$ae^{iqz} + be^{-iqz}, \qquad (6.45)$$

where a and b are constants. This function diverges as $|z| \to \pm\infty$ when $\mathrm{Im}(q) > 0$, and so we must have $a = b = 0$. We are left with the particular integral, which is of the form

$$u(z) = c\exp(iq|z-z'|), \qquad (6.46)$$

where c is a constant. This last step (along with obtaining the value of the constant c) can be verified by substituting it back into the differential equation, taking care to check that the correct behavior is obtained when $z - z' \to 0$ or $\pm\infty$. First, if we denote $q = q_1 + iq_2$ with $q_1 > 0$ and $q_2 > 0$, we have

$$\exp(iq|z-z'|) = \exp(iq_1|z-z'|)\exp(-q_2|z-z'|),$$

and this clearly tends to zero as $z - z' \to \pm\infty$. So the solution is bounded at these points, as required physically. Next, when $z \neq z'$ we can write

$$u = \begin{cases} u^+ = c\exp(iq[z - z']), & z > z' \\ u^- = c\exp(-iq[z - z']), & z < z'. \end{cases} \tag{6.47}$$

It is easily verified that u^+ and u^- satisfy the differential equation in their respective regions. Finally, we examine the behavior as $z - z' \to 0$. Clearly, there is no discontinuity between u^+ and u^-, but there is a slope discontinuity of $2iqc$ because

$$\frac{d}{dz}u^\pm = \pm iqu^\pm \to \pm iqc \quad \text{as} \quad |z - z'| \to 0.$$

Another expression for the slope discontinuity is obtained by integrating the differential equation with respect to z over an infinitesimal range from $z' - \delta$ to $z' + \delta$ (where $\delta \to 0$), giving

$$\int_{z'-\delta}^{z'+\delta} \left(\frac{d^2u}{dz^2} + q^2u\right) dz = -\frac{f_0}{\rho v_L^2 \overline{A}} \int_{z'-\delta}^{z'+\delta} \delta(z - z')dz.$$

This yields

$$\left(\frac{du}{dz}\right)_{z'+\delta} - \left(\frac{du}{dz}\right)_{z'-\delta} = -\frac{f_0}{\rho v_L^2 \overline{A}} = 2iqc.$$

Hence we conclude for the constant that $c = if_0/(2\rho v_L^2 q \overline{A})$ and the particular integral is

$$u(z) = \frac{if_0}{2\rho v_L^2 q \overline{A}} \exp\left(iq\left|z - z'\right|\right). \tag{6.48}$$

Finally, applying Equation (6.22) gives the result for the GF as

$$G(u(z); u(z') \mid \omega) = -\frac{i}{4\pi\rho v_L^2 q \overline{A}} \exp(iq|z - z'|). \tag{6.49}$$

This result illustrates the expected spatial dependence, i.e., it depends on the coordinates only through the separation $|z - z'|$, due to translational invariance. The result looks more familiar, however, when it is reexpressed in terms of the wavevector representation, as follows. If k is the wavenumber in the z direction, then we need

$$G(k, \omega) = \int_{-\infty}^{\infty} e^{-ik(z-z')} G(u(z); u(z') \mid \omega) d(z - z').$$

Denoting $\zeta = z - z'$, the integral can be written as

$$G(k,\omega) = -\frac{i}{4\pi\rho v_L^2 q \overline{A}} \int_{-\infty}^{\infty} e^{-ik\zeta} e^{iq|\zeta|} d\zeta$$

$$= -\frac{i}{4\pi\rho v_L^2 q \overline{A}} \left\{ \int_0^{\infty} e^{-ik\zeta} e^{iq\zeta} d\zeta + \int_{-\infty}^0 e^{-ik\zeta} e^{-iq\zeta} d\zeta \right\}.$$

The remaining integrals can easily be done, and the final result is

$$G(k,\omega) = \frac{1}{4\pi \overline{A}(\omega^2 + i\omega\Gamma - v_L^2 k^2)}, \tag{6.50}$$

where q has been rewritten in terms of ω, assuming the damping to be small. Note that in the zero-damping limit this has poles at $\omega = \pm v_L k$, as expected because $v_L k$ is the angular frequency of the acoustic phonon with wavenumber k.

6.6 Application to the Kubo Formalism

In the present section, we apply the linear response approach to derive a Kubo formula for transport-related phenomena such as the electrical conductivity, thereby providing a connection with correlation functions (and hence GFs). This method represents an alternative to the Boltzmann equation approach in which the concept of a collision time plays an important role (see textbooks on statistical physics, such as [4, 56, 96, 97]). This treatment of electrical conductivity is followed by analogous applications of Kubo-like formulas to the magnetic susceptibility and dielectric response function (see, e.g., [87, 100]).

6.6.1 The Electrical Conductivity

Here we assume a system of electrons (with charges $-e$) at instantaneous positions denoted by $\mathbf{r}_n$, where $n = 1, 2, \ldots, N$. A uniform electric field $\mathbf{E}$ is applied, which causes a flow of the charges giving us a current. We will show how to calculate the electrical conductivity for the system, as an example of how this can be done for transport properties more generally by using the linear response formalism of Section 6.2.

We note that the perturbation Hamiltonian in this case is

$$\mathcal{H}_1 = e \sum_n \mathbf{r}_n \cdot \mathbf{E}. \tag{6.51}$$

Now we employ Equation (6.13), choosing the operator A for this application to be a component of the current density, denoted as J^μ where μ is a Cartesian component. There is no net current flow in the absence of the perturbation, meaning $\langle J^\mu \rangle_0 = 0$, and so

$$\langle J^\mu \rangle = -i \int_{-\infty}^{t} \text{Tr}\left\{ e^{i\mathcal{H}_0(t'-t)}[\mathcal{H}_1, \rho_0] e^{-i\mathcal{H}_0(t'-t)} J^\mu \right\} dt'$$

$$= -i \int_{-\infty}^{t} \text{Tr}\left\{ [\mathcal{H}_1, \rho_0] e^{i\mathcal{H}_0(t-t')} J^\mu e^{-i\mathcal{H}_0(t-t')} \right\} dt', \qquad (6.52)$$

where a cyclic permutation of the operators under the trace has been made to obtain the second line of the equation. Then using the definition as in Equation (6.14) we may write

$$\langle J^\mu \rangle = -i \int_{-\infty}^{t} \text{Tr}\left\{ [\mathcal{H}_1, \rho_0] J^\mu(t - t') \right\} dt'$$

$$= -i \int_{0}^{\infty} \text{Tr}\left\{ \left[e \sum_{n,\nu} r_n^\nu E^\nu, \rho_0 \right] J^\mu(\xi) \right\} d\xi. \qquad (6.53)$$

In the last line we have denoted $\xi = t - t'$ and substituted for $\mathcal{H}_1$ using Equation (6.51). Also ν denotes a Cartesian component. Because the components $\sigma_{\mu\nu}$ of the conductivity tensor are defined by

$$\langle J^\mu \rangle = \sum_\nu \sigma_{\mu\nu} E^\nu, \qquad (6.54)$$

we obtain a formal expression for the electrical conductivity as

$$\sigma_{\mu\nu} = -i \int_{0}^{\infty} \text{Tr}\left\{ \left[e \sum_{n} r_n^\nu, \rho_0 \right] J^\mu(\xi) \right\} d\xi.$$

Noting now that for the unperturbed system of electrons we have the density matrix $\rho_0 = Q_0^{-1} e^{-\beta\mathcal{H}_0}$ with $Q_0 = \text{Tr}(e^{-\beta\mathcal{H}_0})$, the preceding result becomes

$$\sigma_{\mu\nu} = -\frac{i}{Q_0} \int_{0}^{\infty} \text{Tr}\left\{ \left[e \sum_{n} r_n^\nu, e^{-\beta\mathcal{H}_0} \right] J^\mu(\xi) \right\} d\xi. \qquad (6.55)$$

The preceding form of the expression for the conductivity is not very practical because it depends on the instantaneous positions of all the charges and there is no obvious way to evaluate the trace. It turns out to be convenient to use the following identity, which holds for any QM operator A:

$$[A, e^{-\beta\mathcal{H}_0}] = -i e^{-\beta\mathcal{H}_0} \int_{0}^{\beta} e^{\lambda\mathcal{H}_0} \frac{dA}{dt} e^{-\lambda\mathcal{H}_0} d\lambda. \qquad (6.56)$$

For the proof we start by considering the quantity $e^{\beta \mathcal{H}_0}[A, e^{-\beta \mathcal{H}_0}]$ and we formally differentiate this expression with respect to β, giving

$$
\frac{d}{d\beta}\left\{e^{\beta \mathcal{H}_0}[A, e^{-\beta \mathcal{H}_0}]\right\} = \frac{d}{d\beta}\left\{e^{\beta \mathcal{H}_0} A e^{-\beta \mathcal{H}_0} - A\right\}
$$

$$
= e^{\beta \mathcal{H}_0} \mathcal{H}_0 A e^{-\beta \mathcal{H}_0} - e^{\beta \mathcal{H}_0} A \mathcal{H}_0 e^{-\beta \mathcal{H}_0}
$$

$$
= e^{\beta \mathcal{H}_0}[\mathcal{H}_0, A]e^{-\beta \mathcal{H}_0} = -i e^{\beta \mathcal{H}_0}\frac{dA}{dt}e^{-\beta \mathcal{H}_0}.
$$

In the last step, we have used the operator equation of motion with respect to the unperturbed system. On integrating both sides with respect to β we find

$$
e^{\beta \mathcal{H}_0}[A, e^{-\beta \mathcal{H}_0}] = -i \int_0^\beta e^{\lambda \mathcal{H}_0}\frac{dA}{dt}e^{-\lambda \mathcal{H}_0}d\lambda.
$$

This can now be rearranged to give the required result in Equation (6.56).

Next we use this identity with the choice of the operator A being taken as

$$
A = e \sum_n r_n^\nu,
$$

which brings in another current operator through

$$
\frac{dA}{dt} = e \sum_n \frac{dr_n^\nu}{dt} = -J^\nu.
$$

Then, using this result in Equation (6.56) leads us to

$$
\left[e \sum_n r_n^\nu, e^{-\beta \mathcal{H}_0}\right] = -i e^{-\beta \mathcal{H}_0} \int_0^\beta e^{\lambda \mathcal{H}_0} J^\nu e^{-\lambda \mathcal{H}_0}d\lambda
$$

$$
- i e^{-\beta \mathcal{H}_0} \int_0^\beta J^\nu(-i\lambda)\, d\lambda,
$$

where we use the operator definition that $J^\nu(t) = e^{i \mathcal{H}_0 t} J^\nu e^{-i \mathcal{H}_0 t}$, as before, but with a *complex* time label corresponding to $t = -i\lambda$ in this case.

On substituting the preceding result into Equation (6.55), we find that

$$
\sigma_{\mu\nu} = \frac{1}{Q_0} \int_0^\infty d\xi \int_0^\beta d\lambda \, \mathrm{Tr}\left\{e^{-\beta \mathcal{H}_0} J^\nu(-i\lambda) J^\mu(\xi)\right\},
$$

which is just equivalent to

$$
\sigma_{\mu\nu} = \int_0^\infty d\xi \int_0^\beta d\lambda \langle J^\nu(-i\lambda) J^\mu(\xi)\rangle_0. \tag{6.57}
$$

This is known as the *Kubo formula* for the electrical conductivity [99]. It relates a matrix element of the conductivity to a current-current correlation function, which is to be evaluated for the equilibrium system. Our result shows that it is necessary only to calculate a GF of the type $G(J^\mu; J^\nu \mid \omega)$ from which the correlation function may be deduced by using the fluctuation-dissipation theorem. The final step would involve substituting the correlation function into Equation (6.57) and carrying out the integrations.

6.6.2 Other Kubo Formulas

Here we discuss two other applications more briefly by considering how Kubo-type formulas may be obtained for the magnetic susceptibility and for the dielectric response function.

The Magnetic Susceptibility

In this case, we consider a material subjected to a magnetic field and examine the linear response of the system to the field. We assume that the time-varying magnetic field is spatially homogeneous. The perturbation Hamiltonian for this situation is simply the Zeeman energy

$$\mathcal{H}_1 = -\mathbf{m} \cdot \mathbf{B}(t) = -\sum_\nu g\mu_B B^\nu(t) \sum_n S_n^\nu, \tag{6.58}$$

where ν is a Cartesian component, $\mathbf{m}$ is the total magnetic moment of the sample, and $\mathbf{S}_n$ is the spin at any magnetic site n. The magnetization $\mathbf{M}$ (or magnetic moment per unit volume) is defined by

$$\mathbf{M} = \frac{1}{V}\langle \mathbf{m} \rangle = \frac{g\mu_B}{V} \sum_n \langle \mathbf{S}_n \rangle. \tag{6.59}$$

Now we employ Equation (6.17), choosing the operator A for this application to be any component M^μ of the magnetization $\mathbf{M}$, so that at time t we have

$$\Delta M^\mu(t) \equiv \{M^\mu(t) - M^\mu(0)\} = \frac{g\mu_B}{V} \sum_n \{\langle S_n^\mu(t)\rangle - \langle S_n^\mu(0)\rangle\}$$

$$= -\frac{(g\mu_B)^2}{V} \sum_{n,m,\nu} \int_{-\infty}^\infty dt'\, B^\nu(t')\, g(S_n^\mu; S_m^\nu \mid t - t'). \tag{6.60}$$

We note that at least one of the equilibrium components (at $t = 0$) of the magnetization will be nonvanishing in a ferromagnetic material. The preceding equation can be used to define the magnetic susceptibility tensor:

$$\chi_{n,m}^{\mu\nu}(t,t') = -\frac{\mu_0}{V}(g\mu_B)^2 g(S_n^\mu; S_m^\nu \mid t - t'), \tag{6.61}$$

where μ_0 denotes the vacuum permeability. The preceding result, which is an analogue of the Kubo formula applied to the magnetic susceptibility, shows that it is sufficient to calculate a retarded GF of the type $g(S_n^\mu; S_m^\nu \mid t - t')$ to study the magnetic response of the system.

Inserting Equation (6.61) into Equation (6.60) leads to

$$\Delta M^{\mu}(t) = \frac{1}{\mu_0} \sum_{n,m} \sum_{\nu} \int_{-\infty}^{\infty} dt' \chi_{n,m}^{\mu\nu}(t - t') B^{\nu}(t').$$

Alternatively, in terms of frequency Fourier transforms on the right-hand side, we have

$$\Delta M^{\mu}(t) = \frac{1}{\mu_0} \sum_{n,m} \sum_{\nu} \int_{-\infty}^{\infty} d\omega \, e^{-i\omega t} \chi_{n,m}^{\mu\nu}(\omega) \, B^{\nu}(\omega).$$

Two particular forms of the magnetic susceptibility are usually of interest. If z denotes the direction of net magnetization, the longitudinal susceptibility is defined as

$$\chi_{n,m}^{zz}(\omega) = -\frac{\mu_0}{V}(g\mu_B)^2 G(S_n^z; S_m^z \mid \omega), \tag{6.62}$$

and the transverse susceptibility takes the form

$$\chi_{n,m}^{+-}(\omega) = -\frac{\mu_0}{V}(g\mu_B)^2 G(S_n^+; S_m^- \mid \omega). \tag{6.63}$$

The latter expression enters into the FMR response of the system, and we recall that this form of GF gives the SW frequencies (see Section 5.2).

The Dielectric Response Function of an Electron Gas

As another example, we now obtain the dielectric function $\varepsilon(\mathbf{q}, \omega)$ by studying the response function for an electron gas to an external charge density $\rho^{ext}(\mathbf{r}, t)$. A Fourier transform of the external charge density that will take us from position labels to wave-vector labels is defined by

$$\rho^{ext}(\mathbf{r}, t) = \int_{-\infty}^{\infty} d\omega \rho^{ext}(\omega) e^{-i\omega t}. \tag{6.64}$$

The interaction between the electrons and $\rho_{ext}(\mathbf{r}, t)$ is given by the Hamiltonian

$$\mathcal{H}_1(t) = \int_{-\infty}^{\infty} d^3r \int_{-\infty}^{\infty} d^3r' \rho(\mathbf{r}) w(|\mathbf{r} - \mathbf{r}'|) \rho^{ext}(\mathbf{r}', t), \tag{6.65}$$

where $\rho(\mathbf{r})$ is the charge density of the conduction electrons and $w(|\mathbf{r} - \mathbf{r}'|)$ is the electron–electron interaction as in Section 1.4.

Next we reexpress the preceding result in terms of wave vectors. A Fourier transform of $\rho(\mathbf{r})$ is defined by

$$\rho(\mathbf{r}) = \frac{1}{V} \sum_{\mathbf{q}} \rho_{\mathbf{q}} e^{i\mathbf{q}\cdot\mathbf{r}}. \tag{6.66}$$

Here we are assuming that the electrons occupying a volume V may all interact with one another. Therefore, Equation (6.64) becomes

$$\rho^{ext}(\mathbf{r},t) = \frac{1}{V} \int_{-\infty}^{\infty} d\omega \sum_{\mathbf{q}} \rho_{\mathbf{q}}^{ext}(\omega) e^{i\mathbf{q}\cdot\mathbf{r}} e^{-i\omega t}. \tag{6.67}$$

We will adopt a Coulomb interaction between charges, so $w(|\mathbf{r} - \mathbf{r}'|)$ is taken to be as in Equation (1.65) in the absence of screening ($\lambda \to 0$) for simplicity. Then its Fourier components $v(\mathbf{q})$ are given by Equation (1.66). After some straightforward algebraic manipulation the interaction Hamiltonian $\mathcal{H}_1$ takes the form

$$\mathcal{H}_1(t) = \int_{-\infty}^{\infty} d\omega\, e^{-i\omega t} \sum_{\mathbf{q}} v(\mathbf{q}) \rho_{-\mathbf{q}} \rho_{\mathbf{q}}^{ext}(\omega). \tag{6.68}$$

We remark that in the presence of the perturbation we will have a nonzero induced charge ρ^{in}, for which the expectation value can be calculated by linear response theory. The total charge density produced in the electron gas is given by

$$\rho^{tot}(\mathbf{r},t) = \rho^{ext}(\mathbf{r},t) + \rho^{in}(\mathbf{r},t). \tag{6.69}$$

At this stage, we introduce the frequency-dependent dielectric response function $\varepsilon(\mathbf{q},\omega)$ from electromagnetism by

$$\rho_{\mathbf{q}}^{tot}(\omega) = \frac{\rho_{\mathbf{q}}^{ext}(\omega)}{\varepsilon(\mathbf{q},\omega)}, \tag{6.70}$$

where $\rho_{\mathbf{q}}^{tot}(\omega)$ is the Fourier transform of $\rho^{tot}(\mathbf{r},t)$. From the preceding equations we obtain the induced charge density as

$$\rho_{\mathbf{q}}^{in}(\omega) = \left(\frac{1}{\varepsilon(\mathbf{q},\omega)} - 1 \right) \rho_{\mathbf{q}}^{tot}(\omega). \tag{6.71}$$

So far the treatment has been classical, but now we will make use of second quantization by replacing the electron density term $\rho_{-\mathbf{q}}$ in Equation (6.68). Specifically, using Equations (2.77) and (2.88), it is clear that the property $\rho_{\mathbf{q}}^{\dagger} = \rho_{-\mathbf{q}}$ is satisfied. Therefore, $\mathcal{H}_1$ becomes

$$\mathcal{H}_1 = \sum_{\mathbf{q}} \rho_{\mathbf{q}}^{\dagger} f(t,\mathbf{q}), \tag{6.72}$$

where the time-dependent, scalar perturbing term $f(t,\mathbf{q})$ is given by

$$f(t,\mathbf{q}) = v(\mathbf{q}) \int_{-\infty}^{\infty} d\omega\, e^{-i\omega t} \rho_{\mathbf{q}}^{ext}(\omega).$$

We may next examine how the induced charge density ρ^{in} responds to this perturbation. For this purpose we will find $\langle \rho^{in}(\mathbf{r},t) \rangle$ in accordance with the previous linear response theory by writing

$$\langle \rho^{in}(\mathbf{r},t)\rangle = (\langle \rho_{\mathbf{q}}(t)\rangle - \langle \rho_{\mathbf{q}}(0)\rangle)$$

$$= -\int_{-\infty}^{\infty} dt' f(t',\mathbf{q})\, g(\rho_{\mathbf{q}}; \rho_{\mathbf{q}}^\dagger \,|\, t-t').$$

In the last line we have used Equation (6.17) and the overall translational symmetry. After Fourier transforming the preceding equation we obtain

$$\langle \rho_{\mathbf{q}}^{in}(\omega)\rangle = -v(\mathbf{q})\rho_{\mathbf{q}}^{ext}(\omega)G(\rho_{\mathbf{q}}; \rho_{\mathbf{q}}^\dagger \,|\, \omega). \tag{6.73}$$

The dielectric function is now found in terms of a retarded GF because it follows from Equation (6.71) that

$$\frac{1}{\varepsilon(\mathbf{q},\omega)} = 1 - v(\mathbf{q})\, G(\rho_{\mathbf{q}}; \rho_{\mathbf{q}}^\dagger \,|\, \omega). \tag{6.74}$$

This result is a Kubo formula analogous to the expression (6.57) for the electrical conductivity response. To help understand the preceding result we may examine two limiting cases. First, if $\varepsilon(\mathbf{q},\omega)$ is very large, Equations (6.73) and (6.74) give the result that $\langle \rho^{in}\rangle = -\rho^{ext}$. This means that the induced charges in the electron gas produce a complete screening of the perturbation charges. Second, in the limit of $\varepsilon(\mathbf{q},\omega) \to 0$, we must have a singular behavior for the GF. It can then be seen that the poles of the GF correspond to the plasmon frequencies, as considered in Problem 5.11 (see also Section 9.2).

6.7 Inelastic Light Scattering

Here a brief discussion will be presented for the dynamics of the inelastic scattering of light, which technically may be either Raman scattering (RS) or Brillouin light scattering (BLS) depending on details of the experimental technique. The theoretical formalism is essentially the same in either case. In these processes the incoming light (or photon) is scattered out from the sample with a changed frequency and wave vector, due to the interactions that have taken place within the sample. These interactions may involve scattering by the crystal excitations, for example, by phonons due to the thermal disorder in the crystal lattice, or by SWs due to fluctuations in the magnetization, as in Figure 6.2. Some general references on inelastic light scattering are [76, 101].

The interaction of the EM field (light) with the system may be expressed in a general form through the perturbing Hamiltonian

$$\mathcal{H}_1 = \sum_{\mu,\nu}\sum_{\mathbf{r}} E_1^\mu \chi^{\mu\nu}(\mathbf{r}) E_2^\nu.$$

Here $\chi^{\mu\nu}(\mathbf{r})$ may be interpreted as a component of the polarisability (or electric susceptibility) tensor at position $\mathbf{r}$ with μ and ν denoting Cartesian coordinates.

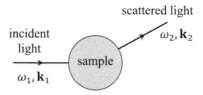

Figure 6.2 Schematic of the inelastic scattering of light, showing the incident light with angular frequency ω_1 and wave vector $\mathbf{k}_1$ together with a component of the scattered light specified by ω_2 and wave vector $\mathbf{k}_2$.

The vectors $\mathbf{E}_1$ and $\mathbf{E}_2$ refer to the incident and scattered electric fields, respectively (assumed to be independent of position inside the sample).

We assume that in the scattering process the system undergoes a transition from a state $|\alpha\rangle$ with energy W_α to another state $|\alpha'\rangle$ with energy $W_{\alpha'}$. From Fermi's golden rule the transition probability per unit time is proportional to

$$|\langle\mathbf{k}_1,\alpha|\mathcal{H}_1|\mathbf{k}_2,\alpha'\rangle|^2\delta(\omega_1 + W_\alpha - \omega_2 - W_{\alpha'}).$$

To arrive at the differential cross section $d^2\sigma/d\Omega d\omega_2$ for the light scattered into an elementary solid angle $d\Omega$ with frequency between ω_2 and $\omega_2 + d\omega_2$, we have to divide the above expression by the incident light flux (proportional to $|\mathbf{E}_1|^2$), then sum over all final states of the system, and finally average over all initial states of the system (assuming a probability distribution that we denote by P_α). All these steps combine to give

$$\frac{d^2\sigma}{d\Omega d\omega_2} \propto \sum_{\alpha,\alpha'} P_\alpha \frac{|\langle\mathbf{k}_1,\alpha|\mathcal{H}_1|\mathbf{k}_2,\alpha'\rangle|^2}{|\mathbf{E}_1|^2\,|\mathbf{E}_2|^2}\delta(\omega + W_\alpha - W_{\alpha'}),$$

where $\omega = \omega_1 - \omega_2$ is the frequency shift of the light. The extra division by $|\mathbf{E}_2|^2$ is just for convenience later, and also it symmetrizes the preceding expression.

On substituting for the interaction term $\mathcal{H}_1$ in the matrix element of states, we have

$$
\begin{aligned}
\langle\mathbf{k}_1,\alpha|\mathcal{H}_1|\mathbf{k}_2,\alpha'\rangle &= \sum_{\mu,\nu} E_1^\mu E_2^\nu \sum_{\mathbf{r}}\langle\mathbf{k}_1,\alpha|\chi^{\mu\nu}(\mathbf{r})|\mathbf{k}_2,\alpha'\rangle \\
&= \sum_{\mu,\nu} E_1^\mu E_2^\nu \frac{1}{N}\sum_{\mathbf{r}}\langle\alpha|\chi^{\mu\nu}(\mathbf{r})e^{i(\mathbf{k}_1-\mathbf{k}_2)\cdot\mathbf{r}}|\alpha'\rangle \\
&= \sum_{\mu,\nu} E_1^\mu E_2^\nu \langle\alpha|\chi^{\mu\nu}(\mathbf{k}_1-\mathbf{k}_2)|\alpha'\rangle,
\end{aligned}
$$

where plane-wave representations for the photon states $|\mathbf{k}_1\rangle$ and $|\mathbf{k}_2\rangle$ have been employed by writing, for example,

$$|\mathbf{k}_1\rangle = \frac{1}{\sqrt{N}}e^{-i\mathbf{k}_1 \cdot \mathbf{r}}, \quad |\mathbf{k}_2\rangle = \frac{1}{\sqrt{N}}e^{-i\mathbf{k}_2 \cdot \mathbf{r}},$$

where N is the number of atoms in the scattering volume. Also we have introduced a wave-vector Fourier transform for the polarisability by

$$\chi^{\mu\nu}(\mathbf{q}) = \frac{1}{N}\sum_{\mathbf{r}}\chi^{\mu\nu}(\mathbf{r})e^{i\mathbf{q}\cdot\mathbf{r}}.$$

We may now express the differential cross section for light scattering as

$$\frac{d^2\sigma}{d\Omega d\omega_2} \propto \sum_{\mu,\nu,\gamma,\delta}\frac{E_1^\mu E_2^\nu E_1^\gamma E_2^\delta}{|\mathbf{E}_1|^2|\mathbf{E}_2|^2}I_{scatt},$$

where we define the factor I_{scatt} for the scattered intensity by

$$I_{scatt} = \sum_{\alpha,\alpha'}P_\alpha\langle\alpha|\chi^{\mu\nu}(\mathbf{k}_1-\mathbf{k}_2)|\alpha'\rangle\langle\alpha'|\{\chi^{\gamma\delta}(\mathbf{k}_1-\mathbf{k}_2)\}^*|\alpha\rangle\,\delta(\omega+W_\alpha-W_{\alpha'}).$$

The preceding expression can be rewritten by using the delta function representation

$$\delta(\omega+W_\alpha-W_{\alpha'}) = \frac{1}{2\pi}\int_{-\infty}^{\infty}dt\,e^{i(\omega+W_\alpha-W_{\alpha'})t}.$$

From this result, and employing the notation $\mathbf{k} = \mathbf{k}_1 - \mathbf{k}_2$ for the change in wave vector, we have

$$I_{scatt} = \frac{1}{2\pi}\int_{-\infty}^{\infty}dt\,e^{i\omega t}\sum_{\alpha,\alpha'}P_\alpha e^{i(W_\alpha-W_{\alpha'})t}\langle\alpha|\chi^{\mu\nu}(\mathbf{k})|\alpha'\rangle\langle\alpha'|\{\chi^{\gamma\delta}(\mathbf{k})\}^*|\alpha\rangle$$

$$= \frac{1}{2\pi}\int_{-\infty}^{\infty}dt\,e^{i\omega t}\sum_{\alpha,\alpha'}P_\alpha\langle\alpha|e^{i\overline{\mathcal{H}}t}\chi^{\mu\nu}(\mathbf{k})e^{-i\overline{\mathcal{H}}t}|\alpha'\rangle\langle\alpha'|\{\chi^{\gamma\delta}(\mathbf{k})\}^*|\alpha\rangle,$$

where $\overline{\mathcal{H}} = \mathcal{H} - \mu\mathcal{N}$ is the Hamiltonian of the system.

On defining, as previously, $A(t) = e^{i\overline{\mathcal{H}}t}Ae^{-i\overline{\mathcal{H}}t}$ for any operator A, we now have

$$I_{scatt} = \frac{1}{2\pi}\int_{-\infty}^{\infty}dt\,e^{i\omega t}\sum_{\alpha,\alpha'}P_\alpha\langle\alpha|\chi^{\mu\nu}(\mathbf{k},t)|\alpha'\rangle\langle\alpha'|\{\chi^{\gamma\delta}(\mathbf{k},0)\}^*|\alpha\rangle.$$

Next we may make use of the completeness property for the states that

$$\sum_{\alpha'}|\alpha'\rangle\langle\alpha'| = 1,$$

which leads to

$$I_{scatt} = \frac{1}{2\pi} \int_{-\infty}^{\infty} dt\, e^{i\omega t} \sum_{\alpha} P_{\alpha} \langle \alpha | \chi^{\mu\nu}(\mathbf{k}, t) \{ \chi^{\gamma\delta}(\mathbf{k}, 0) \}^* | \alpha \rangle$$

$$= \frac{1}{2\pi} \int_{-\infty}^{\infty} dt\, e^{i\omega t} \langle \chi^{\mu\nu}(\mathbf{k}, t) \{ \chi^{\gamma\delta}(\mathbf{k}, 0) \}^* \rangle. \tag{6.75}$$

Putting together all the preceding steps, our final result for the light scattering cross-section can be expressed as

$$\frac{d^2\sigma}{d\Omega d\omega_2} = \mathcal{F} \sum_{\mu, \nu, \gamma, \delta} \hat{e}_1^{\mu} \hat{e}_2^{\nu} \hat{e}_1^{\gamma} \hat{e}_2^{\delta} \int_{-\infty}^{\infty} dt\, e^{i\omega t} \langle \chi^{\mu\nu}(\mathbf{k}, t) \chi^{\gamma\delta}(\mathbf{k}, 0) \}^* \rangle, \tag{6.76}$$

where $\hat{e}_1$ and $\hat{e}_2$ are the unit vectors in the direction the incident and scattered electric fields, $\mathbf{E}_1$ and $\mathbf{E}_2$, respectively. The proportionality factor has been denoted by $\mathcal{F}$ on the right-hand side of Equation (6.76), and according to [76, 101] it is given by

$$\mathcal{F} = \frac{\omega_1 \omega_2^3 n_2 \bar{V}}{16\pi^2 c^4 n_1}, \tag{6.77}$$

where n_1 and n_2 are the values of the refractive index at the incident and scattered light frequencies respectively, and $\bar{V}$ is the scattering volume within the sample.

The usual approach to be followed for evaluating the time-correlation function $\langle \chi^{\mu\nu}(\mathbf{k}, t) \chi^{\gamma\delta}(\mathbf{k}, 0)^* \rangle$, which appears in the integrand of Equation (6.76), is to calculate the corresponding GF of the form $G(\chi^{\mu\nu}(\mathbf{k}); \chi^{\gamma\delta}(\mathbf{k})^* | \omega)$ and then use the fluctuation-dissipation theorem. The light-scattering mechanism varies according to the type of excitation that may be involved. Typically, one may expand $\chi^{\mu\nu}$ as a power series in terms of the generalized coordinate (or the normal-mode variable) Q_j at any site j for the excitation, giving

$$\chi^{\mu\nu} = \chi_0^{\mu\nu} + \sum_j \chi_1^{\mu\nu\lambda} Q_j^{\lambda} + \sum_{j, j'} \chi_2^{\mu\nu\lambda\lambda'} Q_j^{\lambda} Q_{j'}^{\lambda'} + \cdots. \tag{6.78}$$

Here the χ coefficients on the right-hand side represent optical coupling parameters. In the case of light scattering from phonons, for example Q_j may be a component of the strain, which is just a spatial derivative of the lattice displacement $\mathbf{u}_j$ at site j (see, e.g., [56, 101]). To take a specific physical situation, we calculated in Section 6.5 the GF $G(u(z); u(z') | \omega)$ for an infinite elastic medium, considering for simplicity atomic displacements $u(z)$ in the z direction only, i.e., it is the case of a longitudinal elastic wave travelling in the z direction. The component of the strain in this case is $u^{zz} = du^z/dz$, and hence the corresponding strain-strain GF can be found from the result in Equation (6.49) by using

$$G(u^{zz}(z); u^{zz}(z') \mid \omega) = \frac{\partial}{\partial z} \frac{\partial}{\partial z'} G(u(z); u(z') \mid \omega). \tag{6.79}$$

From this and Equation (6.49) we may easily obtain

$$G(u^{zz}(z); u^{zz}(z') \mid \omega) = -\frac{iq}{4\pi \rho v_L^2 \overline{A}} \exp(iq|z - z'|). \tag{6.80}$$

The simplest (lowest-order) GF terms contributing to the light-scattering response are usually those having the form

$$G(Q_j; Q_{j'}^* \mid \omega).$$

In this case, it follows that it is Equation (6.80), multiplied by the appropriate elasto-optic coupling coefficients for the strength of the light scattering [101], that should be used together with Equation (6.76). The resulting description is for light-scattering processes involving the creation or annihilation of a single excitation (one phonon, in this case). The higher-order terms in the expansion in Equation (6.78) correspond to scattering processes that involve several excitations simultaneously.

As a final comment in this section, we note that the cross-section for scattering of particles (such as neutrons or electrons) can be expressed in terms of correlation functions, and hence GFs, by following a very similar procedure to the case of inelastic light scattering described here (e.g., see [102]).

Problems

6.1. By making the substitutions into Equation (6.20) as outlined in Section 6.3, verify that the linear response result quoted in Equation (6.21) is obtained.

6.2. With the interaction Hamiltonian taking the form in Equation (6.16) and using Equation (6.31), show that the average energy dissipation per unit time due to the force specified by Equation (6.30) is as quoted in Equation (6.33).

6.3. Use Equation (6.56) to show that the retarded commutator GF involving any two operators A and B can be expressed as

$$g(A; B \mid t - t') = -\theta(t - t') \int_0^\beta d\lambda \left\langle e^{\lambda \mathcal{H}_0} (dB/dt') e^{-\lambda \mathcal{H}_0} A(t) \right\rangle.$$

6.4. Consider a particle of mass m moving along a fixed direction with velocity v in a viscous fluid. The equation of motion is

$$\frac{dv}{dt} + \frac{\Gamma}{m} v = \frac{1}{m} f(t),$$

where Γ is a damping coefficient and $f(t)$ is the driving force. If the driving force is a delta function $f(t) = \delta(t)$, show that the GF (or the impulse response function) can be expressed as $g(t) = -(1/m)e^{-t/\tau}$ with $\tau = m/\Gamma$.

Suppose now we have a harmonic driving force with $f(t) = f_0(\omega)e^{-i\omega t}$. By using Equation (6.17) to find the expectation value of $\langle v(t) \rangle$ (or otherwise), deduce that the response produced in the velocity is also harmonic with the form $v_0(\omega)e^{-i\omega t}$. Hence show that the generalized susceptibility $\chi(\omega)$ for the system is proportional to

$$\frac{\tau}{m} \left(\frac{1 + i\omega\tau}{1 + \omega^2\tau^2} \right).$$

Sketch the real and imaginary parts of $\chi(\omega)$ as a function of ω.

6.5. Consider a driven harmonic oscillator corresponding to a particle of mass m and angular frequency ω_0 for free oscillations. Assume that the external coupling term has the form $-xf(t)$. Find an expression for the response function of the average displacement and obtain the susceptibility function.

6.6. Suppose a uniform electric field E is applied at time $t = 0$ to the electrons in a metal causing them to accelerate at a rate $dv/dt = eE/m$, with e and m being the charge and the mass of the electrons, respectively. Show that, after a small time interval δt, the electron flow (assumed to be in 1D) corresponds to a current density $J = ne^2 E\delta t/m$, where n is the electron density.

Suppose the impulsive application of the electric field is now over and the amplitude of the current drops off as $e^{-t/\tau}$. Find the conductivity $\sigma = J/E$ of this system.

6.7. Consider the Brownian (random) motion of a particle with mass m and velocity $v(t)$ at time t when it is immersed in a fluid with viscosity Γ. For simplicity, consider the motion in 1D only and assume that the total force on the particle is dominated by a frictional force proportional to the velocity. Find the susceptibility function for the Brownian particle.

6.8. Suppose the susceptibility function for the motion (in 1D) of a particle of mass m immersed in a fluid is given by $\chi(\omega) = (\Gamma - im\omega)^{-1}$, where Γ is a friction term. Verify that the Kramers–Kronig relations (3.36) are satisfied for this susceptibility.

6.9. In a macroscopic theory the semiclassical torque equation of motion for the magnetization in a ferromagnet is

$$d\mathbf{M}/dt = \gamma(\mathbf{M} \times \mathbf{B}),$$

where $\mathbf{M}$ and $\mathbf{B}$ denote the total effective magnetization and magnetic field, respectively, and γ is the gyromagnetic ratio. In the dipole-dipole limit where exchange can be neglected, we can write both of these fields in terms of a static part and a fluctuating part (with angular frequency ω in the form $\mathbf{M} = M_0\hat{\mathbf{i}}_z + \mathbf{m}(\mathbf{r})\exp(-i\omega t)$ and $\mathbf{B} = B_0\hat{\mathbf{i}}_z + \mu_0\mathbf{h}(\mathbf{r})\exp(-i\omega t)$. Here M_0 and B_0 are the static components, both taken to be in the z direction of magnetization (where $\hat{\mathbf{i}}_z$ denotes a unit vector). The fluctuating parts involve small terms $\mathbf{m}(\mathbf{r})$ and $\mathbf{h}(\mathbf{r})$. Show that in the linear response approximation the equation of motion implies the susceptibilty relationship

$$
\begin{pmatrix} m_x \\ m_y \\ m_z \end{pmatrix} = \begin{pmatrix} \chi_a & i\chi_b & 0 \\ -i\chi_b & \chi_a & 0 \\ 0 & 0 & 0 \end{pmatrix} \begin{pmatrix} h_x \\ h_y \\ h_z \end{pmatrix},
$$

where $\chi_a = \omega_m\omega_0/(\omega_0^2 - \omega^2)$ and $\chi_b = \omega_m\omega/(\omega_0^2 - \omega^2)$. We have defined the effective angular frequencies by $\omega_0 = \gamma B_0$ and $\omega_m = \gamma M_0$.

6.10. Typically, in magnetic resonance experiments (e.g., electron spin resonance or nuclear magnetic resonance) a static external magnetic field $\mathbf{B}_0 = B_0\hat{\mathbf{i}}_z$ (in the notation of the previous Problem 6.9) is applied to a paramagnetic system. The static magnetization of the system has the magnitude $M_0 = \chi_0 B_0$, with χ_0 denoting the static susceptibility of the system. A weak transverse oscillating pumping field corresponding to $\mathbf{h}(t) = h_x(t)\hat{\mathbf{i}}_x + h_y(t)\hat{\mathbf{i}}_y$ is also applied to the sample. In this case, the system can be described by the following torque equation of motion:

$$
\frac{d}{dt}M_\alpha = \gamma(\mathbf{M} \times \mathbf{B}(t))_\alpha - \Gamma_\alpha(M_\alpha - M_\alpha^0),
$$

where $\alpha = x, y, z$ are Cartesian components and $\mathbf{B}(t) = \mathbf{B}_0 + \mu_0\mathbf{h}(t)$. Damping is included through the coefficients Γ_α, and M_α^0 is the equilibrium value of the magnetization. Assuming $\Gamma_z = 1/T_1$ and $\Gamma_{x,y} = 1/T_2$, where $T_{1,2}$ are relaxation times, find the responses of the various components of the magnetization to the transverse field when $M_z = M_0$. Deduce the explicit form for $M_x(t)$ when the transverse field is harmonic, taking $\mathbf{h}(t) = h_0\cos(\omega_p t)\hat{\mathbf{i}}_x$.

7

Green's Functions for Localized Excitations

In this chapter we present some further examples of Green's function (GF) calculations, obtained either from the equation-of-motion approach or from linear response theory. The emphasis here will be on the spatial *localization* aspects of the quasiparticles or excitations. In general, localization of the excitations within a system may occur due to some breaking of the translational symmetry that would otherwise exist in an effectively infinite crystal lattice, where there are no boundary effects to be considered explicitly. Mode localization can arise in various circumstances. For example, there may be localization close to isolated impurities or defects in a lattice, at the surfaces of a 3D material, at the lateral edges of a 2D material like graphene, or at the interfaces between different layered materials.

Due to the symmetry breaking as produced in the previously mentioned examples, there may be modifications in the bulk modes or excitations (e.g., through changes induced in their spectral intensities as a function of position) as well as the occurrence of additional spatially decaying or modulated excitations in the vicinity of impurities or near surfaces and interfaces. Some general references covering these properties, typically with regard to the mode frequencies and/or amplitudes, for several different kinds of excitations are [39, 103–108]. By contrast, the topics in this chapter are directed more specifically toward applications of the GF and linear response methods.

7.1 Acoustic Phonons at Surfaces

So far, we have discussed phonons in solids using both a lattice dynamics approach (the infinite linear chain model) in Chapter 2 and a continuum theory of waves in an elastic medium in Chapter 6. Specifically, in Section 6.5 we used linear response theory to calculate GFs of the form $G(u(z); u(z') \mid \omega)$ for the acoustic phonons in an *infinite* elastic medium. We recall that $u(z)$ denotes the longitudinal displacement for an elastic wave traveling in the z direction. Here we will introduce the effects of a planar surface by considering a *semi-infinite* medium that fills the half space $z < 0$, while the region with $z > 0$ is taken to be a vacuum.

Previously, in Section 6.5 we required a solution of the wave equation that was bounded at infinity for both $z \to \pm\infty$. Following the same approach here, it is necessary only that the solution is bounded for $z \to -\infty$ inside the semi-infinite medium, while additionally it satisfies a boundary condition at the surface (which corresponds to the plane $z = 0$). The simplest elastic boundary condition is the requirement for zero stress at the free surface, implying

$$\frac{du}{dz} = 0 \quad \text{at } z = 0. \tag{7.1}$$

At this stage we may refer back to Equations (6.41)–(6.43), which still apply in the present linear response calculation, provided the impulse point with coordinate z' and the response point with coordinate z are both within the semi-infinite medium. For the complementary function, which has the same general form as in Equation (6.45), it is necessary that the constant $a = 0$ for the solution to be bounded as $z \to -\infty$. With the particular integral found as in Equation (6.48), it follows that the complete solution for $u(z)$ can be written as

$$u(z) = b \exp(-iqz) + \frac{if_0}{2\rho v_L^2 q \bar{A}} \exp(iq|z - z'|). \tag{7.2}$$

Finally, we must apply the boundary condition at $z = 0$ stated in Equation (7.1) to determine the constant b. It is easily verified that this gives $b = if_0/(2\rho v_L^2 q \bar{A}) \exp(-iqz')$. Hence the linear response result for the GF of the semi-infinite medium is

$$G(u(z); u(z') \mid \omega) = -\frac{i}{4\pi \rho v_L^2 q \bar{A}} \left\{ \exp\left(iq|z - z'|\right) + \exp\left(-iq(z + z')\right) \right\}. \tag{7.3}$$

The dependence of the preceding GF on the distances $-z$ and $-z'$ from the surface is typical of what is found in many other cases for excitations in a semi-infinite medium. In Equation (7.3) there are two terms, one involving $|z - z'|$ and the other $-(z + z')$. The former quantity represents the distance for direct propagation from the impulse (source) point to the response point, whereas the second term involves the path length for propagation involving a surface, as represented schematically in Figure 7.1.

The preceding calculation of the GF for the semi-infinite medium can be straightforwardly generalized to the case of a finite-thickness film (see Problem 7.1). There are now two surfaces, one at $z = 0$ and the other at $z = -L$, where L is the

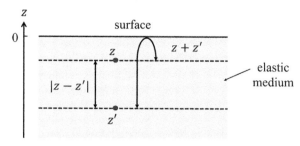

Figure 7.1 Schematic illustration of the dependence on z and z' for the two exponential terms of the GF in Equation (7.3), which applies for the acoustic phonons in a semi-infinite elastic medium.

film thickness. The same mathematical steps may be followed, and it should not be surprising that the GF result is similar to Equation (7.3) but with additional exponential terms such as $\exp(iq(2L + z + z'))$ representing a wave propagating between the points with z and z' involving the lower surface.

Apart from the displacement-displacement type of GF discussed in the preceding text for the phonons, it is often of interest (e.g., for applications to inelastic light scattering [101]) to study the strain-strain GFs such as $G(u^{zz}(z); u^{zz}(z') \mid \omega)$. As mentioned in Section 6.7, the longitudinal strain component is $u^{zz}(z) = du^z(z)/dz \equiv du(z)/dz$ because we are considering only longitudinal displacements in the z direction for this example. The relatively simple connection between the strain-strain GF and the displacement-displacement GF was given by Equation (6.79). Hence, using Equation (7.3) it follows that in the semi-infinite case we have

$$G(u^{zz}(z); u^{zz}(z') \mid \omega) = -\frac{iq}{4\pi \rho v_L^2 A} \left\{ \exp\left(iq|z - z'|\right) - \exp\left(-iq(z + z')\right) \right\}.$$

(7.4)

Applications of the preceding result, by analogy with what was done for the infinite elastic medium in Section 6.5, can be made to include surface effects in inelastic light scattering (see [109]). Also it is a simple matter to deduce the spectra of mean-square displacement $\langle |u(z)|^2 \rangle_\omega$ or mean-square strain $\langle |u^{zz}(z)|^2 \rangle_\omega$ as a function of distance $-z$ from the surface by taking $z' = z$ in Equations (7.3) or (7.4), respectively, and then applying the fluctuation-dissipation theorem as in Equation (3.40). For example, in the case of the mean-square strain we find

$$\langle |u^{zz}(z)|^2 \rangle_\omega = \frac{\{n(\omega) + 1\}}{\pi \rho v_L^2 A} \mathrm{Im}\left[iq\{1 - \exp(-2iqz)\}\right],$$

(7.5)

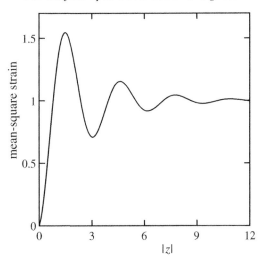

Figure 7.2 Schematic illustration of the dependence of the mean-square strain $\langle |u^{zz}(z)|^2 \rangle_\omega$ when plotted versus distance $|z|$ from the surface of a semi-infinite elastic medium, taking $q_2 = 0.2\, q_1$. The mean-square strain is expressed here in units of its value within an infinite crystal and $|z|$ is in units of π/q_1.

where $n(\omega) = (e^{\beta\omega} - 1)^{-1}$ is the Bose–Einstein thermal population factor for a phonon with frequency ω. If we denote $q = q_1 + iq_2$ with $q_1 > 0$ and $q_2 > 0$, as before in Section 6.5, we have

$$\langle |u^{zz}(z)|^2 \rangle_\omega \simeq \frac{\{n(\omega) + 1\}}{\pi \rho v_L^2 \overline{A}} q_1 \{1 - \exp(2q_2 z) \cos(2q_1 z)\}, \qquad (7.6)$$

assuming weak damping (so that $q_2 \ll q_1$). This quantity is zero at the surface ($z = 0$) and oscillates strongly as a function of $|z|$ with wavelength π/q_1 near the surface. Further into the material, the oscillations decrease in amplitude, and $\langle |u^{zz}(z)|^2 \rangle_\omega$ tends to a constant value (the same value as found for an infinite crystal) when $|z| \gg 2\pi/q_2$. A numerical example to illustrate this behavior is given in Figure 7.2.

7.2 Surface Spin Waves in Ferromagnets

In this next application we will investigate the spectrum of the SWs in a 3D Heisenberg ferromagnet with a planar surface. The objectives will be to study how, by using GFs, the bulk SWs are modified near a surface and also to show that localized surface SWs may exist under certain conditions. Specifically, we consider a b.c.c. ferromagnet such as Ni, and we take the case of a semi-infinite material with

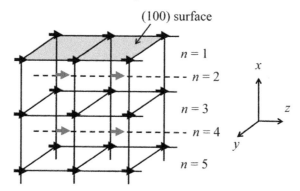

Figure 7.3 Schematic illustration of a semi-infinite b.c.c. Heisenberg ferromagnet with a (100) oriented surface (shown shaded) in the yz plane. The layers of atoms parallel to the surface are labeled with integer n $(= 1, 2, 3, \ldots)$ starting at the surface.

a (100) oriented surface. We have already shown in Subsection 1.5.2 that the bulk SW dispersion relation in an infinite Heisenberg ferromagnet at low temperatures $T \ll T_C$ and in the absence of dipole-dipole interactions is given by Equations (1.93) and (1.94).

The assumed geometry for the semi-infinite ferromagnet is illustrated in Figure 7.3, where we take all the spins to be aligned in the z direction parallel to the surface, which lies in the yz plane. We show the spins at the vertices of a cube (with sides equal to the lattice parameter a) in black and those at the body-centered sites in gray. The scheme for labeling the atomic layers parallel to the surface is indicated, and so the distance of layer n from the surface layer is given by $(n-1)a/2$. Each spin in layer n is coupled by the exchange J to its eight nearest neighbors in layers $n-1$ or $n+1$, with the exception of the spins in layer $n = 1$, which are coupled only to their four nearest neighbors in layer 2.

We will employ the Holstein–Primakoff (HP) transformation from spin operators to boson operators at low temperatures, as was done in Sections 1.5 and 2.6, so the Hamiltonian $\mathcal{H}$ in the site representation is again given by Equation (1.81) where the summations are now restricted to the sites in the semi-infinite ferromagnet. We can Fourier transform to a "mixed" representation for the boson operators in which there is a 2D wave vector $\mathbf{k}_{\parallel} = (k_y, k_z)$ parallel to the surface and a layer number n perpendicular to the surface. Therefore, we choose to study GFs of the form $G(a_n(\mathbf{k}_{\parallel}); a_{n'}^{\dagger}(\mathbf{k}_{\parallel}) \mid \omega)$, where the appropriate form of $\mathcal{H}$ in this representation can be expressed (apart from a constant term) as

$$\mathcal{H} = \sum_{\mathbf{k}_{\parallel}} \sum_{n, n'} A_{n,n'}(\mathbf{k}_{\parallel}) a_n^{\dagger}(\mathbf{k}_{\parallel}) a_{n'}(\mathbf{k}_{\parallel}). \tag{7.7}$$

The summations here are over the layer indices n and n', which range from 1 to ∞. It can be straightforwardly shown for the b.c.c. geometry that (see Problem 7.3) the only nonzero terms in the array $A_{n,n'}(\mathbf{k}_{\parallel})$ are

$$A_{1,1}(\mathbf{k}_{\parallel}) = g\mu_B B_0 + 4SJ,$$
$$A_{n,n}(\mathbf{k}_{\parallel}) = g\mu_B B_0 + 8SJ, \qquad (n > 1),$$
$$A_{n,n+1}(\mathbf{k}_{\parallel}) = A_{n+1,n}(\mathbf{k}_{\parallel}) = -4SJ\gamma(\mathbf{k}_{\parallel}), \qquad (n \geq 1), \qquad (7.8)$$

where we have introduced the 2D structure factor defined by $\gamma(\mathbf{k}_{\parallel}) = \cos\left(\frac{1}{2}k_y a\right)\cos\left(\frac{1}{2}k_z a\right)$. All the other $A_{n,n'}(\mathbf{k}_{\parallel})$ terms are zero because the exchange couples only the nearest-neighbor sites. It follows that when we construct the GF equations of motion for $G_{n,n'} \equiv G(a_n(\mathbf{k}_{\parallel}); a_{n'}^{\dagger}(\mathbf{k}_{\parallel}) \mid \omega)$, keeping n' to be fixed, we obtain the coupled equations

$$(\omega - g\mu_B B_0 - 4SJ)G_{1,n'} - 4SJ\gamma(\mathbf{k}_{\parallel})G_{2,n'} = \delta_{1,n'}/2\pi, \qquad (n = 1),$$
$$(\omega - g\mu_B B_0 - 8SJ)G_{n,n'} - 4SJ\gamma(\mathbf{k}_{\parallel})\left\{G_{n-1,n'} + G_{n+1,n'}\right\}$$
$$= \delta_{n,n'}/2\pi, \qquad (n > 1). \qquad (7.9)$$

The equation for $n = 1$ is different from the others because there are fewer nearest neighbors for a spin in the surface layer.

Next we may divide throughout by a factor $4SJ\gamma(k_{\parallel})$ to rewrite the set of coupled equations (7.9) in a matrix form as

$$\mathbf{M}\mathbf{G}_{n'} = \mathbf{b}_{n'}, \qquad (7.10)$$

Here the matrix elements of the column matrix $\mathbf{G}_{n'}$ are the required GFs for a fixed value of n', and $\mathbf{M}$ is a matrix having the form

$$\mathbf{M} = \begin{pmatrix} d+\Delta & -1 & 0 & 0 & \cdots \\ -1 & d & -1 & 0 & \cdots \\ 0 & -1 & d & -1 & \cdots \\ 0 & 0 & -1 & d & \cdots \\ \vdots & \vdots & \vdots & \vdots & \ddots \end{pmatrix}. \qquad (7.11)$$

The quantity d is related to the excitation frequency ω by

$$d = \frac{g\mu_B B_0 + 8SJ - \omega}{4SJ\gamma(\mathbf{k}_{\parallel})}, \qquad (7.12)$$

whereas Δ depends only on the wave vector through

$$\Delta = -\frac{1}{\gamma(\mathbf{k}_{\parallel})}. \qquad (7.13)$$

Finally, the inhomogeneous term $\mathbf{b}_{n'}$ in Equation (7.10) is a column matrix with matrix elements defined by

$$(\mathbf{b}_{n'})_n = \delta_{n,n'}/[8\pi\, S J \gamma\, (\mathbf{k}_{\parallel})]. \tag{7.14}$$

The formal solution for the GFs from Equation (7.10) is given by $\mathbf{G}_{n'} = \mathbf{M}^{-1}\mathbf{b}_{n'}$, so it becomes necessary to find the inverse of the matrix $\mathbf{M}$. Fortunately, this is relatively straightforward because $\mathbf{M}$ has a special form known as a *tridiagonal matrix* (or TDM), i.e., its only nonzero matrix elements are on the leading diagonal and on the two diagonals on either side. In addition, the matrix elements are the same along the leading diagonal, except at the top left-hand corner. With this in mind, it is convenient to separate $\mathbf{M}$ into two parts as $\mathbf{M}_0 + \mathbf{D}$, where

$$\mathbf{M}_0 = \begin{pmatrix} d & -1 & 0 & \cdots \\ -1 & d & -1 & \cdots \\ 0 & -1 & d & \cdots \\ \vdots & \vdots & \vdots & \ddots \end{pmatrix}, \qquad \mathbf{D} = \begin{pmatrix} \Delta & 0 & 0 & \cdots \\ 0 & 0 & 0 & \cdots \\ 0 & 0 & 0 & \cdots \\ \vdots & \vdots & \vdots & \ddots \end{pmatrix}. \tag{7.15}$$

Unperturbed TDMs having the same form as $\mathbf{M}_0$ are frequently encountered in surface and interface problems when the interactions are short range, such as between nearest neighbors only (see, e.g., [110, 111]). In fact, we shall come across another example involving graphene nanoribbons (GNRs) later in this chapter.

It is known from linear algebra that the inverse of the unperturbed matrix $\mathbf{M}_0$ can be written as

$$(\mathbf{M}_0^{-1})_{i,j} = \frac{x^{i+j} - x^{|i-j|}}{x - x^{-1}}, \tag{7.16}$$

where x is a variable (which may be complex in general) defined by

$$x + x^{-1} = d \tag{7.17}$$

and $|x| \le 1$. In the present case d is related to the frequency ω by Equation (7.12). Although it is no straightforward matter to prove Equation (7.16) from first principles (see [112] for a derivation), it may readily be checked by matrix multiplication that the right-hand side of Equation (7.16) is indeed the inverse. We refer to Problem 7.4 for more details.

Given the preceding result for the inverse of $\mathbf{M}_0$, the inverse of $\mathbf{M}$ can next be deduced by using

$$\mathbf{M}^{-1} = (\mathbf{M}_0 + \mathbf{D})^{-1} = (\mathbf{I} + \mathbf{M}_0^{-1}\mathbf{D})^{-1}\mathbf{M}_0^{-1}, \tag{7.18}$$

where $\mathbf{I}$ is the unit matrix. It is then a simple algebraic exercise to evaluate the terms on the right-hand side of the preceding equation by using the result quoted for $\mathbf{M}_0^{-1}$ together with the fact that $\mathbf{D}$ is a sparse matrix, which has only one nonzero matrix element. In this way it can be verified that

$$(\mathbf{M}^{-1})_{i,j} = \frac{1}{x - x^{-1}} \left[x^{i+j} \left(\frac{1 + x^{-1}\Delta}{1 + x\Delta} \right) - x^{|i-j|} \right]. \tag{7.19}$$

From the formal properties of the matrix $\mathbf{M}$ as described in Equations (7.15)–(7.19), it becomes a matter of simple algebra to obtain the solutions for the GFs in this case. The results are

$$G(a_n(\mathbf{k}_\parallel); a_{n'}^{\dagger}(\mathbf{k}_\parallel) \mid \omega) = \frac{1}{8\pi SJ\gamma(\mathbf{k}_\parallel)(x - x^{-1})} \left[x^{n+n'} \left(\frac{1 + x^{-1}\Delta}{1 + x\Delta} \right) - x^{|n-n'|} \right]. \tag{7.20}$$

We turn now to the physical interpretation of these rather formal expressions for the GFs. First, each GF involves numerator factors of x to an integer power. If the parameter x behaves like a propagating phase factor (i.e., if $x = e^{i\theta}$ with θ being real), we have a description of bulk SW modes. This can be seen, in fact, by putting $x = \exp(i\frac{1}{2}k_x a)$. Then, using Equations (7.12) and (7.17), we find that $d = 2\cos(\frac{1}{2}k_x a)$ and the solution for the mode frequency is $\omega = \omega_B(k_x, \mathbf{k}_\parallel)$. In this context k_x represents the third wave-vector component and

$$\omega_B(k_x, \mathbf{k}_\parallel) = g\mu_B B_0 + 8SJ \left[1 - \cos(\frac{1}{2}k_x a)\gamma(\mathbf{k}_\parallel) \right]. \tag{7.21}$$

Comparison with Equations (1.93) and (1.94) shows that the preceding result is just equivalent, as expected, to the SW dispersion relation found previously for an infinite b.c.c. ferromagnet. More significantly, another excitation frequency may be found from the factor $(1 + x\Delta)$ appearing in the denominator of Equation (7.20) for the GF. The pole corresponds to $x = -1/\Delta = \gamma(\mathbf{k}_\parallel)$, which satisfies the requirement $|x| < 1$ for a localized (spatially decaying) mode, except when $\mathbf{k}_\parallel = 0$. The frequency for the surface SW is found using Equations (7.12) and (7.17) to be $\omega = \omega_S(\mathbf{k}_\parallel)$, where

$$\omega_S(\mathbf{k}_\parallel) = g\mu_B B_0 + 4SJ \left[1 - \gamma^2(\mathbf{k}_\parallel) \right]. \tag{7.22}$$

It can be verified that there are no further poles in the GF Equation (7.20) that give rise to SW excitations. In particular, the denominator factor $(x - x^{-1})$ yields only frequencies that are already included in the bulk SW band. The dispersion relations represented by Equations (7.21) and (7.22) for a semi-infinite b.c.c. ferromagnet are illustrated in Figure 7.4. In this plot of the scaled SW frequency against the 2D in-plane wave vector $|\mathbf{k}_\parallel|$, we have a surface SW branch that splits off from a band (shown shaded) representing the bulk SWs at different k_y values. This bulk band has lower and upper boundaries corresponding to $k_x = 0$ and π/a, respectively.

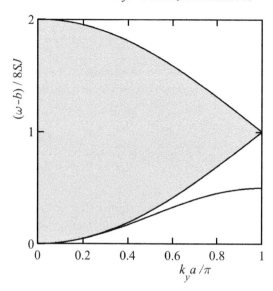

Figure 7.4 Example of SW dispersion relations in a semi-infinite b.c.c. ferromagnet with a (100) surface. The band of bulk SW modes (shaded) and a surface SW branch (split off below the bulk continuum) are shown plotted against the scaled k_y component of the in-plane wave vector (taking the other component $k_z = 0$ for simplicity). The Brillouin zone boundary is at $k_y = \pi/a$. We denote $b = g\mu_B B_0$.

7.3 Edge Modes in Graphene Nanoribbons

The lattice structure for a 2D sheet of graphene in the xy plane was depicted in Figure 2.6, showing the two types of sublattice sites labeled as A and B. If the sheet is terminated to form an edge, there are two high-symmetry directions, described as zig-zag (ZZ) and armchair (AC), in which this may occur. Thus, for example, in Figure 2.6 the vertical sides of the segment of the lattice shown are ZZ edges, while the horizontal parts at the top and bottom are AC edges. For an infinitely extended graphene sheet the bulklike electronic bands were calculated in Section 2.7 using the operator equations of motion, and a generalization of this result to obtain the GFs was given in Section 4.2.

Now we turn our attention to the localized modes that may exist near an edge in a finite graphene sheet. Specifically, we will consider a nanoribbon with parallel ZZ edges as shown in Figure 7.5. The case of AC edges could also be studied in a similar way, but it turns out to be less interesting. For our chosen geometry the nanoribbon is assumed to be infinite in the y direction, so the lattice has translational symmetry in this direction, allowing a Fourier transform from site labels to a 1D wave vector k_y to be made. In the x direction, however, there is no translational symmetry, and we introduce a labeling of the rows according to an integer n ($= 1, 2, 3, \ldots, N$), where N is the (even) total number of rows. Also,

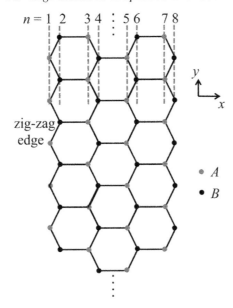

Figure 7.5 Geometry of a graphene nanoribbon with zig-zag (ZZ) longitudinal edges. The nanoribbon is infinitely long in the y direction, and the rows are labeled with an integer n across the width (in the x direction). In this example there are $N = 8$ rows.

we now have the possibility that the nearest-neighbor hopping energy (which was denoted as t in the interior or bulk case) can have a modified value denoted by t_e for nearest neighbors at either edge of the nanoribbon. With these changes we seek to modify the GF equation-of-motion approach used for the infinite sheet in Section 4.2 to apply to the nanoribbon geometry. Instead of having a 2D wave-vector Fourier transform from site labels as previously for a graphene sheet, there is now just the 1D wave vector k_y along the y direction and the row number n instead of the x coordinate. It can be seen from Figure 7.5 that the odd or even n values correspond to all sites being either on the A or B sublattice, respectively.

For simplicity, we discuss here the case in which N is sufficiently large that only one edge needs to be considered (i.e., the case of a semi-infinite nanoribbon). An extension of the calculation to the case of finite-width nanoribbons can be found in [113]. It is convenient for the fermion operators of this system to denote

$$c_n(k_y) = \begin{cases} a_n(k_y) & \text{if } n \text{ is odd} \\ b_n(k_y) & \text{if } n \text{ is even} \end{cases}, \qquad (7.23)$$

enabling us to reexpress the tight-binding Hamiltonian in Equation (2.71) in a form that is convenient for a nanoribbon as

$$\mathcal{H} = -\frac{1}{2} \sum_{k_y, n, n'} \left[\tau_{n,n'}(k_y) c_n^\dagger(k_y) c_{n'}(k_y) + \tau_{n,n'}(-k_y) c_n(k_y) c_{n'}^\dagger(k_y) \right], \qquad (7.24)$$

where the factor of $\frac{1}{2}$ is introduced to avoid double counting. The transformed hopping factor $\tau_{n,n'}(k_y)$ satisfies $\tau_{1,2}(q_y) = \tau_{2,1}(q_y) = t_e\beta(k_y)$ near the ZZ edge, whereas when $n > 1$ and $n' > 1$ we have

$$\tau_{n,n'}(k_y) = t\left[\beta(k_y)\delta_{n,n'\mp1} + \delta_{n,n'\pm1}\right]. \tag{7.25}$$

The upper (lower) signs refer to the cases of n odd (even), and we have defined the structure factor

$$\beta(k_y) = 2\cos\left(\frac{\sqrt{3}}{2}k_y a_0\right). \tag{7.26}$$

The electronic excitations can be investigated by using the GF equation-of-motion method, generalizing what was done for the infinite-sheet GFs of graphene in Section 4.2. From Equations (3.66) and (7.24) we deduce (with no need for any decoupling approximation) that

$$\omega G(c_n(k_y); c_{n'}^\dagger(k_y) \,|\, \omega) = \frac{1}{2\pi}\delta_{n,n'} - \sum_m \tau_{n,m}(k_y)\, G(c_m(k_y); c_{n'}^\dagger(k_y) \,|\, \omega). \tag{7.27}$$

Choosing now $G_{n,n'}$ as a convenient shorthand for $G(c_n(k_y); c_{n'}^\dagger(k_y) \,|\, \omega)$ and taking a fixed value of n', the first few of the preceding coupled equations (starting from the edge at $n = 1$) are given explicitly by

$$\omega\, G_{1,n'} = \frac{1}{2\pi}\delta_{1,n'} - t_e\beta(k_y)\, G_{2,n'}, \qquad (n = 1),$$

$$\omega\, G_{2,n'} = \frac{1}{2\pi}\delta_{2,n'} - t\, G_{3,n'} - t_e\beta(k_y)\, G_{1,n'}, \qquad (n = 2),$$

$$\omega\, G_{3,n'} = \frac{1}{2\pi}\delta_{3,n'} - t\beta(k_y)\, G_{4,n'} - t\, G_{2,n'}, \qquad (n = 3),$$

$$\omega\, G_{4,n'} = \frac{1}{2\pi}\delta_{4,n'} - t\, G_{5,n'} - t\beta(k_y)\, G_{3,n'}, \qquad (n = 4), \tag{7.28}$$

and so on. At this stage we might reasonably expect to be able to follow an approach similar to the TDM method used in the previous section. To do so here, however, we will need to take account of the graphene geometry where odd and even rows consist of sites on A and B sublattices, respectively. This property is easily dealt with because we note that the odd-n equations in (7.28) can be used to eliminate the odd-n GFs from the even-n equations. For example, on substituting the $n = 1$ and $n = 3$ equations in the preceding text into the $n = 2$ equation, we deduce that

$$\left[\omega^2 - t^2 - t_e^2 \beta^2(k_y)\right] G_{2,n'} - t^2 \beta(k_y) G_{4,n'} = \frac{1}{2\pi} \left[\omega \delta_{2,n'} - t_e \beta(k_y) \delta_{1,n'} - t \delta_{3,n'}\right].$$

The process may be continued with other odd and even equations.

The end result, after some straightforward algebra, is that we arrive at a new set of coupled finite-difference GF equations that involve only the even-n GF equations (i.e., those from the sublattice B). These equations can then be recast into a form involving TDMs like in Section 7.2. The coupled equations take on the same matrix form as Equation (7.10) provided we redefine the terms appropriately. Specifically, the matrix elements of the column matrix $\mathbf{G}_{n'}$ consist only of the GFs with n *even*:

$$\mathbf{G}_{n'} = \begin{pmatrix} G_{2,n'} \\ G_{4,n'} \\ G_{6,n'} \\ \vdots \end{pmatrix}. \tag{7.29}$$

The matrix $\mathbf{M}$ can be written in the same form as Equation (7.11), provided we now redefine

$$d = \frac{(\omega/t)^2 - \beta^2(k_y) - 1}{\beta(k_y)}, \tag{7.30}$$

and Δ is a parameter describing the perturbation due to the ZZ edge of the nanoribbon at $n = 1$. It is given by

$$\Delta = \left(1 - \frac{t_e^2}{t^2}\right) \beta(k_y). \tag{7.31}$$

Finally, the column matrix $\mathbf{b}_{n'}$ can be shown to have only very few (in fact, three or less) nonzero elements, depending on the values for n and n'. For example, if $n' = 1$ we have

$$(\mathbf{b}_1)_n = \frac{\omega}{2\pi t^2 \beta(k_y)} \delta_{n,1}. \tag{7.32}$$

Using the preceding results, together with the formal solutions obtained previously in Equations (7.16)–(7.19), it is now straightforward to solve for the individual GFs $G(c_n(k_y); c_{n'}^\dagger(k_y) \mid \omega)$ of the graphene problem. The results found initially will be for even n only, but then the odd-n results can be deduced using the coupled GF expressions in Equation (7.28).

To simplify matters we will quote here only some of the GF results, taking the case in which one of the row labels refers to the ZZ edge with $n' = 1$ (see Problem 7.6 for the derivations). It is found that

$$G_{1,1} = G(c_1(k_y); c_1^\dagger(k_y) \mid \omega) = \frac{1}{2\pi\omega} + \frac{t_e^2 \beta(k_y)}{2\pi t^2 \omega (1 + x\Delta)} x \qquad (n = 1), \qquad (7.33)$$

$$G_{n,1} = G(c_n(k_y); c_1^\dagger(k_y) \mid \omega) = \frac{t_e x^{1/2}[\beta(k_y) + x^{-1}]}{2\pi t\omega(1 + x\Delta)} x^{n/2} \qquad (\text{odd } n > 1), \quad (7.34)$$

$$G_{n,1} = G(c_n(k_y); c_1^\dagger(k_y) \mid \omega) = \frac{-t_e}{2\pi t^2 (1 + x\Delta)} x^{n/2} \qquad\qquad (\text{even } n). \qquad (7.35)$$

Much of the interpretation of these GF results follows analogously to that given for ferromagnets in the previous section, but we will see that there is a novel additional property that is rather special for graphene. First, when $|x| = 1$ (or $x = e^{i\theta}$ with θ being real), the $x^{n/2}$ factors behave like a phase propagation term with respect to distance from the edge. This is characteristic of bulklike propagating modes, by analogy with the previous section. In fact if we put $x = \exp(i3k_x a_0/2)$ and use Equations (7.17) and (7.30) we deduce that the corresponding solution for the frequency is

$$\omega_B(k_x, k_y) = \pm t\sqrt{\beta^2(k_y) + 2\beta(k_y)\cos(3k_x a_0/2) + 1}. \qquad (7.36)$$

This is identical to the standard form of the dispersion relation for the electronic band in an infinite graphene sheet (see Section 2.7), where k_x is a real wave-vector component in the x direction.

Contrasting with the bulk-mode case, any poles of the preceding GFs with $|x| < 1$ must correspond to localized modes, meaning edge modes in this graphene example because the $x^{n/2}$ factors now describe an attenuation with distance from the edge. One obvious way in which this can occur is through the factor $(1 + x\Delta)$ appearing in the denominator for all cases in Equations (7.33)–(7.35), so we conclude that there is an edge mode for $x = -1/\Delta$, but it exists only *provided* that the localization condition that $|\Delta| > 1$ can be satisfied. Solving for the dispersion relation then gives the edge-mode frequency as

$$\omega_E(k_y) = \pm t\sqrt{(t_e^2/t^2)\beta^2(k_y) + [t_e^2/(t_e^2 - t^2)]}. \qquad (7.37)$$

The localization condition for this type of edge mode can be satisfied either if $(t_e/t) < \sqrt{0.5}$ (≈ 0.71), which makes $\Delta > 1$, or if $(t_e/t) > \sqrt{1.5}$ (≈ 1.22), which makes $\Delta < -1$. These cases represent, respectively, the *acoustic* and *optic* edge modes because they come below and above the bulk band of electronic modes, as illustrated by the numerical examples in Figure 7.6. We note, in particular, that there will be no edge modes of this type if $t_e = t$ because Δ then vanishes. Therefore, they arise specifically as a consequence of the edge hopping t_e being sufficiently

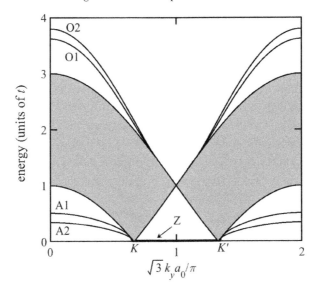

Figure 7.6 Example of the dispersion relations of electronic modes in a semi-infinite graphene nanoribbon. The energies or frequencies for the bulk bands (shaded), two examples of acoustic edge modes (A1 with $t_e/t = 0.3$, A2 with $t_e/t = 0.2$), two examples of optic edge modes (O1 with $t_e/t = 1.7$, O2 with $t_e/t = 1.8$), and the zero edge mode (Z) are shown plotted against scaled wave-vector component k_y in the length direction. Only the positive frequencies are shown in this figure.

modified. Two examples for each type of localized mode (acoustic and optic) are shown in the figure for different t_e/t ratios. The same modes occur at negative energies, i.e., in the valence band. We note that, as t_e/t is decreased (below 0.71), the acoustic edge mode is split off to a greater degree corresponding to increased localization near the edge [113]. An analogous type of behavior holds for the optic edge modes.

There is, however, another edge-mode effect that is rather more subtle and can be seen only from the odd-n GF results quoted in Equations (7.33) and (7.34) for the semi-infinite nanoribbon. It is evident that these particular GFs also have poles (vanishing denominators) at $\omega = 0$, coinciding with the Fermi frequency, and this will be the case even when $t_e = t$. A simple rearrangement of Equations (7.17) and (7.30) leads to an expression for ω^2 in terms of x as

$$\omega^2 = t^2[\beta(k_y) + x][\beta(k_y) + x^{-1}]. \tag{7.38}$$

This implies therefore that $\omega = 0$ occurs either when $x = -\beta(k_y)$ or when $x = -1/\beta(k_y)$. In fact, it is the former condition that correspond to an edge mode, as can be seen if we rewrite Equation (7.34) as

$$G(c_n(k_y); c_1^{\dagger}(k_y) \mid \omega) = \frac{-t_e}{2\pi t^2(1 + x\Delta)} \sqrt{\frac{x\beta(k_y) + 1}{\beta(k_y) + x}} \, x^{n/2} \quad \text{(odd } n > 1\text{)}. \quad (7.39)$$

We may conclude that this type of edge mode exists only on the odd rows of the semi-infinite graphene nanoribbon. Also the localization condition that must be satisfied is $|\beta(k_y)| < 1$, and so k_y must lie in the range $2\pi/3\sqrt{3} < |k_y a_0| < \pi/\sqrt{3}$ in the first Brillouin zone. This excitation is often referred to as the *zero edge mode* because it corresponds to $\omega = 0$ in the semi-infinite GNR (i.e., it occurs at the Fermi frequency). This mode was first predicted by Fujita et al. [114], and it is an example of a topological edge mode. It is seen as the flat line in Figure 7.6 extending from the Dirac points (labeled K and K') to the Brillouin zone boundary. The occurrence of these Dirac points, corresponding to values of the 2D wave vector for which the electronic spectrum is gapless, is one of the properties that makes graphene of such special interest. The wave vectors for K and K' in our notation are $(k_x, k_y) = (2\pi/3a_0, \pm 2\pi/3\sqrt{3}a_0)$.

7.4 Photonic Bands in Multilayer Superlattices

Next, we consider applications to the excitations in periodic multilayer structures, or *superlattices*. A simple structural arrangement is to build up the multilayers by having alternating thin films of two materials, denoted as A and B, to form a $\cdots ABABAB \cdots$ pattern of layer growth (see Figure 7.7) deposited on a substrate material. A feature that makes these structures of special interest occurs when all the A layers have identical properties (for composition, thickness d_A, etc.) to one

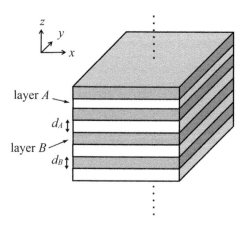

Figure 7.7 Geometry and notation for the calculation of the photonic (optical) band structure of a two-component alternating periodic superlattice with the layers labeled as A and B.

another and likewise for the properties of all the B layers. The overall structure is then formed by repeats of the same basic AB building block, which has thickness $D = d_A + d_B$. If there are many repeats of this basic unit, then it is clear that a new symmetry operation emerges for this artificial structure, namely, there are translation operations in the z direction through a periodicity length D. A consequence is that there is a corresponding new Brillouin zone for excitation wave vectors in the z direction that extends in reciprocal space from $-\pi/D$ to π/D. Because D may be engineered to be (for example) several hundred nanometres and therefore much larger than the atomic-scale lattice parameter of either constituent material, it follows that the artificial Brillouin zone will typically be much smaller than the crystal Brillouin zone. This can be advantageous as regards some experimental techniques for studying the excitations in superlattices, and we give an example in the following text for the optical properties. A general reference covering excitations in superlattices is the book by Cottam and Tilley [39].

In general, the bands for the excitations in superlattices (giving their frequency vs. wave vector relationships) can be very complicated. Recently, the properties for the occurrence and manipulation of such bands have led to the tremendous interest in *band-gap materials*, initially in the *photonics* case for photonic band gaps (PBGs) in optics (see [108] for a thorough review). Also, there have been analogous developments for other excitations, such as SWs or magnons. Applications to devices and materials in *magnonics* are reviewed in, e.g., [115]. Here as an example we present a simple formulation for a 1D photonic band-gap material (a photonic "crystal").

We will consider alternating layers of two materials as in Figure 7.7, where the layer thicknesses d_A and d_B are both assumed to be large compared with the atomic lattice parameters of the materials. Further, it will be assumed that the dielectric functions are ε_A and ε_B, taken for simplicity to be the bulk dielectric constants (treated as constants independent of the excitation angular frequency ω). This is sometimes referred to as the *bulk slab model* of a superlattice. We will employ standard results from electromagnetism for the optical wave propagation in the layers. Two cases arise, depending on whether the waves have s-polarization (meaning that the electric field vector $\mathbf{E}$ is in the y direction) or p-polarization (with $\mathbf{E}$ in the xz plane). A 1D wave-vector component Q enters into the calculations through the analogue of Bloch's theorem (see Chapter 2) in the form

$$\mathbf{E}(z + D) = \exp(i\,QD)\mathbf{E}(z), \tag{7.40}$$

while all the EM field components will be taken to have a dependence on x and t like $\exp(i q_x x - i\omega t)$. We are allowing for the possibility of a wave vector q_x in the x direction parallel to the planar interfaces corresponding to oblique incidence of the optical waves at each interface.

In this example, we will present for brevity just the calculation for s-polarization. With ε denoting the relevant dielectric function and c being the speed of light in a vacuum, the EM wave equation (see, e.g., [7, 8]) for the electric field $\mathbf{E}$ is

$$\varepsilon \frac{\partial^2 \mathbf{E}}{\partial t^2} - c^2 \nabla^2 \mathbf{E} = 0. \tag{7.41}$$

In any layer the solution for the E_y component, as required for s-polarization, will be a superposition of a forward- and a backward-traveling wave in the z direction. Two alternative forms can be written down, depending on whether the phases are expressed relative to the lower (L) or the upper (U) interface of each layer, giving

$$E_y = a_l^L \exp[iq_{Az}(z - lD)] + b_l^L \exp[-iq_{Az}(z - lD)]$$
$$= a_l^U \exp[iq_{Az}(z - lD - d_A)] + b_l^U \exp[-iq_{Az}(z - lD - d_A)]$$

in an A layer (with $lD \leq z \leq lD + d_A$). The amplitude terms are as indicated. Here q_{Az} must satisfy

$$q_{Az}^2 + q_x^2 = \varepsilon_A \omega^2 / c^2. \tag{7.42}$$

For the corresponding electric field in the adjacent layer B in the same cell l (with $lD + d_A \leq z \leq (l+1)D$) and with q_{Bz} defined similarly to Equation (7.42), we have

$$E_y = d_l^L \exp[iq_{Bz}(z - lD - d_A)] + e_l^L \exp[-iq_{Bz}(z - lD - d_A)]$$
$$= d_l^U \exp[iq_{Bz}(z - (l+1)D)] + e_l^U \exp[-iq_{Bz}(z - (l+1)D)].$$

Next, on switching over to a matrix notation for these results applied to the A and B layers, we conclude that the amplitudes in the preceding two equations are related by

$$|u_l^U\rangle = F_A |u_l^L\rangle, \qquad |w_l^U\rangle = F_B |w_l^L\rangle, \tag{7.43}$$

where in the case of an A layer we denote

$$|u_l^{L,U}\rangle = \begin{pmatrix} a_l^{L,U} \\ b_l^{L,U} \end{pmatrix} \quad \text{and} \quad F_A = \begin{pmatrix} f_A & 0 \\ 0 & f_A^{-1} \end{pmatrix} \tag{7.44}$$

with the phase term $f_A = \exp(iq_{Az} d_A)$. There are similar results for the B layer.

The standard EM boundary conditions (see, e.g., [7, 8]) at the $z = lD + d_A$ and $z = (l+1)D$ interfaces include the requirements that E_y and H_x are continuous (and so E_y and $\partial E_y / \partial z$ must be continuous). These conditions lead to additional relationships between the column matrices of the coefficients, and in matrix form the requirements can be stated compactly as

$$X_A |u_l^U\rangle = X_B |w_l^L\rangle \quad \text{and} \quad X_B |w_l^U\rangle = X_A |u_{l+1}^L\rangle, \tag{7.45}$$

where

$$X_i = \begin{pmatrix} 1 & 1 \\ q_{iz} & -q_{iz} \end{pmatrix}, \qquad i = A, B. \tag{7.46}$$

The preceding equations may now be combined to give

$$|u_{l+1}^L\rangle = T|u_l^L\rangle. \tag{7.47}$$

This expression relates the amplitudes in cell l to those at the equivalent part in cell $l + 1$, thereby introducing a new 2×2 matrix T known as the *transfer matrix*. It is given explicitly by

$$T = X_A^{-1} X_B F_B X_B^{-1} X_A F_A. \tag{7.48}$$

The transfer matrix has some useful properties that relate to the spectrum of excitations in the superlattice. First, it satisfies $\det T = 1$, which can be proved from the form of the matrix products in Equation (7.48). Second, Bloch's theorem in Equation (7.40) is equivalent to the condition $|u_{l+1}^L\rangle = \exp(iQD)|u_l^L\rangle$, and therefore we have the property that

$$[T - \exp(iQD)I]|u_l^L\rangle = 0,$$

where I is the unit 2×2 matrix. There is a similar equation obtained by relating $|u_{l-1}^L\rangle$ to $|u_l^L\rangle$, giving

$$[T^{-1} - \exp(-iQD)I]|u_l^L\rangle = 0.$$

Then, by adding the preceding two equations, we deduce that

$$[T + T^{-1} - 2\cos(QD)I]|u_l^L\rangle = 0. \tag{7.49}$$

This result holds for any cell l, and so from the property that $(T + T^{-1}) = \mathrm{Tr}(T)I$, it follows that we must have

$$\cos(QD) = \frac{1}{2}\mathrm{Tr}(T). \tag{7.50}$$

This is an important result because it provides us with an implicit dispersion relation for the excitation frequencies of the superlattice in terms of the Bloch wave vector Q provided that we evaluate the transfer matrix T. An explicit evaluation (see Problem 7.8) shows that in the present example the diagonal matrix elements of T (as required for the trace) are

$$T_{1,1} = \frac{f_A[f_B(q_{Az} + q_{Bz})^2 - f_B^{-1}(q_{Az} - q_{Bz})^2]}{4q_{Az}q_{Bz}},$$

$$T_{2,2} = \frac{f_A^{-1}[f_B^{-1}(q_{Az} + q_{Bz})^2 - f_B(q_{Az} - q_{Bz})^2]}{4q_{Az}q_{Bz}}. \tag{7.51}$$

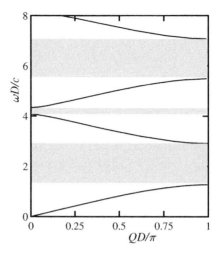

Figure 7.8 Dispersion curves for the frequency (expressed in terms of $\omega D/c$) versus the Bloch wave vector (in terms of QD/π) for normal-incidence optical waves in a 1D photonic crystal consisting of a two-component alternating periodic superlattice. The assumed parameters are $d_A = d_B = 0.5D$, $\varepsilon_A = 3$ and $\varepsilon_B = 2$. The photonic band gaps are shown as the shaded areas.

When these results are substituted into Equation (7.50) we find after some straightforward algebra that the dispersion relation for s-polarization is

$$\cos(QD) = \cos(q_{Az}d_A)\cos(q_{Bz}d_B) - g_s \sin(q_{Az}d_A)\sin(q_{Bz}d_B), \qquad (7.52)$$

where

$$g_s = \frac{1}{2}\left(\frac{q_{Bz}}{q_{Az}} + \frac{q_{Az}}{q_{Bz}}\right).$$

The calculation for the case of p-polarization is very similar to the preceding. Briefly, it is found that Equation (7.52) is still applicable provided g_s is replaced by a different quantity g_p. In fact, the dispersion equation in the general form represented by Equation (7.52) occurs commonly for excitations in periodic structures, and it is sometimes referred to as a *Rytov equation* because its first derivation is attributed to Rytov in the context of acoustic waves in superlattices [116].

We present a numerical example of a superlattice dispersion relation in Figure 7.8 for the special case of normal incidence where $q_x = 0$ and the distinction between s- and p-polarizations vanishes. The plot is in dimensionless units for the excitation frequency ω versus Q. In the reduced-Brillouin-zone scheme used here the dispersion curves are "folded back" and band gaps open up at $Q = 0$ and $Q = \pi/D$. In this example, the band gaps (or stop bands) are relatively large because we have assumed a significant difference, or mismatch, between the dielectric constants of the adjacent layers. If the dielectric constants are taken to be close in value, the gaps shrink to a small value.

7.5 Impurity Modes in Ferromagnets

In this section, we consider the mode localization in an infinite material when there is a single substitutional impurity embedded in the lattice, giving rise to a breaking of the translational (and other) symmetries of the lattice in that vicinity. Additional modes or excitations may occur associated with the impurity, as well as the modes of the pure material. In ferromagnets and antiferromagnets there have been extensive experimental studies by inelastic neutron scattering and Raman scattering (see, e.g., [76, 103] for reviews). While the term "impurity mode" serves as a general description, a distinction of terminology is sometimes made between a *defect mode* occurring outside the band of bulk SWs and a *resonance mode* within the SW band.

To illustrate how a theory for the modes may be developed using a GF formalism, we consider the relatively straightforward case of a simple-cubic (s.c.) Heisenberg ferromagnet with nearest-neighbor exchange interactions. The calculations may, however, be readily extended to other lattice structures and/or to antiferromagnets. The approach described here is analogous to that developed by Wolfram and Callaway [117]. Our first step is to obtain the difference between the Hamiltonian when the impurity is present and that for the pure system (without the impurity). The differences occur only within a small cluster of atomic spin sites situated at and around the actual impurity site. For the situation represented in Figure 7.9

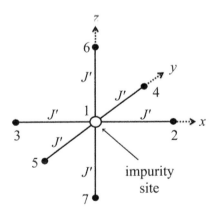

Figure 7.9 Cluster of atomic spin sites formed by an impurity spin (labeled 1) and its six nearest neighbors (labeled 2, 3, ... , 7) in a Heisenberg ferromagnet with s.c. lattice structure. The exchange interaction between the impurity and its nearest neighbors is J'.

we assume that the impurity is located at a site labeled 1: in general, it may be a magnetic impurity with spin quantum number S' and with modified exchange interaction J' to its six nearest neighbors in the pure material. The spin quantum number is denoted by S in the pure material, and the nearest-neighbor exchange is denoted by J as before (in Subsection 1.5.2).

Within a low-temperature bosonic approximation the total Heisenberg Hamiltonian $\mathcal{H}$ is given by the previous Equation (1.88). In the present context this may be separated into two parts as $\mathcal{H} = \mathcal{H}_0 + \mathcal{H}'$, where $\mathcal{H}_0$ is the Hamiltonian for the pure ferromagnet and $\mathcal{H}'$ represents the perturbation due to the impurity. It is straightforward to show that

$$\mathcal{H}' = JS \left\{ \xi a_1^\dagger a_1 + \zeta \sum_{n=2}^{7} a_n^\dagger a_n - \gamma \sum_{n=2}^{7} (a_1^\dagger a_n + a_n^\dagger a_1) \right\}, \tag{7.53}$$

where the summations run over the six nearest neighbors of the impurity site at 1, and we have defined the following quantities measuring the perturbation:

$$\xi = \frac{J'}{J} - 1, \qquad \zeta = \frac{J'S'}{JS} - 1, \qquad \gamma = \frac{J'}{J}\sqrt{\frac{S'}{S}} - 1. \tag{7.54}$$

Next we introduce GFs of the same general form $G(a; a^\dagger \,|\, \omega)$ as we considered previously for ferromagnets (e.g., as in Section 4.4 and again earlier in the present chapter). Here we want the GFs to be in a site-dependent representation to deal with the spatial effects localized around the impurity, and so we define $G_{i,j}^{(0)}(\omega) = G(a_i; a_j^\dagger \,|\, \omega)$ for the pure ferromagnet, while $G_{i,j}(\omega)$ is the corresponding quantity for the impure ferromagnet. The GF for the translationally invariant pure system is already known in the 3D wave-vector representation. Thus, using Equation (4.58) and transforming to the site representation, we have for the retarded GF the expression

$$G_{i,j}^{(0)}(\omega) = \frac{1}{N} \sum_{\mathbf{k}} \exp[i\mathbf{k} \cdot (\mathbf{r}_i - \mathbf{r}_j)] \left(\frac{1}{2\pi}\right) \frac{1}{\omega - E_{\mathbf{k}} + i\eta}, \tag{7.55}$$

where N is the (macroscopically large) number of sites in the ferromagnet and $E_{\mathbf{k}}$ is the dispersion relation for bulk SWs at low temperatures $T \ll T_C$ as given in Equation (1.93).

Our objective now is to find $G_{i,j}(\omega)$ for the ferromagnet with an isolated impurity, and to do so we follow an approach analogous to that introduced in [103, 117] where a generalized matrix formalism is employed for impure systems. Noting that the GF equation of motion in Chapter 3 is an inhomogeneous finite-difference equation in terms of site labels, we write

$$G_{i,j}(\omega) = G_{i,j}^{(0)}(\omega) + \sum_{i',j'} G_{i,i'}^{(0)}(\omega) V_{i',j'} G_{j',j}(\omega).$$

Here $V_{i',j'}$ describes the impurity-host interaction and so is related to the perturbation Hamiltonian $\mathcal{H}'$ as quoted in Equation (7.53). In an obvious matrix shorthand notation (with respect to the site labels) the preceding equation can be reexpressed more transparently as

$$\mathbf{G} = \mathbf{G}^{(0)} + \mathbf{G}^{(0)}\mathbf{V}\mathbf{G}. \tag{7.56}$$

It is relevant to note that we are dealing here with matrices of dimension 7×7 because there are only seven sites (labeled as the cluster in Figure 7.9) involved in the perturbation Hamiltonian $\mathcal{H}'$. After some straightforward rearrangement of Equation (7.56) we may write the formal solution for $\mathbf{G}$ as

$$\begin{aligned} \mathbf{G} &= (\mathbf{I} - \mathbf{G}^{(0)}\mathbf{V})^{-1}\mathbf{G}^{(0)} \\ &= \mathbf{G}^{(0)} + \mathbf{G}^{(0)}\mathbf{V}\mathbf{G}^{(0)} + \mathbf{G}^{(0)}\mathbf{V}\mathbf{G}^{(0)}\mathbf{V}\mathbf{G}^{(0)} + \cdots . \end{aligned} \tag{7.57}$$

The expression in the first line is often referred to as a *Dyson equation* and its expansion (given in the second line) shows that events for multiple scattering of the host magnons by the defect are incorporated. If we now examine the poles of $\mathbf{G}$ using the first line of Equation (7.57), it is evident that any new poles (other than those already present in $\mathbf{G}^{(0)}$ for the pure system) must come from the $(\mathbf{I}-\mathbf{G}^{(0)}\mathbf{V})^{-1}$ factor and therefore correspond to the condition that

$$\det(\mathbf{I} - \mathbf{G}^{(0)}\mathbf{V}) = 0. \tag{7.58}$$

Returning now to the specific example of a single impurity in a s.c. Heisenberg ferromagnet, we deduce from Equation (7.53) for $\mathcal{H}'$ that the form of the host-impurity scattering matrix $\mathbf{V}$ is

$$\mathbf{V} = 2\pi J S \begin{pmatrix} \xi & -\gamma & -\gamma & -\gamma & -\gamma & -\gamma & -\gamma \\ -\gamma & \zeta & 0 & 0 & 0 & 0 & 0 \\ -\gamma & 0 & \zeta & 0 & 0 & 0 & 0 \\ -\gamma & 0 & 0 & \zeta & 0 & 0 & 0 \\ -\gamma & 0 & 0 & 0 & \zeta & 0 & 0 \\ -\gamma & 0 & 0 & 0 & 0 & \zeta & 0 \\ -\gamma & 0 & 0 & 0 & 0 & 0 & \zeta \end{pmatrix}, \tag{7.59}$$

where the impurity-related parameters ξ, ζ, and γ are defined in Equation (7.54). Also from considerations of cubic symmetry it follows that there are only four independent GF matrix elements of $\mathbf{G}^{(0)}$ for the pure system. These can be identified in terms of the *distance apart* of the pairs of sites in the cluster in Figure 7.9. This distance can take the values 0 as in the case of $G_{1,1}^{(0)}$ and similar pairs of sites, a as for $G_{1,2}^{(0)}$, $2a$ as for $G_{2,3}^{(0)}$, and $\sqrt{2}a$ as for $G_{2,4}^{(0)}$. Using Equation (7.55) it follows that

$$G_{1,1}^{(0)}(\omega) = \frac{1}{2\pi N} \sum_{\mathbf{k}} \frac{1}{\omega - E_{\mathbf{k}} + i\eta},$$

$$G_{1,2}^{(0)}(\omega) = \frac{1}{2\pi N} \sum_{\mathbf{k}} \frac{\cos(k_x a)}{\omega - E_{\mathbf{k}} + i\eta},$$

$$G_{2,3}^{(0)}(\omega) = \frac{1}{2\pi N} \sum_{\mathbf{k}} \frac{\cos(2k_x a)}{\omega - E_{\mathbf{k}} + i\eta},$$

$$G_{2,4}^{(0)}(\omega) = \frac{1}{2\pi N} \sum_{\mathbf{k}} \frac{\cos(k_x a)\cos(k_y a)}{\omega - E_{\mathbf{k}} + i\eta}. \tag{7.60}$$

The determinantal condition expressed in Equation (7.58), which leads to the results for the frequency ω of the impurity modes, can now be implemented by substituting the explicit forms discussed in the preceding text for $\mathbf{G}^{(0)}$ and $\mathbf{V}$. After some lengthy but straightforward algebraic manipulation, it is found (see Problem 7.9) that the result may be factorized as

$$\det(\mathbf{I} - \mathbf{G}^{(0)}\mathbf{V}) = [D_A(\omega)]^3 [D_B(\omega)]^2 D_C(\omega). \tag{7.61}$$

Here the individual terms are

$$D_A(\omega) = \zeta[G_{2,3}^{(0)}(\omega) - G_{1,1}^{(0)}(\omega)] + 1/(2\pi J S),$$

$$D_B(\omega) = \zeta[2G_{2,4}^{(0)}(\omega) - G_{1,1}^{(0)}(\omega) - G_{2,3}^{(0)}(\omega)] + 1/(2\pi J S),$$

while $D_C(\omega)$ comes from a 2×2 determinant and can be expressed in terms of the preceding GFs and the impurity parameters.

We turn now to some numerical results. As mentioned earlier in this section it is useful to distinguish between *defect modes*, which are nonresonant with any of the bulk SWs of the pure host material, and *resonance modes*, which occur for ω within the band of bulk SWs. The latter are more complicated because the terms $(\omega - E_{\mathbf{k}} + i\eta)$ appearing in the denominators in Equation (7.60) will effectively vanish for some value(s) of the wave vector $\mathbf{k}$ making the numerical evaluation of the spatial GFs more difficult. For simplicity, therefore, we focus on the defect-mode case for which it is found [117] that solutions do indeed occur corresponding to each of the factors in Equation (7.61) when ω is above the top of the bulk SW band. This quantity corresponds to energy $(g\mu_B B_0 + 12J S)$ in the s.c. geometry.

Some examples of calculations for the defect modes are shown in Figure 7.10. Here we plot (in dimensionless units) the frequencies of the modes versus J'/J for different values of this exchange ratio. Depending on the value of J'/J, there can be three modes, labeled according to convention [117] in the figure as s, p,

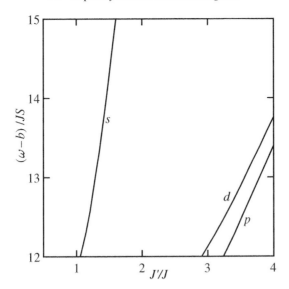

Figure 7.10 Frequencies of the defect modes associated with an isolated magnetic impurity embedded in a s.c. ferromagnet. The figure shows the dimensionless quantity $(\omega - b)/SJ$, where $b = g\mu_B B_0$, plotted versus the exchange ratio J'/J. We have assumed $S' = 2$ and $S = 1$ for the spin quantum numbers of the impurity and host atoms, respectively. The mode frequencies lie above the bulk SW band, which extends up to the value 12 for this plot. The mode labels are as described in the text.

and d, which come from the factors on the right-hand side of Equation (7.61). It can be shown that the s mode corresponds to a mode that is highly localized near the impurity, whereas the p and d modes are more spread out over the nearest neighbors to the impurity.

Problems

7.1. Consider a finite-thickness elastic film with thickness L. Its two planar surfaces correspond to $z = 0$ and $z = -L$. Extend the linear response theory in Section 7.1 to obtain the GFs $G(u(z); u(z') \mid \omega)$ and $G(u^{zz}(z); u^{zz}(z') \mid \omega)$ for the film. You should employ the stress-free boundary condition in Equation (7.1) at both surfaces.

7.2. The boundary condition employed in Section 7.1 was that for a stress-free surface. Suppose, instead of Equation (7.1), a zero-displacement boundary condition is applied, i.e., it is required that $u(z) = 0$ at $z = 0$. Determine how the subsequent calculations given in Section 7.1 for the GFs $G(u(z); u(z') \mid \omega)$ and $G(u^{zz}(z); u^{zz}(z') \mid \omega)$ will be modified for this case.

7.3. Verify that the nonzero coefficients $A_{n,n'}(\mathbf{k}_\parallel)$ have the form quoted in Equation (7.8) for the calculation of the surface SWs in a semi-infinite b.c.c. ferromagnet.

7.4. Verify that the inverse of the matrix $\mathbf{M}_0$ is as quoted in Equation (7.16). In other words, prove that when the matrix given by Equation (7.16) is either premultiplied or postmultiplied by the matrix $\mathbf{M}_0$ in Equation (7.15) the result is the unit matrix.

7.5. In Section 7.2 we found that for a b.c.c. Heisenberg ferromagnet a surface SW is predicted decaying with distance away from the (100) surface of a semi-infinite material. Now consider the same situation, but for a Heisenberg ferromagnet with a s.c. structure and nearest-neighbor exchange only. Show that there cannot be a localized surface SW in this case when all nearest-neighbor exchange interactions have the bulk value J.

7.6. A set of GFs that describes correlations between any two sites in the semi-finite graphene lattice with a ZZ edge is given by Equation (7.27). The GFs can be written in a matrix form $\mathbf{G}_{n'} = \mathbf{M}^{-1}\mathbf{b}_{n'}$ with the definitions being given in Section 7.3. Derive explicit expression for the GFs to verify the results for the special cases that are quoted in Equations (7.33)–(7.35).

7.7. Extend the analysis used in the previous Problem 7.6 for a semi-infinite graphene lattice with a ZZ edge to obtain expressions for the GFs denoted as $G_{2,2}$ and $G_{4,2}$. Note that you will now need to take $n' = 2$ for part of your derivation. Verify that there is no contribution from these GFs to the spectral intensity of the zero mode.

7.8. For the example of a periodic multilayer (or superlattice) considered in Section 7.4, use Equation (7.48) and the preceding expressions to verify that the diagonal elements of the transfer matrix T are as quoted in Equation (7.51). Then show that the Rytov Equation (7.52) can be deduced as an implicit expression for the mode frequencies.

7.9. For the model of an isolated impurity in a s.c. Heisenberg ferromagnet (as in Section 7.5) use the explicit forms for $\mathbf{V}$ in Equation (7.59) and the matrix elements of $\mathbf{G}^{(0)}$ in Equation (7.60) to construct $(\mathbf{I} - \mathbf{G}^{(0)}\mathbf{V})$ in the 7×7 matrix representation. Then verify the result quoted in Equation (7.61) for the impurity modes.

7.10. Consider a ferromagnetic medium occupying the half space $x < 0$, so there is a single surface in the yz-plane corresponding to $x = 0$. The in-plane magnetization M_0 is taken to be in the z direction and the external medium (where $x > 0$) is a vacuum. Calculate the magnetic excitations (SWs) of this

system in the dipolar limit (neglecting exchange) by extending Problem 6.9
as follows. The relevant Maxwell's equations for the fluctuation terms $\mathbf{m}(\mathbf{r})$
and $\mathbf{h}(\mathbf{r})$ in the magnetostatic case are $\nabla \times \mathbf{h} = 0$ and $\nabla \cdot (\mathbf{h} + \mathbf{m}) = 0$. The first
of these equations is automatically satisfied if we introduce the magnetostatic
scalar potential ψ by $\mathbf{h} = \nabla \psi$. Show that the second equation implies

$$(1 + \chi_a) \left(\frac{\partial^2 \psi}{\partial x^2} + \frac{\partial^2 \psi}{\partial y^2} \right) + \frac{\partial^2 \psi}{\partial z^2} = 0,$$

using the susceptibility relation from Problem 6.9. This holds in all space,
including the vacuum region (where $\chi_a = 0$). Next, assume SWs with the
in-plane wave vector along the y direction, i.e., $\mathbf{q}_\| = (q_y, 0)$, and look for
solutions of the above equation with the form

$$\psi(x, y, z) = \begin{cases} a_1 \exp(-|q_y| x) \exp(i q_y y), & \text{for } x > 0, \\ a_2 \exp(-i q_x x) \exp(i q_y y), & \text{for } x < 0. \end{cases}$$

Here q_x is a wavenumber (complex, in general) in the x direction. Show
that there are two possible SW modes: one corresponds to the frequency
$\pm[\omega_0(\omega_0 + \omega_m)]^{1/2}$ and the other to $[\omega_0 + \frac{1}{2}\omega_m]$. The first is a bulk mode and
the second is a localized surface mode (called the Damon–Eshbach mode)
which exists only if $q_y < 0$.

8

Diagrammatic Perturbation Methods

Starting in this chapter, a perturbative Green's function (GF) technique will be established in a form that is applicable to interacting boson and fermion many-body systems in second quantization. The method will be expressed in terms of the imaginary-time (or Matsubara) GFs, which were introduced already in Section 3.5, rather than the real-time GFs that we have mainly employed so far. The results will eventually be formulated in terms of a diagrammatic representation, which offers many advantages over following purely algebraic procedure. The theory will be developed here for finite temperatures, incorporating the standard results from equilibrium statistical physics.

We shall see that, on the one hand, the method has the considerable advantage (compared with the previous GF equation-of-motion and linear response methods) that approximations can be introduced systematically through a rigorous expansion parameter in a controlled fashion by including contributions up to a particular chosen order. In this way, the diagrammatic method avoids the arbitrariness sometimes involved in previous decoupling approximations or linearization procedures. Therefore, it is often advantageous for higher-order calculations. On the other hand, a drawback potentially is that the method is harder to learn and can sometimes be more complicated to apply.

Detailed descriptions of the standard diagrammatic perturbation formalism, as applied to bosons and fermions, interacting through a scalar pairwise interaction potential, can be found in various text books on the many body theory of condensed matter systems (see, e.g., [48, 49, 51, 52, 118, 119]). Some similar material also occurs in books on the quantum field theory of particles (e.g., [120–122]), where the treatment is often relativistic and/or restricted to a zero-temperature formalism.

8.1 The Grand Partition Function

For developing the perturbation technique it is convenient to write the previous Hamiltonian for a system of interacting bosons or fermions in second quantization as

$$\mathcal{H} = \mathcal{H}_0 + \mathcal{H}_1. \tag{8.1}$$

Here $\mathcal{H}_0$ denotes the Hamiltonian for the noninteracting particles, which can be expressed as

$$\mathcal{H}_0 = \sum_{\mathbf{k}} E_{\mathbf{k}} a_{\mathbf{k}}^\dagger a_{\mathbf{k}}, \tag{8.2}$$

with typically $E_{\mathbf{k}} = (k^2/2m) - \mu$. The perturbation term $\mathcal{H}_1$ is due to the interactions between particles and will be taken to have the same form as in Equation (1.62):

$$\mathcal{H}_1 = \frac{1}{2} \sum_{\mathbf{k}_1, \mathbf{k}_2, \mathbf{q}} v(\mathbf{q}) a_{\mathbf{k}_1}^\dagger a_{\mathbf{k}_2}^\dagger a_{\mathbf{k}_2+\mathbf{q}} a_{\mathbf{k}_1-\mathbf{q}}. \tag{8.3}$$

We recall that $v(\mathbf{q})$ is the Fourier transform with respect to wave vector of the pairwise interaction potential, and it depends on the nature and strength of the interaction. The operators satisfy $[a_{\mathbf{k}}, a_{\mathbf{k}'}^\dagger]_\varepsilon = \delta_{\mathbf{k}, \mathbf{k}'}$, etc., where ε is equal to 1 (for commutators) or -1 (for anticommutators) in the case of bosons and fermions, respectively.

The equilibrium thermal averages are defined as before. Hence, for an operator A the average with respect to the full Hamiltonian can be written as

$$\langle A \rangle = \frac{1}{Q} \mathrm{Tr} \left(A e^{-\beta \mathcal{H}} \right) \quad \text{with} \quad Q = \mathrm{Tr} \left(e^{-\beta \mathcal{H}} \right). \tag{8.4}$$

If we know Q we can in principle calculate all other thermodynamic quantities of the system in equilibrium, so this will be our goal for now. Notice here that we have conveniently absorbed the effect of the chemical potential μ into $E_{\mathbf{k}}$, so we are continuing to use a grand canonical ensemble and Q is the grand partition function.

8.1.1 Thermal Averages for the Unperturbed System

Suppose, as a preliminary step, we wish to calculate the corresponding grand partition function Q_0 for the *unperturbed* system. We have

$$Q_0 = \mathrm{Tr}\{ \exp(-\beta \mathcal{H}_0) \} = \mathrm{Tr} \left\{ \exp \left(-\beta \sum_{\mathbf{k}} E_{\mathbf{k}} a_{\mathbf{k}}^\dagger a_{\mathbf{k}} \right) \right\}$$

$$= \mathrm{Tr} \left\{ \prod_{\mathbf{k}} \exp \left(-\beta E_{\mathbf{k}} a_{\mathbf{k}}^\dagger a_{\mathbf{k}} \right) \right\},$$

where $a_{\mathbf{k}}^\dagger a_{\mathbf{k}}$ represents the number operator. For the case of fermions the number operator can only have eigenvalues 0 and 1, and so

$$Q_0 = \prod_{\mathbf{k}} \left(1 + e^{-\beta E_{\mathbf{k}}}\right). \tag{8.5}$$

For the case of bosons the number operator has eigenvalues $0, 1, 2, 3, \ldots$, and so

$$Q_0 = \prod_{\mathbf{k}} \left(1 + e^{-\beta E_{\mathbf{k}}} + e^{-2\beta E_{\mathbf{k}}} + e^{-3\beta E_{\mathbf{k}}} + \cdots\right)$$
$$= \prod_{\mathbf{k}} \left(\frac{1}{1 - e^{-\beta E_{\mathbf{k}}}}\right). \tag{8.6}$$

We can then use the preceding results for Q_0 to evaluate simple averages for the unperturbed system, such as $\langle a_{\mathbf{k}}^\dagger a_{\mathbf{k}}\rangle_0$ because

$$\langle a_{\mathbf{k}}^\dagger a_{\mathbf{k}}\rangle_0 = \frac{1}{Q_0}\mathrm{Tr}\left\{a_{\mathbf{k}}^\dagger a_{\mathbf{k}} \exp\left(-\beta \sum_{\mathbf{k}'} E_{\mathbf{k}'} a_{\mathbf{k}'}^\dagger a_{\mathbf{k}'}\right)\right\}$$
$$= \frac{1}{Q_0}\frac{\partial}{\partial(-\beta E_{\mathbf{k}})}\mathrm{Tr}\left\{\exp\left(-\beta \sum_{\mathbf{k}'} E_{\mathbf{k}'} a_{\mathbf{k}'}^\dagger a_{\mathbf{k}'}\right)\right\}$$
$$= -\frac{1}{Q_0}\frac{\partial Q_0}{\partial(\beta E_{\mathbf{k}})}.$$

After substituting for Q_0 this gives (consistent with our earlier notation)

$$n_{\mathbf{k}} \equiv n(E_{\mathbf{k}}) = \langle a_{\mathbf{k}}^\dagger a_{\mathbf{k}}\rangle_0 = \frac{1}{\exp(\beta E_{\mathbf{k}}) - \varepsilon}, \tag{8.7}$$

as expected, for the Bose–Einstein ($\varepsilon = 1$) and Fermi–Dirac ($\varepsilon = -1$) distribution functions.

By extension, it can be shown for other unperturbed averages involving pairs of operators that

$$\langle a_{\mathbf{k}}^\dagger a_{\mathbf{k}'}\rangle_0 = \frac{\delta_{\mathbf{k},\mathbf{k}'}}{\exp(\beta E_{\mathbf{k}}) - \varepsilon}, \qquad \langle a_{\mathbf{k}}^\dagger a_{\mathbf{k}'}^\dagger\rangle_0 = \langle a_{\mathbf{k}} a_{\mathbf{k}'}\rangle_0 = 0.$$

8.1.2 The S-Matrix Expansion

Corresponding to the full Hamiltonian of the interacting system we have by definition $Q = \mathrm{Tr}\{e^{-\beta\mathcal{H}}\}$ for the grand partition function. Then, to relate full averages to the unperturbed averages, we introduce the so-called S-matrix operator $S(\beta)$, which is defined as having the property that [49, 52, 119]

$$e^{-\beta\mathcal{H}} = e^{-\beta\mathcal{H}_0} S(\beta). \tag{8.8}$$

If we can evaluate $S(\beta)$ then we can in principle evaluate the partition function for the interacting system. We note that

$$Q = \text{Tr}\{e^{-\beta\mathcal{H}}\} = \text{Tr}\{e^{-\beta\mathcal{H}_0}S(\beta)\} = Q_0 \frac{1}{Q_0}\text{Tr}\{e^{-\beta\mathcal{H}_0}S(\beta)\},$$

and so

$$Q = Q_0 \langle S(\beta)\rangle_0. \tag{8.9}$$

It follows that we need a way to solve for $S(\beta)$ and then to take its unperturbed average. This can be achieved by differentiating Equation (8.8) with respect to β to give initially

$$-\mathcal{H}e^{-\beta\mathcal{H}} = -\mathcal{H}_0 e^{-\beta\mathcal{H}_0} S + e^{-\beta\mathcal{H}_0} \frac{\partial S}{\partial \beta},$$

from which it follows after rearranging that

$$\begin{aligned}
\frac{\partial S}{\partial \beta} &= -e^{\beta\mathcal{H}_0}\mathcal{H}e^{-\beta\mathcal{H}} + e^{\beta\mathcal{H}_0}\mathcal{H}_0 e^{-\beta\mathcal{H}_0} S \\
&= -e^{\beta\mathcal{H}_0}(\mathcal{H}_0 + \mathcal{H}_1)e^{-\beta\mathcal{H}_0}S + e^{\beta\mathcal{H}_0}\mathcal{H}_0 e^{-\beta\mathcal{H}_0} S \\
&= -e^{\beta\mathcal{H}_0}\mathcal{H}_1 e^{-\beta\mathcal{H}_0} S.
\end{aligned}$$

This result can be rewritten compactly as

$$\frac{\partial S}{\partial \beta} = -\mathcal{H}_1(\beta)\, S, \tag{8.10}$$

where we have introduced the notation for any operator A that

$$A(\beta) = e^{\beta\mathcal{H}_0} A\, e^{-\beta\mathcal{H}_0}. \tag{8.11}$$

The initial condition (or boundary condition) on the differential Equation (8.10) is that $S = 1$ when $\beta = 0$, which follows from the definition of S.

Integrating both sides of Equation (8.10) now gives us a formal solution for the S-matrix as

$$S(\beta) = 1 - \int_0^\beta \mathcal{H}_1(\tau)S(\tau)d\tau. \tag{8.12}$$

By repeatedly iterating for S on the right-hand side we get the series

$$\begin{aligned}
S(\beta) = 1 &- \int_0^\beta d\tau_1 \mathcal{H}_1(\tau_1) + \int_0^\beta d\tau_1 \int_0^{\tau_1} d\tau_2 \mathcal{H}_1(\tau_1)\mathcal{H}_1(\tau_2) - \cdots \\
&+ (-1)^n \int_0^\beta d\tau_1 \int_0^{\tau_1} d\tau_2 \cdots \int_0^{\tau_{n-1}} d\tau_n \mathcal{H}_1(\tau_1)\mathcal{H}_1(\tau_2)\cdots \mathcal{H}_1(\tau_n) + \cdots,
\end{aligned}$$

$$\tag{8.13}$$

where the nth order term in the expansion is displayed. In this form the result is not particularly convenient because the limits of integration involve the τ-labels. At this stage it is helpful to point out a similarity with the analysis given in Subsection 2.1.2 for the expansion of the time evolution operator in a Dyson series. Specifically, the preceding Equation (8.13) is similar in form to Equation (2.19). Accordingly, we now rewrite Equation (8.13) in a more useful way by defining the ordering operator $\hat{T}_W$ for the τ-labels. As before, the ordering operator rearranges the operators in an order with the associated τ labels increasing from right to left. Also, it introduces a overall factor of $(-1)^A$ where the integer A is the number of interchanges of *fermion* operators to achieve the reordering. For boson operators we will simply have $A = 0$. By following now the same procedures as in Subsection 2.1.2 for the Dyson series (including the geometric arguments presented as in Figure 2.1), we can rewrite the second-order term in the expansion in Equation (8.13) as

$$\frac{1}{2} \int_0^\beta d\tau_1 \int_0^\beta d\tau_2 \hat{T}_W \left\{ \mathcal{H}_1(\tau_1) \mathcal{H}_1(\tau_2) \right\}.$$

We note that in the preceding case we always have $(-1)^A = 1$, even for fermion systems, because $\mathcal{H}_1$ contains an even number of operators. Generalizing the preceding argument, we obtain the S-matrix expansion in the form

$$S(\beta) = 1 + \sum_{n=1}^\infty \frac{(-1)^n}{n!} \int_0^\beta d\tau_1 \int_0^\beta d\tau_2 \cdots \int_0^\beta d\tau_n$$
$$\times \hat{T}_W \left\{ \mathcal{H}_1(\tau_1) \mathcal{H}_1(\tau_2) \cdots \mathcal{H}_1(\tau_n) \right\}. \tag{8.14}$$

In a similar way to Equation (2.21), a shorthand form for the preceding result can be introduced as

$$S(\beta) = \hat{T}_W \left[\exp \left\{ -\int_0^\beta d\tau \, \mathcal{H}_1(\tau) \right\} \right]. \tag{8.15}$$

We must keep in mind that the exponential expansion must always be made *before* doing the integration, otherwise the role of the $\hat{T}_W$ operator is not meaningful.

8.2 Wick's Theorem

It follows from Equations (8.9) and (8.14) that to calculate Q for the interacting system we need to be able to calculate unperturbed averages like $\langle \hat{T}_W \{ \mathcal{H}_1(\tau_1) \mathcal{H}_1(\tau_2) \cdots \mathcal{H}_1(\tau_n) \} \rangle_0$. Because each $\mathcal{H}_1$ contains an even number of creation or annihilation operators, we need to calculate expressions like

$$\langle \hat{T}_W \{ b_1 b_2 b_3 \cdots b_m \} \rangle_0,$$

where each b_i denotes an a_k or $a_k^\dagger$ operator and m is an *even* positive integer. Each b_i operator has a label τ associated with it. Wick's theorem is an important algebraic result that eventually enables us to reexpress the preceding average over a product of m operators in terms of averages over products of *pairs of operators*, like

$$\left\langle \hat{T}_W \left\{ b_i b_j \right\} \right\rangle_0 .$$

When (or if) this can be done, and bearing in mind that we already know the results for unperturbed averages over pairs of operators, it becomes possible to make a complete evaluation of Q in terms of a perturbation expansion [123, 124]. Ultimately, it is this property that will lead us to a convenient representation of the expansion terms using Feynman diagrams.

We first need a few mathematical results and definitions involving products of operators and their thermal averages. Proofs of the main steps in establishing Wick's theorem are provided here, but in a first reading of this material it may be preferable to focus on the results and return to the proofs later.

The τ-Dependence of Operators

The τ-dependence of the individual operators corresponds to

$$a_k(\tau) = a_k e^{-E_k \tau}, \quad a_k^\dagger(\tau) = a_k^\dagger e^{E_k \tau}. \tag{8.16}$$

Proof For the $a_k(\tau)$ operator we have from the earlier definition

$$a_k(\tau) = e^{\mathcal{H}_0 \tau} a_k e^{-\mathcal{H}_0 \tau} = e^{E_k \tau a_k^\dagger a_k} a_k e^{-E_k \tau a_k^\dagger a_k}. \tag{8.17}$$

We now use the following operator identity (known as the Baker–Campbell–Hausdorff identity [14]) to make an expansion in terms of a series of nested commutators:

$$e^X Y e^{-X} = Y + [X, Y] + \frac{1}{2!}[X, [X, Y]] + \frac{1}{3!}[X, [X, [X, Y]]] + \cdots . \tag{8.18}$$

We came across a special case of this identity earlier (see Problem 1.4), but the general result here can be verified (for example) by expanding both sides in powers of X and Y and equating terms. Choosing $X = E_k \tau a_k^\dagger a_k$ and $Y = a_k$, we have

$$[X, Y] = E_k \tau [a_k^\dagger a_k, a_k] = -E_k \tau a_k, \tag{8.19}$$

which holds for bosons and fermions. This result leads to

$$a_k(\tau) = a_k - (E_k \tau) a_k + \cdots + \frac{(-1)^n}{n!} (E_k \tau)^n a_k + \cdots$$

$$= a_k e^{-E_k \tau}, \tag{8.20}$$

as required. The proof for the creation operator is similar. Note that the Hermitian conjugate of $a_k(\tau)$ is $a_k^\dagger(-\tau)$, and *not* $a_k^\dagger(\tau)$.

Reduction of Operator Products

Unperturbed averages that involve a product of m operators, where m is an even integer greater than 2, can be rewritten in terms of unperturbed averages over products of $(m-2)$ operators, using

$$\langle b_1 b_2 b_3 \cdots b_m \rangle_0 = \Phi(\beta E_\mathbf{k}) \{ \langle [b_1, b_2]_\varepsilon b_3 \cdots b_m \rangle_0 + \varepsilon \langle b_2 [b_1, b_3] b_4 \cdots b_m \rangle_0$$
$$+ \cdots + \varepsilon^{m-2} \langle b_2 b_3 b_4 \cdots [b_1, b_m] \rangle_0 \}, \tag{8.21}$$

where we define

$$\Phi(\beta E_\mathbf{k}) = \begin{cases} (1 - \varepsilon e^{-\beta E_\mathbf{k}})^{-1} & \text{if} \quad b_1 = a_\mathbf{k} \\ (1 - \varepsilon e^{\beta E_\mathbf{k}})^{-1} & \text{if} \quad b_1 = a_\mathbf{k}^\dagger \end{cases}. \tag{8.22}$$

Proof To be specific, we will take the case of $b_1 = a_\mathbf{k}$ (but the other case when $b_1 = a_\mathbf{k}^\dagger$ follows in a similar fashion). Therefore, we have

$$\langle a_\mathbf{k} b_2 b_3 \cdots b_m \rangle_0 = \text{Tr} \{ a_\mathbf{k} b_2 b_3 \cdots b_m \rho_0 \},$$

where we write $\rho_0 = Q_0^{-1} e^{-\beta \mathcal{H}_0}$ for the equilibrium (unperturbed) density matrix. Now we can reexpress the product $a_\mathbf{k} b_2$ using

$$a_\mathbf{k} b_2 = [a_\mathbf{k}, b_2]_\varepsilon + \varepsilon b_2 a_\mathbf{k}.$$

It follows, on using this result repeatedly for $a_\mathbf{k}$, that we may obtain

$$a_\mathbf{k} b_2 b_3 \cdots b_m \rho_0 = [a_\mathbf{k}, b_2]_\varepsilon b_3 \cdots b_m \rho_0 + \varepsilon b_2 a_\mathbf{k} b_3 \cdots b_m \rho_0$$
$$= [a_\mathbf{k}, b_2]_\varepsilon b_3 \cdots b_m \rho_0 + \varepsilon b_2 [a_\mathbf{k}, b_3]_\varepsilon b_4 \cdots b_m \rho_0 + \cdots$$
$$+ \varepsilon^{m-2} b_2 b_3 \cdots b_{m-1} [a_\mathbf{k}, b_m]_\varepsilon \rho_0$$
$$+ \varepsilon^{m-1} b_2 b_3 \cdots b_m a_\mathbf{k} \rho_0.$$

For the operator product $a_\mathbf{k} \rho_0$ in the last term we also have

$$a_\mathbf{k} \rho_0 = Q_0^{-1} a_\mathbf{k} e^{-\beta \mathcal{H}_0} = Q_0^{-1} e^{-\beta \mathcal{H}_0} e^{\beta \mathcal{H}_0} a_\mathbf{k} e^{-\beta \mathcal{H}_0}$$
$$= \rho_0 a_\mathbf{k}(\beta) = e^{-\beta E_\mathbf{k}} \rho_0 a_\mathbf{k}.$$

Substituting this into the preceding result and taking the Tr operation of both sides of the equation, we find

$$\text{Tr} \{ a_\mathbf{k} b_2 b_3 \cdots b_m \rho_0 \} = \text{Tr} \{ [a_\mathbf{k}, b_2]_\varepsilon b_3 \cdots b_m \rho_0 \}$$
$$+ \varepsilon \text{Tr} \{ b_2 [a_\mathbf{k}, b_3]_\varepsilon b_4 \cdots b_m \rho_0 \} + \cdots$$
$$+ \varepsilon^{m-2} \text{Tr} \{ b_2 b_3 \cdots b_{m-1} [a_\mathbf{k}, b_m]_\varepsilon \rho_0 \}$$
$$+ \varepsilon^{m-1} e^{-\beta E_\mathbf{k}} \text{Tr} \{ b_2 b_3 \cdots b_m \rho_0 a_\mathbf{k} \}.$$

We may now use the cyclic invariance property of Tr to conclude that the last term on the right-hand side is just the same as

$$\varepsilon e^{-\beta E_{\mathbf{k}}} \mathrm{Tr}\{a_{\mathbf{k}}b_2 b_3 \cdots b_m \rho_0\}.$$

Putting these results together and rearranging, we find

$$\left(1 - \varepsilon e^{-\beta E_{\mathbf{k}}}\right)\langle a_{\mathbf{k}}b_2 b_3 \cdots b_m \rangle_0 = \langle [a_{\mathbf{k}}, b_2]_\varepsilon b_3 \cdots b_m \rangle_0$$
$$+ \varepsilon \langle b_2 [a_{\mathbf{k}}, b_3]_\varepsilon b_4 \cdots b_m \rangle_0 + \cdots$$
$$+ \varepsilon^{m-2} \langle b_2 b_3 \cdots b_{m-1}[a_{\mathbf{k}}, b_m]_\varepsilon \rangle_0.$$

After further rearrangement this gives us the required result.

We notice that all the terms on the right-hand side involve a commutator (or anticommutator), which gives either a scalar factor or zero. Therefore, the surviving terms have two fewer operators in the product. The result obtained here will be useful because, when it is applied successively, we can eventually break down averages over products with a large number of operators into products with just two operators.

Definition of Limited Contractions

The right-hand side of Equation (8.21) contains terms involving a commutator or anticommutator (depending on ε) of the form

$$[a_{\mathbf{k}}(\tau_1), b_i(\tau_i)]_\varepsilon \quad \text{or} \quad [a_{\mathbf{k}}^\dagger(\tau_1), b_i(\tau_i)]_\varepsilon.$$

In the first case, if $b_i = a_{\mathbf{k}'}$ the result is zero, whereas if $b_i = a_{\mathbf{k}'}^\dagger$ we have

$$[a_{\mathbf{k}}(\tau_1), a_{\mathbf{k}'}^\dagger(\tau_i)]_\varepsilon = \exp\{E_{\mathbf{k}}(\tau_i - \tau_1)\}[a_{\mathbf{k}}, a_{\mathbf{k}'}^\dagger]_\varepsilon$$
$$- \exp\{E_{\mathbf{k}}(\tau_i - \tau_1)\}\delta_{\mathbf{k},\mathbf{k}'}. \tag{8.23}$$

This leads us to define a quantity that we shall (for the time being) refer to as a *limited contraction* of the annihilation operator $a_{\mathbf{k}}$ with any other boson or fermion operator as being equal to

$$\frac{[a_{\mathbf{k}}(\tau_1), b_i(\tau_i)]_\varepsilon}{1 - \varepsilon e^{-\beta E_{\mathbf{k}}}},$$

where the denominator term comes from Equation (8.22). In other words, for a limited contraction we take the operator $a_{\mathbf{k}}(\tau_i)$, form its (anti)commutator with $b_i(\tau_i)$, and divide by the denominator shown. We shall denote this operation algebraically in a shorthand by

$$\langle \overline{a_{\mathbf{k}}b_2 b_3 \cdots b_i} \cdots b_m \rangle_0 \equiv \frac{1}{1 - \varepsilon e^{-\beta E_{\mathbf{k}}}} \langle b_2 b_3 \cdots [a_{\mathbf{k}}(\tau_1), b_i(\tau_i)]_\varepsilon \cdots b_m \rangle_0,$$
$$\tag{8.24}$$

where the overhead line connects the two operators (a_k and b_i in this case) that are involved in the limited contraction.

In the second case that arises from the preceding discussion, the result is zero if $b_i = a_{k'}^\dagger$, whereas if $b_i = a_{k'}$ the result is

$$[a_k^\dagger(\tau_1), a_{k'}(\tau_i)]_\varepsilon = \exp\{-E_k(\tau_i - \tau_1)\}[a_k^\dagger, a_{k'}]_\varepsilon$$
$$= -\varepsilon \exp\{-E_k(\tau_i - \tau_1)\}\delta_{k,k'}. \qquad (8.25)$$

Then, by analogy with the previous case we define the *limited contraction* of a creation operator $a_k^\dagger$ with any other operator to be

$$\frac{[a_k^\dagger(\tau_1), b_i(\tau_i)]_\varepsilon}{1 - \varepsilon e^{\beta E_k}},$$

and we use the same overhead line notation as previously. Notice, however, that the denominator factor is different in the two cases.

To summarize the outcome, it follows that the results in Equations (8.21) and (8.22) can be expressed symbolically in a convenient shorthand as

$$\langle b_1 b_2 b_3 \cdots b_m \rangle_0 = \overline{\langle b_1 b_2} b_3 \cdots b_m \rangle_0 + \varepsilon \overline{\langle b_1 b_2 b_3} b_4 \cdots b_m \rangle_0$$
$$+ \cdots + \varepsilon^{m-2} \overline{\langle b_1 b_2 b_3 \cdots b_m} \rangle_0. \qquad (8.26)$$

We see that the initial unperturbed thermal average of m operators can be formally evaluated by taking a limited contracted of the first operator with each of the other operators in turn. Because the result of taking a limited contraction is a scalar quantity, each nonzero term on the right-hand side has $(m - 2)$ operators remaining.

Products of Limited Contractions

By extension of the previous result we shall now show that

$$\langle b_1 b_2 b_3 \cdots b_m \rangle_0 = \sum \left[(-1)^A \times \left\{ \begin{array}{c} \text{Product of all limited} \\ \text{contractions taken in pairs} \end{array} \right\} \right], \qquad (8.27)$$

where A is the number of interchanges of *fermion* operators needed to bring the contracted operators together (i.e., to be positioned adjacently in the product).

Proof The simplest way is to use the mathematical method of induction in terms of the integer label m. We consider first the case of $m = 2$, for which we know already that the only two nonzero possibilities are

$$\langle a_k^\dagger(\tau_1) a_{k'}(\tau_2) \rangle_0 = \langle a_k^\dagger a_k \rangle_0 \delta_{k,k'} e^{E_k(\tau_1 - \tau_2)} = \frac{\delta_{k,k'} e^{E_k(\tau_1 - \tau_2)}}{e^{\beta E_k} - \varepsilon}, \qquad (8.28)$$

where we have used the earlier result obtained in Section 8.1 for the thermal distribution function, and similarly

$$\langle a_{\mathbf{k}}(\tau_1)a_{\mathbf{k'}}^{\dagger}(\tau_2)\rangle_0 = \langle a_{\mathbf{k}}a_{\mathbf{k}}^{\dagger}\rangle_0 \, \delta_{\mathbf{k},\mathbf{k'}} \, e^{E_{\mathbf{k}}(\tau_2-\tau_1)}$$
$$= \{1 + \varepsilon\langle a_{\mathbf{k}}^{\dagger}a_{\mathbf{k}}\rangle_0\} \, \delta_{\mathbf{k},\mathbf{k'}} \, e^{E_{\mathbf{k}}(\tau_2-\tau_1)}$$
$$= \frac{\delta_{\mathbf{k},\mathbf{k'}} \, e^{E_{\mathbf{k}}(\tau_2-\tau_1)}}{1 - \varepsilon e^{-\beta E_{\mathbf{k}}}}. \tag{8.29}$$

On comparing the right-hand sides of Equations (8.28) and (8.29) with the definitions given of limited contractions, we see that these are just the results as specified, so it has been established that Equation (8.27) is true for the case of $m = 2$.

Next it follows directly from Equation (8.26) that, if the result in Equation (8.27) holds for $m = M$ (with M denoting any positive even integer), then it must be so for $m = M + 2$. Therefore, by induction, it holds for any positive even integer, as stated.

Definition of Contractions

The remaining requirement for Wick's theorem is to include the effect of the ordering operator $\hat{T}_W$, which we defined earlier. This will be achieved through a minor generalization of our previous definition of a limited contraction.

With this in mind, we now define in symbols a *contraction* of two operators $b_i(\tau_i)$ and $b_j(\tau_j)$ by

$$\overbrace{b_i(\tau_i) \cdots b_j(\tau_j)} = \begin{cases} \overline{b_i(\tau_i) \cdots b_j(\tau_j)} & \text{if} \quad \tau_i > \tau_j \\ \varepsilon \overline{b_j(\tau_j) \cdots b_i(\tau_i)} & \text{if} \quad \tau_i < \tau_j \end{cases},$$

where we use ⌢‾‾‾ to denote a contraction and ⋯⋯ to denote a limited contraction as defined before. The preceding definition is sufficient except in the special case in which the τ-labels are equal for two of the contracted operators. We will need to *choose* a convention to deal with this particular case, and we shall adopt the following rule: if $\tau_i = \tau_j$, which gives rise to $\overbrace{b_i(\tau_i) \cdots b_j(\tau_i)}$, then we interpret this as being

$$\lim_{\eta \to 0} \overbrace{b_i(\tau_i + \eta) \cdots b_j(\tau_i)}, \tag{8.30}$$

i.e., our chosen convention is to add a positive infinitesimal η to the first operator (the one on the left).

Wick's Theorem and Examples

The full form of Wick's theorem, which is basically just a minor extension of the result in Equation (8.27), can now be stated as

$$\langle \hat{T}_W \{b_1 b_2 b_3 \cdots b_m\}\rangle_0 = \sum \left[(-1)^A \times \left\{ \begin{array}{l} \text{Product of all } \hat{T}_W\text{-ordered} \\ \text{contractions taken in pairs} \end{array} \right\} \right].$$

$$(8.31)$$

The definition of the integer A is as before.

Proof This result follows immediately as a generalization of Equation (8.27), where we had an analogous result in terms of limited contractions. The only difference comes because the $\hat{T}_W$-ordering of the products has been included in the present result, both on the left-hand side and for the contractions on the right-hand side. Also the sign factors (the powers of ε needed in the fermion case) are automatically taken care of by being absorbed into the chosen definition of a contraction.

As a simple example, taking the case of $m = 4$ and fermion operators, the application of Wick's theorem gives us the result

$$\langle \hat{T}_W \{b_1 b_2 b_3 b_4\}\rangle_0 = \langle \hat{T}_W \{b_1 b_2\}\rangle_0 \langle \hat{T}_W \{b_3 b_4\}\rangle_0$$
$$+ (-1)\langle \hat{T}_W \{b_1 b_3\}\rangle_0 \langle \hat{T}_W \{b_2 b_4\}\rangle_0$$
$$+ (-1)^2 \langle \hat{T}_W \{b_1 b_4\}\rangle_0 \langle \hat{T}_W \{b_2 b_3\}\rangle_0. \qquad (8.32)$$

In this case, there are only three distinct ways to form pairs of operators (for the contractions), and each -1 factor comes from an interchange of the left-to-right order of the fermion operators. A similar result holds for bosons when $m = 4$, except that all the terms on the right-hand side would have $+$ signs. Of course, some of the terms appearing in Equation (8.32) may be equal to zero because the contractions are nonzero only when one operator is an $a_\mathbf{k}$ and the other is an $a_\mathbf{k}^\dagger$ with the same value of the wave vector $\mathbf{k}$.

The complexity of the algebraic expressions arising from Wick's theorem increases rapidly for larger m values. Thus, when $m = 8$, we have $7 \times 5 \times 3 = 105$ terms when the contractions in pairs are formed. This fact strongly points the way to developing an approach to implement Wick's theorem that is not purely algebraic. Eventually, we will be led to a diagrammatic (or graphical) representation of the terms as being more practical.

8.3 The Unperturbed Imaginary-Time Green's Function

For a product of just two operators ($m = 2$) we have already shown that the only nonzero limited contractions (and hence the only nonzero contractions) are those for an $a_\mathbf{k}(\tau_1)$ operator with an $a_\mathbf{k}^\dagger(\tau_2)$ operator, or vice versa.

We now use this property to define a GF for the *unperturbed* system by writing

$$g^0_{M,\mathbf{k}}(\tau_1 - \tau_2) = -\langle \hat{T}_W a_\mathbf{k}(\tau_1) a_\mathbf{k}^\dagger(\tau_2)\rangle_0. \qquad (8.33)$$

We note that this is just identifiable as being same as the imaginary-time (or Matsubara) GF defined previously in Equation (3.50) of Chapter 3, but in the case of the thermal average being taken with respect to the unperturbed Hamiltonian $\mathcal{H}_0$. From Wick's theorem we can also reexpress this imaginary-time GF equivalently in terms of the contraction notation as

$$g^0_{M,\mathbf{k}}(\tau_1 - \tau_2) = -\overline{a_{\mathbf{k}}(\tau_1)a^\dagger_{\mathbf{k}}(\tau_2)}. \tag{8.34}$$

Therefore, from our previous expressions for nonzero contractions it follows that if $\tau_1 > \tau_2$ we have

$$g^0_{M,\mathbf{k}}(\tau_1 - \tau_2) = -\langle a_{\mathbf{k}}a^\dagger_{\mathbf{k}}\rangle_0 e^{E_{\mathbf{k}}(\tau_2-\tau_1)} = -(1 + \varepsilon n_{\mathbf{k}})e^{E_{\mathbf{k}}(\tau_2-\tau_1)}, \tag{8.35}$$

whereas if $\tau_1 < \tau_2$ we have

$$g^0_{M,\mathbf{k}}(\tau_1 - \tau_2) = -\varepsilon\langle a^\dagger_{\mathbf{k}}a_{\mathbf{k}}\rangle_0 e^{E_{\mathbf{k}}(\tau_2-\tau_1)} = -\varepsilon n_{\mathbf{k}}e^{E_{\mathbf{k}}(\tau_2-\tau_1)}. \tag{8.36}$$

We recall that the unperturbed distribution function $n_{\mathbf{k}}$ is defined in Equation (8.7) for bosons and fermions.

In principle, we can now put everything together to evaluate the terms like $\langle \hat{T}_W\{\mathcal{H}_1(\tau_1)\mathcal{H}_1(\tau_2)\cdots\mathcal{H}_1(\tau_n)\}\rangle_0$ as required for the S-matrix expansion in Section 8.1. From Equation (8.14) we would then have for the unperturbed average of the S matrix

$$\langle S(\beta)\rangle_0 = 1 + \sum_{n=1}^{\infty} \frac{(-1)^n}{n!} \int_0^\beta d\tau_1 \int_0^\beta d\tau_2 \cdots \int_0^\beta d\tau_n$$
$$\times \langle \hat{T}_W\{\mathcal{H}_1(\tau_1)\mathcal{H}_1(\tau_2)\cdots\mathcal{H}_1(\tau_n)\}\rangle_0. \tag{8.37}$$

In a final step we would use Equation (8.9) to obtain the partition function Q for the interacting system. Rather than summing over all orders of perturbation (for n ranging from 1 to ∞), we would normally in practice introduce an approximation by including all terms up to a chosen finite value of n.

8.4 Diagrammatic Representation

Instead of proceeding algebraically (which soon becomes tedious in terms of applying Wick's theorem and dealing with the preceding integrations over the τ-labels), it is convenient to relate the GF formalism to a diagrammatic representation that makes it easier to keep track of all the terms. We shall see in the next few pages how this can be done in a manner that systematically deals with all the products that come about when applying Wick's theorem. The diagrams are often referred to as *Feynman diagrams* following the pioneering work by R. P. Feynman, who

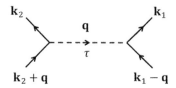

Figure 8.1 The diagrammatic representation of a $\mathcal{H}_1(\tau_i)$ interaction vertex, showing the four lines either emerging from (as a creation operation) or terminating at (as an annihilation operation) the vertex.

introduced a diagram representation in term of lines (representing a GF or "propagator" evolving in real time) and interaction vertices [125–127]. In the formalism to be presented here we shall, in fact, be using the imaginary-time GF, so there is no useful notion in the same sense of forward and backward propagation in time.

As a starting point, we consider the nth order term in the expansion of $\langle S(\beta)\rangle_0$ in Equation (8.37). This term involves a product of n of the $\mathcal{H}_1(\tau_i)$ factors, each of which contains products like

$$\frac{1}{2}v(\mathbf{q})a^{\dagger}_{\mathbf{k}_1}a^{\dagger}_{\mathbf{k}_2}a_{\mathbf{k}_2+\mathbf{q}}a_{\mathbf{k}_1-\mathbf{q}}. \tag{8.38}$$

We will choose to represent each $\mathcal{H}_1(\tau_i)$ as in Figure 8.1. The interaction $v(\mathbf{q})$ is drawn as a broken line carrying the wave-vector (or momentum) label $\mathbf{q}$, and it is also labeled with the appropriate imaginary time τ. The full lines leaving and entering the vertex represent the creation and annihilation operators, respectively, at that vertex. In this chapter, we will be attaching a quantitative significance to the broken and full lines, along with their labels, rather than the schematic representation that was previously depicted in Figure 1.3.

A complete diagram is formed by drawing a vertex as described in the preceding text for each $\mathcal{H}_1(\tau_i)$, and then joining up the full lines in *all* possible ways (consistent with the direction of arrows and the wave-vector conservation). For the simplest case of a diagram with just one interaction vertex (with label τ_1) there are only two possibilities, as shown in Figure 8.2(a). The correspondence between Wick's theorem and those diagrams in Figure 8.2(a) can be illustrated by considering Equation (8.32) when applied specifically to the terms of the interaction Hamiltonian Equation (8.38) in first order:

$$\langle\hat{T}_W v(\mathbf{q})a^{\dagger}_{\mathbf{k}_1}a^{\dagger}_{\mathbf{k}_2}a_{\mathbf{k}_2+\mathbf{q}}a_{\mathbf{k}_1-\mathbf{q}}\rangle_0 = v(\mathbf{q})\langle\hat{T}_W a^{\dagger}_{\mathbf{k}_1}a^{\dagger}_{\mathbf{k}_2}\rangle_0\langle\hat{T}_W a_{\mathbf{k}_2+\mathbf{q}}a_{\mathbf{k}_1-\mathbf{q}}\rangle_0$$
$$- v(\mathbf{q})\langle\hat{T}_W a^{\dagger}_{\mathbf{k}_1}a_{\mathbf{k}_2+\mathbf{q}}\rangle_0\langle\hat{T}_W a^{\dagger}_{\mathbf{k}_2}a_{\mathbf{k}_1-\mathbf{q}}\rangle_0$$
$$+ v(\mathbf{q})\langle\hat{T}_W a^{\dagger}_{\mathbf{k}_1}a_{\mathbf{k}_1-\mathbf{q}}\rangle_0\langle\hat{T}_W a^{\dagger}_{\mathbf{k}_2}a_{\mathbf{k}_2+\mathbf{q}}\rangle_0. \tag{8.39}$$

These are just the three terms expected as in Equation (8.32). The first term here involves averages $\langle\hat{T}_W a^{\dagger}_{\mathbf{k}_1}a^{\dagger}_{\mathbf{k}_2}\rangle_0$ and $\langle\hat{T}_W a_{\mathbf{k}_2+\mathbf{q}}a_{\mathbf{k}_1-\mathbf{q}}\rangle_0$ that are identically zero (their contraction yields zero), while the other two terms involve contractions between an

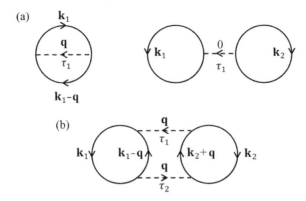

Figure 8.2 The diagrammatic representation: (a) the only two first-order diagrams with no external lines; (b) an example of one of the diagrams in second order.

$a^\dagger$ and an a operator. The latter quantities are just the imaginary time GFs, which can be represented as the solid lines in Figure 8.2(a). After taking account of the wave-vector conservation, the nonzero terms on the right-hand side of the preceding equation reduce to (for fermions)

$$-v(\mathbf{q})\langle\hat{T}_W a^\dagger_{\mathbf{k}_1} a_{\mathbf{k}_1}\rangle_0 \langle\hat{T}_W a^\dagger_{\mathbf{k}_1-\mathbf{q}} a_{\mathbf{k}_1-\mathbf{q}}\rangle_0 + v(0)\langle\hat{T}_W a^\dagger_{\mathbf{k}_1} a_{\mathbf{k}_1}\rangle_0 \langle\hat{T}_W a^\dagger_{\mathbf{k}_2} a_{\mathbf{k}_2}\rangle_0,$$

where the two terms are in a one-to-one correspondence with the two diagrams shown in Figure 8.2(a).

More generally, in the nth order of perturbation there will be n vertices, and the lines must then be connected in all possible ways, giving a complete set of Feynman diagrams. As an example, in Figure 8.2(b) just one of the second-order diagrams is shown. We will return to discuss this process further at a later stage.

With each full line in any diagram that begins at a vertex labeled τ_2 and ends at a vertex τ_1, it follows that we must associate a factor

$$g^0_{M,\mathbf{k}}(\tau_1 - \tau_2),$$

where $\mathbf{k}$ is the wave-vector label for the line. Typically, the τ-labels will refer to different vertices, as in Figure 8.2(b). In some cases, however, the τ-labels may be the same, as in Figure 8.2(a), and the labeling convention adopted in Equation (8.30) must be put into practice, as we describe in later examples.

Finally, to complete the evaluation of the total contribution of any diagram to $\langle S(\beta)\rangle_0$, it is also necessary to carry out several other steps that consist of

- Including the $(-1)^A$ factor (from Wick's theorem);
- Including the $(-1)^n/n!$ factor (from the expansion of $S(\beta)$);
- Summing over all wave-vector labels (from the expression for $\mathcal{H}_1$); and
- Integrating with respect to each τ_i label from 0 to β (from the expansion of $S(\beta)$).

Before attempting to go through these steps, it is convenient to discuss some ways in which carrying them out can be simplified in the diagrammatic context.

The Linked-Cluster Theorem

The various diagrams may be classified as being either *linked* or *unlinked,* depending on whether they form one connected entity (between vertices) or if they have several unconnected parts. The first-order diagrams are (by neccessity) all linked, but the diagrams in second order (and higher) are not necessarily so. As a simple exercise, it is instructive to try drawing some second- and/or third-order diagrams, finding examples of linked and unlinked diagrams. We remark that the diagram in Figure 8.2(b) is a linked diagram, because there is a network of connections (by GF lines) between all the vertices (dashed lines). By contrast, the second-order diagram formed by drawing the two first-order diagrams in Figure 8.2(a) alongside one another as a single entity would be an unlinked diagram. The alternative terms *connected* and *unconnected* are also used in this context, instead of linked and unlinked.

From Equation (8.9) and our proof of Wick's theorem in Section 8.2, we have established for the partition function Q that

$$\frac{Q}{Q_0} = \langle S(\beta) \rangle_0 = \sum \left[\text{All distinct diagrams with no external lines} \right]. \qquad (8.40)$$

Here "all" means that we must include both linked and unlinked diagrams (without making any distinction), and "with no external lines" means that all the GF lines have to join up (i.e., they correspond to all possible contractions being made in pairs). We emphasize this latter point here because later in this chapter we shall come across other diagrams with external lines.

By detailed counting arguments based on the number of possible contractions giving either linked or unlinked diagrams and also by using permutation theory, it can be proved that

$$\ln \left(\frac{Q}{Q_0} \right) = \sum \left[\begin{array}{c} \text{All distinct } \textit{linked} \text{ diagrams} \\ \text{with no external lines} \end{array} \right]. \qquad (8.41)$$

This result is known as the *linked-cluster theorem*. The proof involves some specialized mathematics and will not be given here, but very good discussions can be found in [50, 51]. The use of this theorem simplifies the analysis by allowing us to restrict attention to the evaluation of a smaller number of diagrams (i.e., only the linked or connected ones). This procedure then leads to the natural logarithm of Q/Q_0, instead of Q/Q_0.

The appearance of a logarithmic term on the left-hand side of Equation (8.41) is very convenient in view of some standard definitions in thermodynamics. Because we are working, in general, in a grand canonical ensemble (see, e.g., [3]) it is convenient to make use of the so-called grand potential Ω, which is one of the thermodynamic potentials. It applies for systems that are held at the same temperature and chemical potential as the environment, but the particle number can vary. It satisfies the defining relationship to the grand partition function that

$$\ln(Q) = -\beta\Omega, \tag{8.42}$$

where again $\beta = 1/k_B T$. By forming various partial derivatives of Ω we may deduce other thermodynamic quantities, e.g., the mean number of particles corresponds to

$$\langle N \rangle = -\left(\frac{\partial\Omega}{\partial\mu}\right)_{T,V}.$$

With the introduction of the thermodynamic potential we see that Equation (8.41) becomes

$$\Delta\Omega \equiv (\Omega - \Omega_0) = -\frac{1}{\beta}\sum \left[\begin{array}{c} \text{All distinct } \textit{linked} \text{ diagrams} \\ \text{with no external lines} \end{array} \right]. \tag{8.43}$$

Here $\Delta\Omega$ is the correction to the thermodynamic potential due to the perturbation terms, and Ω_0 is the unperturbed part of the thermodynamic potential given by $(-1/\beta)\ln Q_0$.

The Factor $(-1)^A$

We recall the definition of the integer A, which appears in Wick's theorem in Equation (8.31), as being the number of interchanges of fermion operators required to bring about the set of contractions for Wick's theorem. For a boson system, we have simply $A = 0$ and so the $(-1)^A$ factor is unity. For the fermion case, however, we need to know how to interpret this factor diagrammatically. Fortunately, this turns out to be very simple.

In terms of the corresponding diagrams, it is possible to prove that *each closed fermion loop corresponds to an odd number of interchanges of fermion operators.* Therefore, we may associate a factor of -1 with each closed (or complete) fermion loop. It is necessary only to count up how many such loops there are in any particular diagram, and the $(-1)^A$ algebraic factor coming from Wick's theorem can be replaced by the simple diagrammatic rule:

"For any diagram, we include an overall factor of $(-1)^L$, where L is the number of closed fermion loops."

As an exercise, try finding the $(-1)^A$ values corresponding to the first- and second-order diagrams in Figure 8.2 for the fermion case. Another case for consideration is included in Problem 8.1.

Symmetry Factors

There is a rather subtle property concerning the proper way in which the different contractions that correspond to Wick's theorem should be counted in terms of the

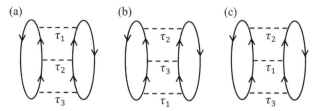

Figure 8.3 Some examples of third-order free-energy diagrams used for consideration of symmetry factors.

diagrammatic representations. It eventually leads to the concept of a symmetry factor to be associated with any diagram.

We will explore these ideas by considering some examples, starting with the third-order diagrams shown in Figure 8.3. It is obvious that they must all eventually lead to the same contribution on evaluation (because they differ only by a permutation of the τ-labels, and the τ-labels are dummy variables that are later integrated from 0 to β).

In general, we shall get the same contribution from sets of diagrams that differ only in that their τ-labels are permuted. Such diagrams are said to be *topologically the same*. As a consequence, it would seem to be sufficient to consider only one member of the set, and to correct for this by introducing a factor $n!$ in the case of nth order diagrams. This would seem to cancel out nicely with the $1/n!$ coming from the S-matrix expansion as in Equation (8.37). However, this would represent an oversimplification, which can lead to an incorrect counting of diagrams. The reason is that some of the $n!$ diagrams might not be distinct diagrams (i.e., they might correspond to the same set of contractions in applying Wick's theorem, and so they should be counted once). One way to think about this situation is as follows. Suppose we have some diagrams that have their τ-labels inserted, so that the vertices are labeled with $\{\tau_1, \tau_2, \ldots, \tau_n\}$ in a nth-order diagram. If any one of these diagrams can be "deformed" (through a repositioning of the vertices and the connecting GF lines, without breaking any of the lines or connections) into another one of the diagrams, then those two diagrams come from the same set of contractions and are *not* distinct.

A specific example to illustrate the preceding argument is provided by the three labeled third-order diagrams shown in Figure 8.3, which all have three vertices that connect the two different GF loops. Clearly, there are $3! = 6$ such diagrams in total (from considering all possible permutations of the three τ-labels). There is an important distinction to be made, however, for the three examples shown. It is evident that diagram (a) can be continuously deformed into diagram (b), but it cannot be deformed into diagram (c). This is because diagrams (a) and (b) correspond to even permutations (those with the same cyclic order) of the labels 1, 2, and 3, whereas diagram (c) corresponds to an odd permutation.

In this case, the diagrams that differ only in having their labels in the same cyclic order are *not* to be regarded as distinct diagrams because they correspond to the same set of contractions in Wick's theorem, as explained. The only distinct cases here are those corresponding to the odd and the even permutations. Hence, the overall counting factor here is $(1/3!) \times 2 = 1/3$. In a more general case, we will evaluate only one of the diagrams of a particular (topological) type, and we introduce a *symmetry factor* denoted generally as $1/p$ to allow for the correct counting (of distinct contractions). It follows that in the preceding example, we have $p = 3$. We shall see further examples of the use of symmetry factors in subsequent diagrammatic calculations.

Transformation to Matsubara Frequencies

Rather than working in terms of the τ-labels that are associated with the vertices and therefore also with the GF lines between vertices, it is convenient to transform to a "frequency" representation by using the following identity for the unperturbed GFs:

$$g^0_{M,\mathbf{k}}(\tau) = \frac{1}{\beta} \sum_m \frac{\exp(-i\omega_m \tau)}{i\omega_m - E_\mathbf{k}}, \tag{8.44}$$

where the ω_m are the discrete Matsubara frequencies that were defined in Equation (3.54). As discussed, they take a different form for bosons and fermions. A proof of the important result quoted in Equation (8.44) is presented in the following text. First we remark, on comparing the preceding result with Equation (3.53), that the Fourier-transformed imaginary-time GF $G^0_\mathbf{k}(i\omega_m)$ for the unperturbed system is given by

$$G^0_\mathbf{k}(i\omega_m) = \frac{1}{i\omega_m - E_\mathbf{k}}. \tag{8.45}$$

This result is consistent, as expected, with the Matsubara GF expression deduced in Subsection 4.1.3 for noninteracting systems.

Proof First we will take the case of fermions and $\tau < 0$ for Equation (8.44). It follows from Equation (8.36) that the appropriate expression for the unperturbed GF is

$$g^0_{M,\mathbf{k}}(\tau) = \frac{e^{-E_\mathbf{k}\tau}}{e^{\beta E_\mathbf{k}} + 1}. \tag{8.46}$$

Next we consider a related contour integral in the complex plane that corresponds to

$$\int_C \frac{e^{-\omega\tau} d\omega}{(\omega - E_\mathbf{k})(e^{\beta\omega} + 1)}, \tag{8.47}$$

where the contour C is taken to be as shown in Figure 8.4(a) with the closure being at $\pm\infty$ on the imaginary axis. The poles enclosed by this contour are those

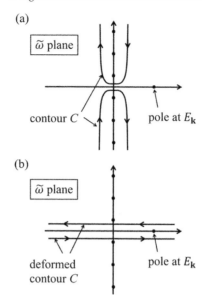

(a)

(b)

Figure 8.4 Contours and poles for the contour integration used in Fourier transforming the unperturbed GF in Section 8.3, showing (a) the original contour and (b) the deformed contour.

associated with the term $(e^{\beta\omega} + 1)$ in the denominator of the contour integral. They occur when $e^{\beta\omega} = -1$, which means

$$\omega = \frac{(2m + 1)i\pi}{\beta} \equiv i\omega_m.$$

This result provides us with the connection to the Matsubara frequencies for fermions defined in Equation (3.54). It is now a straightforward step to use the residue theorem in complex analysis to prove that

$$\int_C \frac{e^{-\omega\tau}\,d\omega}{(\omega - E_{\mathbf{k}})(e^{\beta\omega} + 1)} = 2\pi i\frac{1}{\beta}\sum_m \frac{\exp(-i\omega_m\tau)}{i\omega_m - E_{\mathbf{k}}}. \tag{8.48}$$

The preceding result follows because the contour encloses an infinite sequence of poles (for ω) at discrete points along the imaginary axis, as shown, whereas the other pole on the real axis at $\omega = E_{\mathbf{k}}$ is not enclosed and does not contribute.

Next, we continuously deform the contour C by "opening it out" so that it becomes as shown in Figure 8.4(b). The form of the exponential factors allows the modified contour to be completed at $\pm\infty$ on the real axis without there being any contribution to the integral from this part. Now the only pole enclosed is the one on the real axis at $\omega = E_{\mathbf{k}}$. We use the residue theorem again to evaluate the same integral, but with the modified contour. This gives

$$\int_C \frac{e^{-\omega\tau}\,d\omega}{(\omega - E_{\mathbf{k}})\left(e^{\beta\omega} + 1\right)} = 2\pi i \frac{e^{-E_{\mathbf{k}}\tau}}{e^{\beta E_{\mathbf{k}}} + 1} = 2\pi i g_{M,\mathbf{k}}^0(\tau). \qquad (8.49)$$

By comparing the right-hand side of Equation (8.49) with that in Equation (8.48) we see that the required result stated in Equation (8.44) for $\tau < 0$ is obtained. With only minor changes the proof can be carried out for fermions when $\tau > 0$.

There is a very similar proof for the case of bosons, again treating $\tau < 0$ and $\tau > 0$ separately. The only essential difference is that the poles now associated with a term $(e^{\beta\omega} - 1)$ appearing in the denominator of the contour integral occur for $e^{\beta\omega} = 1$, which means that

$$\omega = \frac{2mi\pi}{\beta} \equiv i\omega_m$$

for the boson case, consistent with Equation (3.54).

8.4.1 The τ-Dependence of a Vertex

One advantage of the previous result, in which we obtained a "frequency" representation for the GFs, is that we arrive at an easy way of dealing with the τ-dependence of any vertex in a diagram. It follows from Equation (8.44) that the overall τ-dependence of a vertex comes from

$$\exp\left[\tau(i\omega_{m_1} + i\omega_{m_2} - i\omega_{m_3} - i\omega_{m_4})\right].$$

When the τ-label is integrated over, with the limits of integration being taken from 0 to β as in Equation (8.37), there are two cases depending on whether $\Delta\omega_m \equiv (\omega_{m_1} + \omega_{m_2} - \omega_{m_3} - \omega_{m_4})$ is zero or otherwise. First, if $\Delta\omega_m \neq 0$, we can perform the integration as specified to obtain

$$\frac{\exp\left[\beta(i\omega_{m_1} + i\omega_{m_2} - i\omega_{m_3} - i\omega_{m_4})\right] - 1}{(i\omega_{m_1} + i\omega_{m_2} - i\omega_{m_3} - i\omega_{m_4})}. \qquad (8.50)$$

This result can immediately be simplified because, using the definitions for the Matsubara frequencies, we see that

$$i\omega_{m_1} + i\omega_{m_2} - i\omega_{m_3} - i\omega_{m_4} = \frac{2i\pi}{\beta}(m_1 + m_2 - m_3 - m_4)$$

$$\equiv \frac{2i\pi}{\beta}M, \qquad (8.51)$$

where M denotes a nonzero integer, irrespective of whether we have boson or fermion operators. This is equivalent to stating that the sums or differences of an *even* number of Matsubara frequencies is always formally the same as a boson frequency. Then we see by using Equation (8.51) that the exponential term in the numerator of Equation (8.50) is unity, implying a zero result for the τ-integration. However, when $\Delta\omega_m = 0$, we have simply an integration over unity leading to a factor of β. Hence, in general we conclude that in the frequency representation each vertex gives rise to a factor of $\beta\delta_{m_1+m_2,m_3+m_4}$. This Kronecker delta just corresponds to a frequency conservation condition at each vertex.

It can therefore be seen that there is a great advantage in using the frequency representation in which $i\omega_m$ labels are associated with each GF line because the τ-integration required for each vertex is automatically taken care of when we make use of the frequency conservation condition at that vertex.

8.4.2 Rules for Perturbation Summation

We can now put together all the results for evaluating diagrams obtained so far and summarize them in terms of a straightforward set of rules as follows:

Draw all the topologically distinct linked diagrams with no external lines, and then calculate the contribution of each diagram according to the rules stated in the following text. The sum for the set of diagrams gives their contribution to $\Delta\Omega = \Omega - \Omega_0$.

1. Label the diagrams (both full lines and vertex lines) with wave vector and frequency so that these quantities are conserved at each end of the vertex.
2. Associate a factor $\frac{1}{2}\beta v(\mathbf{q})$ with each vertex line, where $\mathbf{q}$ is the wave vector along the vertex.
3. For each full line associate a factor $(1/\beta)\mathcal{G}_{\mathbf{k}}^0(i\omega_m)$, where $\mathbf{k}$ and $i\omega_m$ are the wave vector and frequency along the full line.
4. Include an extra factor of $(-1/\beta)$.
5. Include a factor $(-1)^L/p$, where L is the number of closed fermion loops and p is the symmetry number.
6. Include a factor $(-1)^n$, where n is the number of vertices in the diagram.
7. Finally, sum over all wave-vector and frequency labels within the conservation restrictions imposed by rule 1.

It turns out that there is an additional diagram rule needed, which we will introduce later as required.

8.4.3 Examples of Evaluating Diagrams

We take the case of fermion systems in the following examples. The simplest diagrams are the first-order diagrams shown earlier in Figure 8.2(a), where each one has been labeled with wave vector and frequency as described earlier. We start with the diagram on the right for which it is necessarily the case that $\mathbf{q} = 0$ along the vertex. Also, we must have $p = 1$ because this is a first-order diagram. By applying the set of diagrammatic rules we find

$$\sum_{\mathbf{k}_1,\mathbf{k}_2}\sum_{m_1,m_2} \frac{1}{2}\beta v(0) \left(-\frac{1}{\beta}\right)\left(\frac{1}{\beta}\right)^2 \frac{(-1)^2}{1}(-1)^1 \mathcal{G}^0_{\mathbf{k}_1}(i\omega_{m_1})\mathcal{G}^0_{\mathbf{k}_2}(i\omega_{m_2})$$

$$= \frac{v(0)}{2\beta^2}\sum_{\mathbf{k}_1,m_1}\mathcal{G}^0_{\mathbf{k}_1}(i\omega_{m_1})\sum_{\mathbf{k}_2,m_2}\mathcal{G}^0_{\mathbf{k}_2}(i\omega_{m_2}). \tag{8.52}$$

The two sets of double summations in the last line are independent of one another. Suppose we consider either one of them, say

$$\sum_{\mathbf{k}_1,m_1}\mathcal{G}^0_{\mathbf{k}_1}(i\omega_{m_1}) = \sum_{\mathbf{k}_1}\left\{\sum_{m_1}\frac{1}{i\omega_{m_1}-E_{\mathbf{k}_1}}\right\}.$$

We could in principle evaluate the frequency summation by a contour integral method similar to that used for the proof of Equation (8.44). In its preceding form, however, it does not converge properly at infinity in the complex ω plane. We can deal with this difficulty by recalling the convention adopted earlier in Equation (8.30) for a contraction in Wick's theorem when the GF line begins and ends at the same τ label, as is the case here. It involved adding an infinitesimal to τ in the first label. The consequence now is that it leads us to an extra rule, which can be stated as follows:

Extra diagram rule: If a line is self-contracted (meaning that it starts and ends on the same interaction vertex), then we include a convergence factor of $\exp(i\omega_m\eta)$, where $i\omega_m$ is the frequency label and η denotes a positive infinitesimal.

Then, using Equation (8.44) and setting $\tau = -\eta$, we have

$$\sum_{\mathbf{k}_1,m_1}\exp(i\omega_{m_1}\eta)\mathcal{G}^0_{\mathbf{k}_1}(i\omega_{m_1}) = \sum_{\mathbf{k}_1}\frac{\beta}{e^{\beta E_{\mathbf{k}_1}}+1} = \beta\sum_{\mathbf{k}_1}n_{\mathbf{k}_1}, \tag{8.53}$$

from which it follows that the contribution from the diagram on the right in Figure 8.2(a) is simply

$$\frac{1}{2}\sum_{\mathbf{k}_1,\mathbf{k}_2}v(0)n_{\mathbf{k}_1}n_{\mathbf{k}_2}.$$

In a similar way the contribution from the other diagram in Figure 8.2(a) is found to be given by

$$\sum_{\mathbf{k}_1,\mathbf{q}}\sum_{m_1,m_2}\frac{1}{2}\beta v(\mathbf{q})\left(-\frac{1}{\beta}\right)\left(\frac{1}{\beta}\right)^2\frac{(-1)^1}{1}(-1)^1$$

$$\times\, \exp(i\omega_{m_1}\eta)\mathcal{G}^0_{\mathbf{k}_1}(i\omega_{m_1})\exp(i\omega_{m_2}\eta)\mathcal{G}^0_{\mathbf{k}_1-\mathbf{q}}(i\omega_{m_2})$$

$$= -\frac{1}{2\beta^2}\sum_{\mathbf{k}_1,\mathbf{q}}v(\mathbf{q})\left[\sum_{m_1}\exp(i\omega_{m_1}\eta)\mathcal{G}^0_{\mathbf{k}_1}(i\omega_{m_1})\right]$$

$$\times \left[\sum_{m_2} \exp(i\omega_{m_2}\eta) \mathcal{G}^0_{\mathbf{k}_1-\mathbf{q}}(i\omega_{m_2}) \right]$$

$$= -\frac{1}{2} \sum_{\mathbf{k}_1,\mathbf{q}} v(\mathbf{q}) n_{\mathbf{k}_1} n_{\mathbf{k}_1-\mathbf{q}}.$$

The two diagrams appearing in Figure 8.2(a) are the only topologically distinct first-order diagrams. On adding the two contributions just found in the preceding text, we conclude (after some relabeling of wave vectors as dummy summation variables) that the total result for $\Delta\Omega$ in first order of perturbation is

$$\Delta\Omega = \frac{1}{2} \sum_{\mathbf{k}_1,\mathbf{q}} \{v(0) - v(\mathbf{q})\} n_{\mathbf{k}_1} n_{\mathbf{k}_1-\mathbf{q}}. \tag{8.54}$$

The preceding result was derived for fermions, but it is easy to show that the corresponding result for the boson case is simply

$$\Delta\Omega = \frac{1}{2} \sum_{\mathbf{k}_1,\mathbf{q}} \{v(0) + v(\mathbf{q})\} n_{\mathbf{k}_1} n_{\mathbf{k}_1-\mathbf{q}}, \tag{8.55}$$

where the distribution functions are now those for Bose–Einstein statistics.

In the second order of perturbation theory ($n = 2$), where each diagram has two interaction vertices, there are several diagrams to evaluate for $\Delta\Omega$, even when we remember that only the linked (or connected) diagrams need to be considered. One of these diagrams is shown in Figure 8.2(b). It is instructive (see Problem 8.1) to draw all the topologically distinct linked diagrams in second order and to evaluate one of them (e.g., the diagram in Figure 8.2(b)).

8.5 The Interacting Imaginary-Time Green's Function

In the diagrammatic context as used so far, we have employed only the non-interacting (or unperturbed) GFs. These featured as the solid lines in the Feynman diagrams, having been formally defined in Equation (8.33) and taking the values found in Equations (8.35) and (8.36). For the *interacting* GF case we already introduced the corresponding imaginary-time GF following Equation (3.50) as

$$g_{M,\mathbf{k}}(\tau_1 - \tau_2) = -\langle \hat{T}_W \breve{a}_{\mathbf{k}}(\tau_1) \breve{a}^\dagger_{\mathbf{k}}(\tau_2) \rangle, \tag{8.56}$$

where the transformed operators $\breve{a}(\tau)$ and $\breve{a}^\dagger(\tau)$ at any label τ are related to the corresponding a and $a^\dagger$ operators and the full Hamiltonian $\mathcal{H}$ according to Equation (3.49).

Recalling now the definition of the S-matrix in Equation (8.8) and using the definition for an equilibrium thermal average, we can eventually reexpress the GF definition as

$$g_{M,\mathbf{k}}(\tau_1 - \tau_2) = -\frac{\langle \hat{T}_W a_{\mathbf{k}}(\tau_1) a_{\mathbf{k}}^{\dagger}(\tau_2) S(\beta) \rangle_0}{\langle S(\beta) \rangle_0}. \tag{8.57}$$

The algebraic proof of the preceding result, which can be omitted in a first reading, proceeds as follows. We introduce a new operator S defined in terms of τ labels by

$$S(\tau_1, \tau_2) = \hat{T}_W \left[\exp\left\{ -\int_{\tau_2}^{\tau_1} d\tau \mathcal{H}_1(\tau) \right\} \right]. \tag{8.58}$$

This is a generalization of the shorthand result for the S-matrix in Equation (8.15), and it must be interpreted in the same fashion. Clearly, we have the identity that $S(\beta) = S(\beta, 0)$. Also, for any τ' satisfying $\tau_1 < \tau' < \tau_2$, we have

$$S(\tau_1, \tau_2) = \hat{T}_W \left[\exp\left\{ -\int_{\tau_2}^{\tau'} d\tau \mathcal{H}_1(\tau) - \int_{\tau'}^{\tau_1} d\tau \mathcal{H}_1(\tau) \right\} \right]$$

$$= \hat{T}_W \left[\exp\left\{ -\int_{\tau_2}^{\tau'} d\tau \mathcal{H}_1(\tau) \right\} \exp\left\{ -\int_{\tau'}^{\tau_1} d\tau \mathcal{H}_1(\tau) \right\} \right]$$

$$= S(\tau_1, \tau') S(\tau', \tau_2).$$

We may note that, due to the effect of the $\hat{T}_W$ operator, the steps in the derivation are not as trivial as it seems at first. Next it follows from the above property that when $\tau_1 > \tau_2$ we have other similar relationships like $S(\tau_1, \tau_2) = S(\tau_1, 0)\big(S(\tau_2, 0)\big)^{-1}$ and $S(\beta, \tau) = S(\beta)\big(S(\tau, 0)\big)^{-1}$. From these results (see Problem 8.5), taken together with the definition of the S-matrix in Equation (8.8), we may arrive at the GF expression quoted in Equation (8.57).

It is now evident that both the numerator and denominator parts of the right-hand side in Equation (8.57) include factors of $S(\beta)$. Therefore, it follows from the S-matrix expansion in Equation (8.13) that we again have thermal averages of the form $\langle \hat{T}_W \{ b_1 b_2 b_3 \cdots b_m \} \rangle_0$, where m denotes an *even* integer. Hence, we can use Wick's theorem, as before, to express the results in terms of summations over all contractions of the operators in pairs.

The outcome from the previously mentioned discussion is that we can again set up a *diagrammatic representation* for the GF terms in Equation (8.57) in a very similar manner to what was done previously for contributions to the thermodynamic potential Ω. The only essential difference arises from the extra operator terms that appear in the numerator of Equation (8.57), as well as the $S(\beta)$ in the denominator. Specifically, the extra $a_{\mathbf{k}}^{\dagger}$ operator in the numerator of the interacting GF gives an external line (meaning one that does not begin at a vertex) and the extra $a_{\mathbf{k}}$ operator gives another external line (one that does not end at a vertex). Hence, we have two external lines in any diagram for the GF, so in general they look schematically like Figure 8.5 where the shaded area represents any other allowed diagrammatic

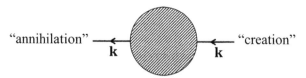

"annihilation" $\quad$ k $\qquad$ k $\quad$ "creation"

Figure 8.5 The general form of a diagrammatic contribution to the interacting GF, as discussed in the text. By contrast with the previous closed diagrams, there are now two external GF lines representing the role of the extra creation and annihilation operators, $a_{\mathbf{k}}^{\dagger}(\tau_2)$ and $a_{\mathbf{k}}(\tau_1)$, respectively, in the numerator of Equation (8.57).

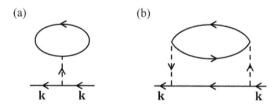

(a) $\qquad$ (b)

k $\quad$ k $\qquad$ k $\qquad$ k

Figure 8.6 Some examples of contributions to the interacting GF in (a) the first order ($n = 1$) and (b) the second order ($n = 2$) of perturbation.

structure of lines and interaction vertices, drawn as before. By contrast, the previous diagrams for the thermodynamic potential were closed diagrams with no external lines. Some simple examples of diagrams for the interacting GF are shown in Figure 8.6.

Fortunately, there are simplifications that arise, just as before for the closed diagrams for the thermodynamic potential. It can again be proved that a linked cluster theorem applies. What happens in this case is that the end result of restricting attention to only the linked (or connected) diagrams has the effect of cancelling out exactly the $\langle S(\beta) \rangle_0$ term in the denominator of Equation (8.57). Hence, the formal result for evaluating the GF becomes rather conveniently

$$g_{M,\mathbf{k}}(\tau_1 - \tau_2) = \sum \left[\begin{array}{c} \text{All distinct linked diagrams} \\ \text{with two external lines} \end{array} \right]. \qquad (8.59)$$

Again, just as before, it is very cumbersome to evaluate the diagrams directly using the τ-labels because these have to be integrated over. The solution again is to define a transformation to a frequency representation by analogy with the result for the noninteracting GF. Following Equation (3.53) we write the GF as

$$g_{M,\mathbf{k}}(\tau) = \frac{1}{\beta} \sum_{m=-\infty}^{\infty} e^{-i\omega_m \tau} \mathcal{G}_{\mathbf{k}}(i\omega_m), \qquad (8.60)$$

where the imaginary frequency $i\omega_m$ is defined for the boson and fermion cases by the same expressions as before. At any interaction vertex (with two lines entering and two leaving) we will have conservation of frequency, as before, but the external lines will carry a fixed imaginary frequency $i\omega_m$ (as well as the wave vector $\mathbf{k}$). We note that, in the case of the noninteracting GF, the previous result was that quoted in Equation (8.44).

We can now write down a set of diagrammatic rules for calculating the interacting GF that are closely analogous to those used for the thermodynamic potential Ω.

Diagrammatic Rules for Evaluating $\mathcal{G}_{\mathbf{k}}(i\omega_m)$

Draw all the topologically distinct linked diagrams with one external line $(\mathbf{k}, i\omega_m)$ entering and another external line $(\mathbf{k}, i\omega_m)$ leaving. Calculate the contribution of each diagram to $\mathcal{G}_{\mathbf{k}}(i\omega_m)$ according to the rules given in the following text.

1. Label the diagrams so that the wave vector and frequency are conserved at each vertex.
2. Associate a factor $\beta v(\mathbf{q})$ with each vertex, where $\mathbf{q}$ is the wave vector along the vertex.
3. For each full line associate a factor $(1/\beta)\mathcal{G}_{\mathbf{k}}^0(i\omega_m)$, where $\mathbf{k}$ and $i\omega_m$ are the wave vector and frequency along the line. Also include an extra factor of $\exp(i\omega_m\eta)$ if the line is self-contracted.
4. Include a factor $(-1)^L/p$, where L is the number of closed fermion loops and p is the symmetry number.
5. Include a factor $(-1)^n$, where n is the number of vertices in the diagram.
6. Finally, sum over all wave-vector and frequency labels within the restrictions imposed by rule 1.

It is important to comment that we do *not* sum over the external $\mathbf{k}$ and $i\omega_m$ because these are fixed labels. Note that most of these rules are the same or similar to the previous case of those used for Ω, where there were no external lines.

Proper Self-Energy and Dyson's Equation

We now introduce the concept of self-energy (and *proper* self-energy, in particular). First, we note that some examples (in no special order) of simple diagrams for $\mathcal{G}_{\mathbf{k}}(i\omega_m)$ are shown in Figure 8.7.

It becomes immediately apparent by inspection that, even using just the two simple loop structures chosen here, there are many diagrams that can be formed, and the process of selecting them appears to be rather haphazard. Therefore, it is useful if we can develop a more *systematic* procedure for writing down a series of contributions to $\mathcal{G}_{\mathbf{k}}(i\omega_m)$. One approach, which can be followed quite generally, is to classify the sum of all diagrams involving a particular structure of vertices (e.g., the loop structures that occur in some parts of Figure 8.7). This process is represented by the diagrammatic series shown in Figure 8.8.

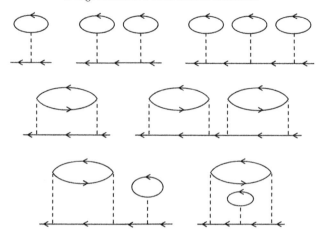

Figure 8.7 Some diagrammatic structures that contribute to the interacting GF $\mathcal{G}_{\mathbf{k}}(i\omega_m)$ in low orders of perturbation.

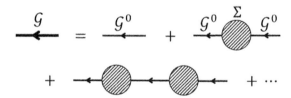

Figure 8.8 A diagrammatic series leading to Dyson's equation. There are GF lines $\mathcal{G}^0$ joining the shaded parts, which are labeled as Σ and explained in the text.

Here the interacting GF $\mathcal{G}_{\mathbf{k}}(i\omega_m)$ is denoted by a heavy arrowed line, while the noninteracting GF $\mathcal{G}_{\mathbf{k}}^0(i\omega_m)$ is denoted by a (lighter) arrowed line, as before. By the conservation properties, all the lines mentioned in the preceding text carry the same labels $(\mathbf{k}, i\omega_m)$. The shaded parts denote any diagrammatic structure (with an arbitrary number of interactions) having the special property that it cannot be separated into two parts by breaking one GF line. Some examples of proper self-energy diagrams are shown in Figure 8.9(a); the diagrams in Figure 8.9(b) are not proper self-energies because the breaking of one line would separate them into two parts.

The summation of the series of diagrams represented by Figure 8.8 can be expressed as

$$\mathcal{G}_{\mathbf{k}}(i\omega_m) = \mathcal{G}_{\mathbf{k}}^0(i\omega_m) + \mathcal{G}_{\mathbf{k}}^0(i\omega_m)(1/\beta)\Sigma(\mathbf{k}, i\omega_m)\mathcal{G}_{\mathbf{k}}^0(i\omega_m)$$
$$+ \mathcal{G}_{\mathbf{k}}^0(i\omega_m)(1/\beta)\Sigma(\mathbf{k}, i\omega_m)\mathcal{G}_{\mathbf{k}}^0(i\omega_m)(1/\beta)\Sigma(\mathbf{k}, i\omega_m)\mathcal{G}_{\mathbf{k}}^0(i\omega_m) + \cdots .$$

It can be seen by a process of iteration with the last term on the right-hand side of the preceding expression that this is just the same as

$$\mathcal{G}_{\mathbf{k}}(i\omega_m) = \mathcal{G}_{\mathbf{k}}^0(i\omega_m) + \mathcal{G}_{\mathbf{k}}^0(i\omega_m)(1/\beta)\Sigma(\mathbf{k}, i\omega_m)\mathcal{G}_{\mathbf{k}}(i\omega_m). \tag{8.61}$$

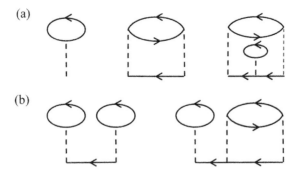

Figure 8.9 Some examples of self-energy diagrams that (a) are proper and (b) are not proper.

The preceding result is known as *Dyson's equation*. It gives a connection between $\mathcal{G}$ and $\mathcal{G}^0$ for any choice of the proper self-energy Σ. A separate calculation must be made for Σ using the diagrammatic rules.

Leaving aside the evaluation of Σ until later, we can easily rearrange Dyson's equation into an alternative form as

$$[\mathcal{G}_{\mathbf{k}}^0(i\omega_m)]^{-1} = [\mathcal{G}_{\mathbf{k}}(i\omega_m)]^{-1} + (1/\beta)\Sigma(\mathbf{k}, i\omega_m). \tag{8.62}$$

If Equation (8.45) for the noninteracting GF is substituted into the preceding equation, the result for the interacting GF can easily be obtained more explicitly as

$$\mathcal{G}_{\mathbf{k}}(i\omega_m) = \frac{1}{i\omega_m - E_{\mathbf{k}} - (1/\beta)\Sigma(\mathbf{k}, i\omega_m)}. \tag{8.63}$$

In other words, the interacting GF is obtained from the noninteracting GF by making the formal replacement

$$E_{\mathbf{k}} \rightarrow E_{\mathbf{k}} + (1/\beta)\Sigma(\mathbf{k}, i\omega_m). \tag{8.64}$$

Because any pole of the interacting GF corresponds to the vanishing of its energy denominator, we have arrived at a procedure to find the poles in any order of perturbation (for Σ). It is necessary to replace the discrete imaginary frequency $i\omega_m$ by a (usually complex) frequency label $\tilde{\omega}$ through an analytic continuation as in Chapter 3, and then we look for the self-consistent solution(s) of

$$\tilde{\omega} - E_{\mathbf{k}} - (1/\beta)\Lambda(\mathbf{k}, \tilde{\omega}) - i(1/\beta)\Gamma(\mathbf{k}, \tilde{\omega}) = 0. \tag{8.65}$$

Here the proper self-energy Σ, which might be complex in general, has been expressed in terms of its real and imaginary parts as $\Sigma = \Lambda + i\Gamma$.

Although Equation (8.65) should be solved self-consistently, it is often possible to find approximate solution(s). For example, if the interaction terms are weak (implying that the real and imaginary parts of Σ are small), the solution for $\tilde{\omega}$

will still be approximately real and close to $E_{\mathbf{k}}$ and so we can consider replacing $\Lambda(\mathbf{k}, \tilde{\omega})$ and $\Gamma(\mathbf{k}, \tilde{\omega})$ by $\Lambda(\mathbf{k}, E_{\mathbf{k}})$ and $\Gamma(\mathbf{k}, E_{\mathbf{k}})$, respectively. Hence, an *approximate* solution of Equation (8.65) is

$$\tilde{\omega} = E_{\mathbf{k}} + (1/\beta)\Lambda(\mathbf{k}, E_{\mathbf{k}}) + i(1/\beta)\Gamma(\mathbf{k}, E_{\mathbf{k}}). \tag{8.66}$$

The physical interpretation of the preceding result is as follows. The *real* part of the preceding expression shows that the energy $E_{\mathbf{k}}$ for a free particle is modified, or *renormalized*, to become $E_{\mathbf{k}} + \Delta E_{\mathbf{k}}$, where

$$\Delta E_{\mathbf{k}} = (1/\beta)\Lambda(\mathbf{k}, E_{\mathbf{k}}). \tag{8.67}$$

The occurrence of an *imaginary* part (usually known as the *damping*) in the pole of the GF is inversely related to a decay time (or lifetime) of the particle. If we write the complex solution for the frequency as $\tilde{\omega} = \omega - i\omega''$, then it is clear for any Fourier component that

$$\exp(-i\tilde{\omega}t) = \exp\{-i(\omega - i\omega'')t\} = \exp(-i\omega t)\exp(-\omega''t). \tag{8.68}$$

This confirms that the real part ω is related to the renormalized frequency, as deduced in the preceding text, while the lifetime $T_{\mathbf{k}}$ of the excitation is given by

$$T_{\mathbf{k}} = \frac{1}{\omega''} = -\beta\{\Gamma(\mathbf{k}, E_{\mathbf{k}})\}^{-1}. \tag{8.69}$$

Examples for the evaluation of self-energy diagrams to calculate the interacting GFs, as well as the renormalized energy and damping of the excitations, will be given in the next chapter. In some simple theories, such as those where the proper self-energy is calculated only to some relatively low order in the perturbation expansion, we might find that Σ is real. In these cases, we have a renormalization for the energy of the excitation but the damping is zero.

Problems

8.1. One of the Feynman diagrams contributing to the thermodynamic potential Ω in second order of perturbation is depicted in Figure 8.2(b). Add to this by drawing all the other topologically distinct linked diagrams that occur in second order.

8.2. Draw *four* examples of third-order diagrams that contribute to the thermodynamic potential Ω for an interacting boson or fermion system. Keep in mind that only the linked, or connected, diagrams with the appropriate number of interaction vertices are to be considered.

8.3. The Feynman diagram shown in Figure 8.10 represents a second-order contribution to the thermodynamic potential Ω. In terms of the labeling scheme

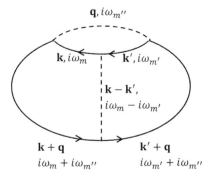

Figure 8.10 A diagram representing a second-order contribution to the thermodynamic potential (see Problem 8.3). A labeling scheme is included.

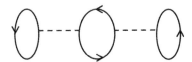

Figure 8.11 Another diagram representing a second-order contribution to the thermodynamic potential (see Problem 8.4).

shown in the figure, employ the diagrammatic rules to obtain an expression for the contribution to Ω for the case of a fermion system. Do *not* in this case carry out the summations over the internal frequencies or wave vectors, but indicate where they should be applied.

8.4. The Feynman diagram shown in Figure 8.11 represents another second-order contribution to the thermodynamic potential Ω. Insert wave-vector and frequency labels as appropriate, and employ the diagrammatic rules to obtain an expression for the contribution to Ω for the case of a boson system. Now carry out the summations over the internal frequencies (noting that a double pole arises in one of the summation).

8.5. By using the properties described in the text for the generalized quantity $S(\tau_1, \tau_2)$, which was defined in Equation (8.58), prove that Equation (8.56) can be rewritten in terms of unperturbed thermal averages as stated in Equation (8.57). It will help to make use of some of the properties of the trace (Tr) operator and the definition in Equation (8.8).

8.6. Evaluate the contribution to the proper self-energy for a boson system arising from the Feynman diagram shown in Figure 8.12(a). Next use your result to deduce the contribution (if any) to the renormalized energy and the damping of a boson due to this self-energy term.

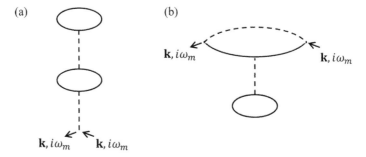

Figure 8.12 Two diagrammatic contributions to the self-energy of the interacting GF $\mathcal{G}_{\mathbf{k}}(i\omega_m)$ (see Problems 8.6 and 8.7).

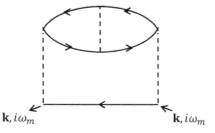

Figure 8.13 A contribution to the proper self-energy $\Sigma(\mathbf{k}, i\omega_m)$ (see Problem 8.8).

8.7. Repeat the previous Problem 8.6, but taking in this case the proper self-energy contribution of a boson system to be represented by the Feynman diagram shown in Figure 8.12(b).

8.8. Consider the diagrammatic contribution to the proper self-energy $\Sigma(\mathbf{k}, i\omega_m)$ shown in Figure 8.13 for an interacting fermion system. Label this diagram with appropriate wave vectors and frequencies (consistent with the conservation properties at vertices). Then write down a formal expression for its total contribution in terms of the different factors that arise from the application of the diagram rules. Do *not* attempt to carry out any of the summations.

8.9. In Section 8.5 it was shown that the calculation for the modified poles of a renormalized GF requires finding the self-consistent solution of Equation (8.65), and we gave an approximate solution in Equation (8.66). Now obtain, as follows, an improved solution in the special case where the self-energy is real (i.e., there is no damping). Instead of simply replacing the frequency label by $E_{\mathbf{k}}$ in the self-energy, suppose we now make a Taylor series expansion about the point with $\omega = E_{\mathbf{k}}$ on the real axis. Keeping only the lowest-order terms, deduce the leading-order correction to the result given in Equation (8.66).

9

Applications of Diagrammatic Methods

In this final chapter, we present some applications and examples of the diagrammatic perturbation methods that were developed in Chapter 8. Sometimes it will be the case that certain topics or systems that were analyzed in the earlier chapters of this book using either the Green's function (GF) equation-of-motion method (typically in conjunction with a decoupling approximation) or the linear response approach can also be revisited in terms of the diagrammatic perturbation theory and developed further. Some examples of this occur with Hartree–Fock (HF) theory and with plasmon excitations (both in an interacting electron gas). In some other cases, we will take advantage of the perturbation approach to study some higher-order effects that might not be reliably taken into account using decoupling approximations or linear response methods (e.g., interactions between SWs in ferromagnets).

In addition, we also consider some applications involving other types of diagrammatic formalisms. These include the scattering of particles by random distributions of static impurities and unconventional diagrammatic techniques applied directly to spin operators. The latter case requires a special treatment because the spin operators behave neither as bosons or fermions, so our proof of Wick's theorem given in Chapter 8 does not apply.

9.1 Hartree–Fock Theory for Fermions

The simplest contributions to the proper self-energy for the interacting GF $\mathcal{G}_{\mathbf{k}}(i\omega_m)$ in the perturbation treatment are those arising in first order, i.e., they come from diagrams with just *one* interaction vertex. The only diagrams of this type for $\Sigma(\mathbf{k}, i\omega_m)$ are those shown in Figure 9.1, where the external points are understood to have the labels $\{\mathbf{k}, i\omega_m\}$ entering and leaving, and we will sum the contributions from the two diagrams to get the total proper self-energy in first order of perturbation.

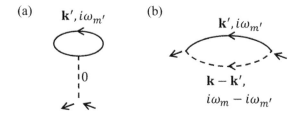

Figure 9.1 The two first-order Feynman diagrams for the proper self-energy $\Sigma(\mathbf{k}, i\omega_m)$, corresponding to Hartree–Fock theory for an interacting fermion gas.

First we consider diagram (a), which can be labeled with wave vector and frequency as shown. On applying the diagram rules given in Section 8.5 of Chapter 8 for this case we obtain

$$\Sigma^{(a)}(\mathbf{k}, i\omega_m) = \sum_{\mathbf{k}', m'} \beta v(0) \left(\frac{1}{\beta}\right) \frac{(-1)^2}{1} \exp(i\omega_{m'}\eta) \mathcal{G}^0_{\mathbf{k}'}(i\omega_{m'})$$

$$= \sum_{\mathbf{k}'} v(0) \sum_{m'} \exp(i\omega_{m'}\eta) \mathcal{G}^0_{\mathbf{k}'}(i\omega_{m'}) = \beta \sum_{\mathbf{k}'} v(0) n_{\mathbf{k}'}, \qquad (9.1)$$

where the frequency summation in the final stage has been carried out as before in Chapter 8, taking account of the occurrence of a convergence factor $\exp(i\omega_{m'}\eta)$ due to a self-contracted GF line. In the final expression $n_{\mathbf{k}'}$ denotes the Fermi–Dirac (FD) distribution function. Similarly, for diagram (b) we find

$$\Sigma^{(b)}(\mathbf{k}, i\omega_m) = \sum_{\mathbf{k}', m'} \beta v(\mathbf{k} - \mathbf{k}') \left(\frac{1}{\beta}\right) \frac{(-1)}{1} \exp(i\omega_{m'}\eta) \, \mathcal{G}^0_{\mathbf{k}'}(i\omega_{m'})$$

$$= -\beta \sum_{\mathbf{k}'} v(\mathbf{k} - \mathbf{k}') n_{\mathbf{k}'}. \qquad (9.2)$$

On summing these two diagrammatic contributions, we find that the *total* proper self-energy in first order is

$$\Sigma(\mathbf{k}, i\omega_m) = \beta \sum_{\mathbf{k}'} \left\{ v(0) - v(\mathbf{k} - \mathbf{k}') \right\} n_{\mathbf{k}'}. \qquad (9.3)$$

We notice that, in the present first-order perturbation approximation, Σ is real and is also independent of the external frequency label $i\omega_m$. It follows therefore (e.g., from the replacement scheme given in Equation (8.64)) that the renormalized energy is

$$E_{\mathbf{k}} + \sum_{\mathbf{k}'} \left\{ v(0) - v(\mathbf{k} - \mathbf{k}') \right\} n_{\mathbf{k}'}. \qquad (9.4)$$

This result is just the diagrammatic equivalent of the HF theory, which we discussed earlier in Chapter 5 in terms of a decoupling approximation. Comparison can be made, for example, with Equation (5.10). Clearly there is no contribution to the damping in this order of perturbation.

An advantage of the diagrammatic technique is that it presents a means to improve on HF theory. To do so, we would need to evaluate all diagrams for the proper self-energy that have two interaction vertices. Suppose we now denote the total proper self-energy up to second order as $\Sigma^{(1)} + \Sigma^{(2)}$, where $\Sigma^{(1)}$ is the first-order contribution made up of the two diagrams that we already evaluated. It is a useful exercise to draw the set of distinct diagrams contributing to the proper self-energy in second order. You should be able to find six (see Problem 9.1). It turns out (as can be directly established by evaluating each of them) that four of these diagrams give *real* contributions to $\Sigma^{(2)}$. Therefore they renormalize the energy, but they do not contribute to the damping. There are two of the second-order diagrams, however, that give a complex contribution to $\Sigma^{(2)}$ and, therefore, they may lead to a damping of the excitations, as well as to a renormalization of the energy.

As an example, we will evaluate here one of these diagrams, specifically the one shown in Figure 9.2. Applying the diagram rules in stages, we obtain

$$\sum_{\mathbf{q}, m'} \beta^2 v^2(\mathbf{q}) \left(\frac{1}{\beta}\right) \frac{1}{i\omega_m - i\omega_{m'} - E_{\mathbf{k}-\mathbf{q}}} F^0(\mathbf{q}, i\omega_{m'}), \tag{9.5}$$

where $F^0(\mathbf{q}, i\omega_{m'})$ denotes the contribution from the GF loop, which is defined in Figure 9.3 and is part of the diagram in Figure 9.2. We will evaluate this loop

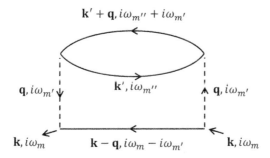

Figure 9.2 One of the diagrams contributing to the proper self-energy of $\mathcal{G}_{\mathbf{k}}(i\omega_m)$ in second order of perturbation. It is evaluated in Section 9.1 as an example.

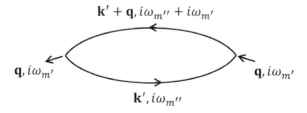

Figure 9.3 The diagrammatic definition of the GF loop $F^0(\mathbf{q}, i\omega_{m'})$ for fermions as used in Section 9.1.

diagram separately because it will provide a useful result for later, as well as being needed here. It is given by

$$F^0(\mathbf{q}, i\omega_{m'}) = (-1)\left(\frac{1}{\beta}\right)^2 \sum_{\mathbf{k'},m''} \frac{1}{(i\omega_{m''} + i\omega_{m'} - E_{\mathbf{k'}+\mathbf{q}})(i\omega_{m''} - E_{\mathbf{k'}})}$$

$$= -\frac{1}{\beta^2}\sum_{\mathbf{k'}}\frac{1}{(i\omega_{m'} + E_{\mathbf{k'}} - E_{\mathbf{k'}+\mathbf{q}})}\sum_{m''}\exp(i\omega_{m''}\eta)$$

$$\times \left[\frac{1}{(i\omega_{m''} - E_{\mathbf{k'}})} - \frac{1}{(i\omega_{m''} + i\omega_{m'} - E_{\mathbf{k'}+\mathbf{q}})}\right],$$

where partial fractions have been used and then the $\exp(i\omega_{m''}\eta)$ factor has been introduced to take care of the convergence of the individual terms for the frequency summation. The preceding frequency summations over index m'' are now of the same type as encountered previously and lead to

$$F^0(\mathbf{q}, i\omega_{m'}) = -\frac{1}{\beta}\sum_{\mathbf{k'}}\frac{n_{\mathbf{k'}} - n_{\mathbf{k'}+\mathbf{q}}}{(i\omega_{m'} + E_{\mathbf{k'}} - E_{\mathbf{k'}+\mathbf{q}})}. \qquad (9.6)$$

In obtaining this form of the result, we have been able to simplify one of the thermal factors by using the property that

$$\frac{1}{\exp[\beta(E_{\mathbf{k'}+\mathbf{q}} - i\omega_{m'})] + 1} = \frac{1}{\exp(\beta E_{\mathbf{k'}+\mathbf{q}}) + 1} = n_{\mathbf{k'}+\mathbf{q}}.$$

This is obtained on realizing that the internal label $i\omega_{m'}$ behaves like a *boson* frequency because it is the difference between two fermion frequencies.

It follows that the contribution to the proper self-energy of the complete diagram in Figure 9.2 is now given by

$$\sum_{\mathbf{k'},\mathbf{q},m'}\frac{-v^2(\mathbf{q})\{n_{\mathbf{k'}} - n_{\mathbf{k'}+\mathbf{q}}\}}{(i\omega_m - i\omega_{m'} - E_{\mathbf{k}-\mathbf{q}})(i\omega_{m'} + E_{\mathbf{k'}} - E_{\mathbf{k'}+\mathbf{q}})}. \qquad (9.7)$$

Finally, the m' summation can be made (e.g., after using partial fractions once again). The final result is straightforwardly found (see Problem 9.2) to be of the form

$$\beta\sum_{\mathbf{k'},\mathbf{q}}\frac{f(\mathbf{k},\mathbf{k'},\mathbf{q})}{(E_{\mathbf{k'}+\mathbf{q}} - E_{\mathbf{k'}} + E_{\mathbf{k}-\mathbf{q}} - i\omega_m)}, \qquad (9.8)$$

where the function $f(\mathbf{k},\mathbf{k'},\mathbf{q})$ in the numerator depends on the wave-vector labels and also on β (through the thermal population factors).

To find the contribution to the damping of the quasiparticle excitation $E_{\mathbf{k}}$ from this diagram, the procedure identified in Section 8.5 can be followed. Thus we replace $i\omega_m$ by its approximate solution $E_{\mathbf{k}} - i\eta$ (where η is a positive infinitesimal as before), giving for the inverse lifetime

$$\frac{1}{T_{\mathbf{k}}} = -\frac{1}{\beta} \Gamma(\mathbf{k}, E_{\mathbf{k}} - i\eta) = -\frac{1}{\beta} \mathrm{Im} \Sigma(\mathbf{k}, E_{\mathbf{k}} - i\eta)$$

$$= \pi \sum_{\mathbf{k}', \mathbf{q}} f(\mathbf{k}, \mathbf{k}', \mathbf{q}) \delta(E_{\mathbf{k}'+\mathbf{q}} - E_{\mathbf{k}'} + E_{\mathbf{k}-\mathbf{q}} - E_{\mathbf{k}}), \qquad (9.9)$$

where the imaginary part has been taken using the identity in Equation (3.31). It is seen that the delta function $\delta(E_{\mathbf{k}'+\mathbf{q}} - E_{\mathbf{k}'} + E_{\mathbf{k}-\mathbf{q}} - E_{\mathbf{k}})$ ensures the conservation of both energy and wave vector. Physically, the preceding damping term describes a scattering process in which an incoming fermion excitation of energy $E_{\mathbf{k}}$ scatters from another (thermally excited) fermion of energy $E_{\mathbf{k}'}$ to produce a pair of outgoing fermions with energies $E_{\mathbf{k}'+\mathbf{q}}$ and $E_{\mathbf{k}-\mathbf{q}}$.

9.2 Density Fluctuations in an Electron Gas

We have considered this topic already in terms of an operator equation-of-motion method with a decoupling approximation in Section 2.8 (see also Problem 5.11). Eventually, it led to a description of plasmons. The same problem will now be reexamined in terms of the diagrammatic method, and we will show that the choice of a particular type of simple diagrammatic chain structure leads to the same formal result as previously.

As before, we introduce the operator $\rho_{\mathbf{q}}^{\dagger}(\mathbf{k}) = a_{\mathbf{k}+\mathbf{q}}^{\dagger} a_{\mathbf{k}}$, which creates an electron of wave vector $\mathbf{k} + \mathbf{q}$ and destroys an electron of wave vector $\mathbf{k}$. The net effect is the formation of an electron-hole pair with total wave vector $\mathbf{q}$. We will also make use of the operator $\rho_{\mathbf{q}}^{\dagger}$, which is obtained by summing over all $\mathbf{k}$ as in Equation (2.88). Then a new imaginary-time GF can be defined by

$$g_M(\rho_{\mathbf{q}}^{\dagger}; \rho_{\mathbf{q}} \mid \tau_1 - \tau_2) = -\langle \hat{T}_w \rho_{\mathbf{q}}^{\dagger}(\tau_1) \rho_{\mathbf{q}}(\tau_2) \rangle. \qquad (9.10)$$

The frequency Fourier components of this GF will be denoted by $F(\mathbf{q}, i\omega_m)$, where $i\omega_m$ is a *boson* frequency label (because sums and differences of two fermion frequencies are involved). We can now represent the diagrammatic contributions to $F(\mathbf{q}, i\omega_m)$ in terms of the noninteracting GF lines G^0 for fermions and the usual interaction vertices. The diagrams for $F(\mathbf{q}, i\omega_m)$ will have a total of four external fermion lines (two associated with the first operator in the definition, and two more associated with the second operator). Therefore, they will have the general form as shown in Figure 9.4, where the shaded part denotes an allowed structure made up of any number of G^0 lines and interaction vertices. It is a simple exercise to try drawing some examples of diagrams for $F(\mathbf{q}, i\omega_m)$ that are of first order and second order in the perturbation (see Problem 9.3).

In fact, the lowest-order contribution for $F(\mathbf{q}, i\omega_m)$ is the zeroth-order diagram that has *no* interaction vertices, and consists only of a single loop of GF lines.

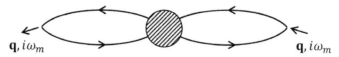

Figure 9.4 The general form of a diagrammatic contribution to $F(\mathbf{q}, i\omega_m)$ with its four external GF lines, as discussed in the text.

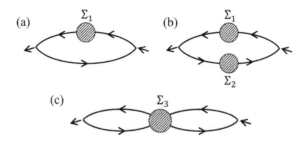

Figure 9.5 Structure of the three categories of diagrams contributing to $F(\mathbf{q}, i\omega_m)$ in first order and higher (see the text).

In other words, this is just the loop diagram drawn in Figure 9.3 and evaluated earlier as a component of a diagram for the proper self-energy. Hence, in the non-interacting case, the result is just the expression already denoted by $F^0(\mathbf{q}, i\omega_{m'})$ and quoted in Equation (9.6). We note that the poles of this quantity occur at $i\omega_{m'} = E_{\mathbf{k}+\mathbf{q}} - E_{\mathbf{k}}$, which is as expected because this is just the noninteracting pair energy discussed previously in Section 2.8. For a fixed value of $\mathbf{q}$ (but variable $\mathbf{k}$) it gives the continuum band of energy states as described previously.

Next, we consider the inclusion of the interaction vertices in the calculation of the GF to study the *plasmon* properties. We may distinguish three topologically different types of diagrammatic contributions, as sketched in Figure 9.5. Cases (a) and (b) correspond just to the renormalization of one or two, respectively, of the *individual* fermions. They will give an energy shift (and eventually a damping) of the individual fermions, and the overall effect is a modification of the continuum of energy states predicted in the noninteracting limit. Case (c) is more interesting and represents a renormalization of the electron-hole *pair*, and so it might be expected to lead to new physical effects.

We start by considering, as an example, the sequence of diagrams represented in Figure 9.6. These consist of "chains" formed by linking F^0 loops and interaction vertices in an alternating fashion. They correspond to a geometric series that can be summed to give

$$F(\mathbf{q}, i\omega_m) = F^0(\mathbf{q}, i\omega_m) - F^0(\mathbf{q}, i\omega_m)\beta v(\mathbf{q}) F^0(\mathbf{q}, i\omega_m) + \cdots$$

$$= \frac{F^0(\mathbf{q}, i\omega_m)}{1 + \beta v(\mathbf{q}) F^0(\mathbf{q}, i\omega_m)}.$$

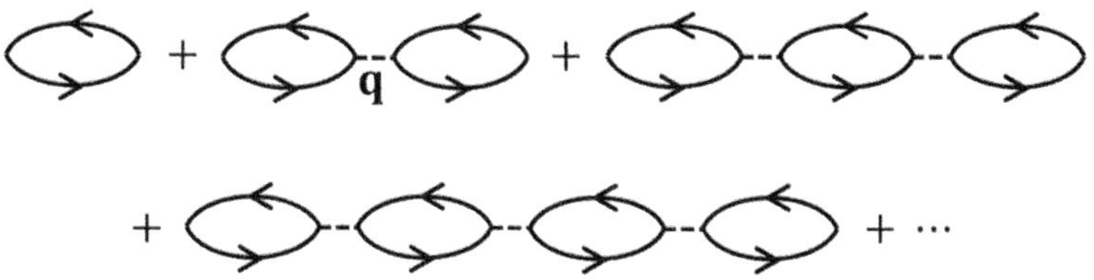

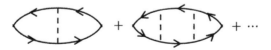

Figure 9.6 A sequence of chain diagrams contributing to $F(\mathbf{q}, i\omega_m)$.

Figure 9.7 Some ladder diagrams contributing to $F(\mathbf{q}, i\omega_m)$.

Therefore, in addition to the pole contained within F^0 (which we have already seen gives rise to an electron-hole continuum of states), the renormalized GF in this case has addition poles that correspond to the condition $\beta v(\mathbf{q}) F^0(\mathbf{q}, i\omega_m) = -1$. Using Equation (9.6) this can be rearranged and written as

$$\sum_{\mathbf{k}} \frac{n_{\mathbf{k}} - n_{\mathbf{k+q}}}{i\omega_m + E_{\mathbf{k}} - E_{\mathbf{k+q}}} = \frac{1}{v(\mathbf{q})}. \tag{9.11}$$

The preceding result can now be recognized as the equivalent of the plasmon dispersion relation in Equation (2.90) that was obtained previously by using an operator equation-of-motion and decoupling method. The analogous calculation based on the GF equation of motion was introduced in Problem 5.11. The connection was also pointed out to the Lindhard function, which is essentially the left-hand side of Equation (9.11) with the analytic continuation $i\omega_m \rightarrow \omega + i\eta$.

Thus, we conclude that the selection of the "chain" diagrams in the present case is just related to the RPA-type of decoupling used before. However, the inclusion of other diagrams within our present formalism, like the "ladder" diagrams shown in Figure 9.7, would represent additional effects previously neglected. It can be shown explicitly that the ladder series does not give rise to plasmon effects and mostly the diagrams have the same types of poles as $F^0(\mathbf{q}, i\omega_m)$. The diagrams that give rise to the renormalization of the plasmon excitations (through plasmon-plasmon scattering, etc.) are more complicated and will not be described here.

9.3 Electron–Phonon Interactions

Here we discuss the application of the diagrammatic methods to electron-phonon interactions, as a first example of interacting fields (one fermion and one boson field). There are several reasons why electron–phonon interactions are of interest.

For example, an effect on the electrons is that they can be scattered to different states and hence there is a contribution to the electrical conductivity. However, an effect on the phonons is to provide a mechanism for the damping of acoustic waves in a metal or semiconductor. Also, in conventional superconductors the electron–phonon interaction provides an explanation for the formation of Cooper pairs of electrons, as required in the BCS theory of superconductivity (see Section 5.6). Hence there has been a strong focus on electron–phonon interactions in the literature (see, e.g., review accounts in [9, 49, 128]).

For a system of interacting electrons and phonons we may write the total Hamiltonian in a second-quantized form as

$$\mathcal{H} = \sum_{\mathbf{k}} E_{\mathbf{k}} a_{\mathbf{k}}^{\dagger} a_{\mathbf{k}} + \sum_{\mathbf{q}} \omega_{\mathbf{q}} \left\{ b_{\mathbf{q}}^{\dagger} b_{\mathbf{q}} + \frac{1}{2} \right\} + \mathcal{H}_d, \tag{9.12}$$

where $E_{\mathbf{k}} = (k^2/2m^*) - \mu$ denotes the electron energy measured relative to the chemical potential and m^* is an effective mass. The a and b operators refer to the electrons and phonons, respectively. Also, $\omega_{\mathbf{q}}$ is a phonon frequency as in the model Hamiltonian in Equation (2.50) for a longitudinal acoustic phonon. More generally, this term may have a summation over several phonon branches (acoustic or optic, and longitudinal or transverse). The mechanism for the interaction term $\mathcal{H}_d$ between the electrons and phonons is typically based on the concept of the deformation potential. In essence, if there is a phonon propagating in an ionic crystal (or metal), there will be a displacement of the ions that will modulate the interionic wave functions (typically taken as Bloch functions) of the ideal lattice of ions. The leading-order effect is similar to a linear modulation of the electron-band energy (e.g., see the book by Rickayzen [49] for a particularly elegant derivation), which eventually gives

$$\mathcal{H}_d = \sum_{\mathbf{k}, \mathbf{q}} f(\mathbf{k}, \mathbf{q}) a_{\mathbf{k}}^{\dagger} a_{\mathbf{k}+\mathbf{q}} (b_{\mathbf{q}} - b_{-\mathbf{q}}^{\dagger}). \tag{9.13}$$

Here $f(\mathbf{k}, \mathbf{q})$ is a strength factor for the electron–phonon interaction, and it takes a form that depends on the electronic band structure of the material and the type of phonon being considered. In some cases, for example with longitudinal acoustic phonons (see [9, 128]), we can approximate by writing

$$f(\mathbf{k}, \mathbf{q}) = i \left(\frac{C_1 |\mathbf{q}|}{\sqrt{2d\omega_{\mathbf{q}}}} \right), \tag{9.14}$$

where C_1 is an elastic constant and d is the density of the material. The $\mathbf{k}$-dependence is neglected for f in this model. In other cases, the dependence of f on $\mathbf{q}$ may be quite different, and we might simply approximate by writing $f = iD$, where D is a real constant. We note that the quoted form of Equation (9.13) is applicable for so-called Normal (N) processes, where the total wave vector for the creation operators is equal to that for the annihilation operators. More generally, Umklapp (U) processes may also occur in which the two wave-vector quantities differ by a reciprocal lattice vector.

The next step is to introduce a diagrammatic formalism. We define unperturbed GFs for the electron and phonon systems, respectively, in the Matsubara frequency representation by

$$\mathcal{G}_{el}^0(\mathbf{k}, i\omega_m) \equiv \mathcal{G}^0(a_\mathbf{k}; a_\mathbf{k}^\dagger \,|\, i\omega_m) = \frac{1}{i\omega_m - E_\mathbf{k}}, \tag{9.15}$$

$$\mathcal{G}_{ph}^0(\mathbf{q}, i\omega_m) \equiv \mathcal{G}^0(\{b_\mathbf{q} - b_{-\mathbf{q}}^\dagger\}; \{b_\mathbf{q}^\dagger - b_{-\mathbf{q}}\} \,|\, i\omega_m)$$
$$= \frac{2\omega_\mathbf{q}}{(i\omega_m)^2 - (\omega_\mathbf{q})^2}. \tag{9.16}$$

The definition of the phonon GF given here reflects the combination in which the phonon operators appear in Equation (9.13). The final form of the expression in Equation (9.16) can be deduced following Problem 9.4. The preceding GFs will be represented by full lines and wavy lines, respectively. We note that $i\omega_m$ are fermion and boson imaginary frequencies in the electron and phonon cases, respectively. The interaction vertices corresponding to the terms in $\mathcal{H}_d$ are shown in Figure 9.8(a) and (b) where a phonon is either being emitted or absorbed, respectively. The black circle represents a factor for the interaction strength $f(\mathbf{k}, \mathbf{q})$.

At first sight it might seem that the previous proof of Wick's theorem in Chapter 8 is not applicable here. This is because $\mathcal{H}_d$ contains an *odd* number of boson or fermion operators, rather than an even number as assumed previously. However, it will be seen that all the diagrammatic contributions of relevance involve the

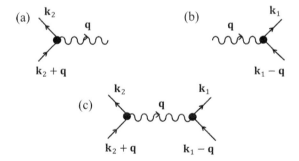

Figure 9.8 The individual interactions contained in the Hamiltonian $\mathcal{H}_d$ for electron–phonon interactions representing (a) phonon emission and (b) phonon absorption. The second-order diagram (c) represents the virtual process in which a phonon is emitted and reabsorbed, resulting in a scattering of a pair of electrons.

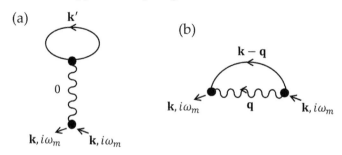

Figure 9.9 The proper self-energy diagrams considered for the electronic GF due to the electron–phonon interactions.

electron–phonon interactions *in pairs*, so the requirements for Wick's theorem are satisfied. Next we present two applications involving electron–phonon interactions.

9.3.1 Polarons

In the absence of electron–phonon interactions, the unperturbed electronic GF $\mathcal{G}_{el}^0(\mathbf{k}, i\omega_m)$ has a simple pole at energy $E_{\mathbf{k}}$ as in Equation (9.15). An effect of electron–phonon interactions will be to modify (renormalize) $E_{\mathbf{k}}$ through processes involving the emission and subsequent reabsorption of phonons. We therefore arrive at a picture of an electron being surrounded by its "cloud" of phonons, as the electron moves through the lattice causing distortions. The composite entity, which is known as a *polaron*, has been a topic of extensive studies (see, e.g., [129–131] for reviews).

A note of caution here is that the electron–phonon interaction is not necessarily weak. It can be quite strong in some materials and/or for some phonon branches, so a perturbation expansion in terms of $\mathcal{H}_d$ vertices will not then be appropriate. In the following, we will consider situations in which a diagrammatic expansion using $\mathcal{H}_d$ is valid. Then the lowest-order proper self-energy contributions to the electronic GF come from the two diagrams shown in Figure 9.9.

We note first that diagram (a) will be zero in many cases. For example, if the interaction strength f has the form given by Equation (9.14), then it vanishes for $\mathbf{q} = 0$. Even when f does not take this simple form, the diagrammatic contribution may be unimportant [49]. Turning now to diagram (b) we find that the proper self-energy is given formally by

$$\Sigma_{el}(\mathbf{k}, i\omega_m) = \frac{1}{\beta} \sum_{\mathbf{q}, m'} |f(\mathbf{k}, \mathbf{q})|^2 \, \mathcal{G}_{el}^0(\mathbf{k} - \mathbf{q}, i\omega_m - i\omega_{m'}) \mathcal{G}_{ph}^0(\mathbf{q}, i\omega_{m'})$$

$$= \frac{2}{\beta} \sum_{\mathbf{q}, m'} \frac{\omega_{\mathbf{q}} |f(\mathbf{k}, \mathbf{q})|^2}{(i\omega_m - i\omega_{m'} - E_{\mathbf{k} - \mathbf{q}})\left[(i\omega_{m'})^2 - (\omega_{\mathbf{q}})^2\right]}. \tag{9.17}$$

It is a straightforward, but lengthy, task to carry out the summation over the imaginary boson frequency $i\omega_{m'}$ (see Problem 9.5) by using the same methods as described in the previous sections of this chapter. The poles for $i\omega_{m'}$ in the preceding equation occur at $(i\omega_m - E_{k-q})$ and at $\pm\omega_q$, from which it can be concluded that the expression for $\Sigma_{el}(\mathbf{k}, i\omega_m)$ has energy denominators of the form $(i\omega_m - E_{k-q} \pm \omega_q)$. Further evaluation would depend on the detailed form of $f(\mathbf{k}, \mathbf{q})$ and the unperturbed electron and phonon dispersion relations.

9.3.2 Cooper Pairs in Superconductivity

At this point, we refer back to the BCS theory of superconductivity covered in Section 5.6. There we mentioned that the electron–phonon interaction was responsible for the attractive component of the effective interaction between two electrons in the formation of Cooper pairs specified by $(\mathbf{k}, \uparrow)$ and $(-\mathbf{k}, \downarrow)$, or vice versa. We now present some results to justify this assumption, which was summarized in Equations (5.74) and (5.75).

We turn our attention to Figure 9.8(c), which describes an interaction (a scattering event) between two electrons in which a phonon of wave vector $\mathbf{q}$ is emitted and then reabsorbed. In a full QM description we also include the virtual process in which the creation of a phonon with wave vector $\mathbf{q}$ is equated to annihilation of a phonon with wave vector $-\mathbf{q}$, and vice versa. These processes are of second order in the electron–phonon vertex (the black circles in the figure) and involve a wavy phonon GF line for $\mathcal{G}_{ph}^0(\mathbf{q}, i\omega_m)$. The effective interaction that acts in combination with the Coulomb (or screened Coulomb) interaction, denoted by $v(\mathbf{q})$ in Equation (1.64), is obtained from $|f(\mathbf{k}_1, \mathbf{q})|^2 \mathcal{G}_{ph}^0(\mathbf{q}, i\omega_m)$. Using Equation (9.16) this leads us to

$$\frac{2\omega_q |f(\mathbf{k}_1, \mathbf{q})|^2}{(i\omega_m)^2 - (\omega_q)^2}. \tag{9.18}$$

We notice that the preceding result depends on the imaginary frequency $i\omega_m$. As in the discussion given in Section 8.5 for the self-energy, the appropriate approximate procedure is make the analytic continuation of $i\omega_m$ to a real quantity corresponding to the net excitation energy. In the present context this energy is just the pair energy corresponding to the external electronic GF lines, and so can be written as $(E_{k_1} - E_{k_1-q})$. Therefore, we arrive at an expression for the effective interaction energy in the form

$$\frac{2\omega_q |f(\mathbf{k}_1, \mathbf{q})|^2}{(E_{k_1} - E_{k_1-q})^2 - (\omega_q)^2}. \tag{9.19}$$

This quantity is seen to be negative, representing an attraction between the electrons, whenever $|E_{k_1} - E_{k_1-q}| < \omega_q$; otherwise the contribution represents a repulsion. If the electron–phonon interaction is sufficiently strong, the attractive term may overcome the repulsive effects of the (screened) Coulomb interaction. The condition for $|E_{\mathbf{k}_1} - E_{\mathbf{k}_1 - \mathbf{q}}|$ to be smaller than the phonon energy is usually satisfied by having both electron energies lying close to the Fermi energy ϵ_F, specifically within a range from about $\epsilon_F - \omega_D$ to $\epsilon_F + \omega_D$ where ω_D denotes the Debye energy for the phonons. This leads us to the condition quoted in Equation (5.75).

9.4 Boson Expansion Methods for Spin Waves

We now use the diagrammatic methods for further discussion of SWs in Heisenberg ferromagnets at low temperatures and, in particular, to investigate the nature of the *interactions* between SWs. These interactions are expected to lead to a renormalization of the SW energy and (in a sufficiently high order of perturbation) to a SW damping. The Heisenberg Hamiltonian from Chapter 1, when rewritten in component form, is just

$$\mathcal{H} = -\frac{1}{2} \sum_{i,j} J_{i,j} \left\{ \frac{1}{2}(S_i^+ S_j^- + S_i^- S_j^+) + S_i^z S_j^z \right\} - b \sum_i S_i^z, \qquad (9.20)$$

where we employ the shorthand $b = g\mu_B B_0$, with B_0 denoting the applied magnetic field taken in the z direction.

It should be emphasized at this point that the diagrammatic method as developed in Chapter 8 does *not* apply directly to spin systems. For example, the proof outlined for Wick's theorem was valid specifically for operators with the boson commutation relations or the fermion anticommutation relations. The spin operators do satisfy commutation relations, as discussed in Chapter 1, but the result is either zero or another spin operator and not a scalar. The available options to circumvent this difficulty for the spin systems are typically the following:

- *Either* we seek to develop a modified version of Wick's theorem and a new diagrammatic method specifically for spin systems;
- *Or* we transform from the spin operators to a representation in terms of boson or fermion operators, and then we may use the standard techniques (although Wick's theorem might still need generalizing in some cases).

We will give examples of both methods, but in the present section we will employ the second approach, by making use of the Holstein–Primakoff (HP) transformation from spin to boson operators as described in Chapter 1, where the full transformation is quoted in Equation (1.85). As before, we will focus on studying the excitations in Heisenberg ferromagnets at low temperatures ($T \ll T_C$). In this case, the spins are well aligned, so $S_i^z \approx S$ for each spin, implying $a_i^\dagger a_i \ll S$, as explained before. We once again use this property to approximate the square root terms in the HP transformation, but now we go to higher terms in the binomial expansion so that the interaction effects previously ignored can be included. We have

$$\left(1 - \frac{a_i^\dagger a_i}{2S}\right)^{1/2} \approx 1 - \frac{a_i^\dagger a_i}{4S} - \frac{a_i^\dagger a_i a_i^\dagger a_i}{32S^2} + \cdots, \tag{9.21}$$

and so it follows that

$$S_i^+ = \sqrt{2S} \left[a_i - \frac{a_i^\dagger a_i a_i}{4S} - \cdots\right],$$

$$S_i^- = \sqrt{2S} \left[a_i^\dagger - \frac{a_i^\dagger a_i^\dagger a_i}{4S} - \cdots\right], \tag{9.22}$$

with $S_i^z = S - a_i^\dagger a_i$ as before. When these expressions are substituted into the Hamiltonian in Equation (9.20) we may write the result as

$$\mathcal{H} = \mathcal{E}_0 + \mathcal{H}_0 + \mathcal{H}_1 + \cdots, \tag{9.23}$$

where $\mathcal{E}_0$ is a scalar constant, corresponding to the ground-state energy as quoted in Equation (1.92). The next two terms $\mathcal{H}_0$ and $\mathcal{H}_1$ are, respectively, quadratic and quartic in the boson operators (because there are no odd-order terms in $\mathcal{H}$ for the Heisenberg Hamiltonian). We shall neglect the other higher-order terms, which should be justified provided $T \ll T_C$.

First, considering only the $\mathcal{H}_0$ term, we have as before

$$\mathcal{H}_0 = -\frac{1}{2}S \sum_{i,j} J_{i,j}\left\{u_i u_j^\dagger + a_i^\dagger a_j \quad a_i^\dagger a_i - a_j^\dagger a_j\right\} + h \sum_i a_i^\dagger a_i, \tag{9.24}$$

which on Fourier transforming to a wave-vector representation, as in Equations (1.89) and (1.90), becomes

$$\mathcal{H}_0 = \sum_{\mathbf{k}} E_{\mathbf{k}} a_{\mathbf{k}}^\dagger a_{\mathbf{k}}. \tag{9.25}$$

This is the quasiparticle form of the Hamiltonian, and $E_{\mathbf{k}}$ is the SW energy at low temperatures, as given in Equation (1.93). Hence we have the result $E_{\mathbf{k}} = b + S\{J(0) - J(\mathbf{k})\}$ and we recall that $J(\mathbf{k})$ was defined as the wave-vector Fourier transform of the exchange interaction.

Next, we include the effects of the Hamiltonian term $\mathcal{H}_1$ as a perturbation. This will provide the leading-order description of any interactions between the SWs and is given by

$$\mathcal{H}_1 = \frac{1}{2} \sum_{i,j} J_{i,j} \left\{ \frac{1}{4} (a_j^\dagger a_j^\dagger a_i a_j + a_i^\dagger a_j^\dagger a_i a_i + a_i^\dagger a_j^\dagger a_j a_j + a_i^\dagger a_i^\dagger a_i a_j) - a_i^\dagger a_j^\dagger a_i a_j \right\}.$$

Here we have used the fact that i and j always refer to different sites, allowing us to interchange the order of some operators. Then, on transforming to the wave-vector representation, we eventually obtain

$$\mathcal{H}_1 = \frac{1}{2} \sum_{\mathbf{k},\mathbf{k}',\mathbf{q}} \left[\frac{1}{4} \{ J(\mathbf{k}) + J(\mathbf{k}') + J(\mathbf{k}+\mathbf{q}) + J(\mathbf{k}'-\mathbf{q}) \} - J(\mathbf{q}) \right] a_{\mathbf{k}}^\dagger a_{\mathbf{k}'}^\dagger a_{\mathbf{k}'+\mathbf{q}} a_{\mathbf{k}-\mathbf{q}}.$$

$$(9.26)$$

It can be seen that this result is formally very similar to our standard form of the interaction (perturbation) term in a boson or fermion gas, except for the replacement of the interaction energy, which previously depended only on $\mathbf{q}$, with a more complicated quantity that depends on $\mathbf{k}$ and $\mathbf{k}'$ as well as $\mathbf{q}$. In other words, we have the replacement $v(\mathbf{q}) \rightarrow \bar{v}(\mathbf{k}, \mathbf{k}', \mathbf{q})$ for the interaction vertex, where

$$\bar{v}(\mathbf{k}, \mathbf{k}', \mathbf{q}) = \frac{1}{4} \{ J(\mathbf{k}) + J(\mathbf{k}') + J(\mathbf{k}+\mathbf{q}) + J(\mathbf{k}'-\mathbf{q}) \} - J(\mathbf{q}). \qquad (9.27)$$

In particular, we see that when all the wave-vector labels are put equal to zero we have the property that $\bar{v}(0,0,0) = 0$, so the interaction is weak at small wave vectors. Nevertheless, apart from the differences mentioned in the preceding text, we can use all the previous formulation of the diagrammatic perturbation method. Thus the noninteracting GF defined with respect to $\mathcal{H}_0$ is

$$\mathcal{G}^0(\mathbf{k}, i\omega_m) = \frac{1}{i\omega_m - E_{\mathbf{k}}}, \qquad (9.28)$$

where $i\omega_m$ is a boson frequency.

To carry out the renormalization of this GF in first order of perturbation theory, we need the proper self-energy diagrams. These are formally the same as those shown in Figure 9.1 for the HF theory, except that now we are dealing with boson statistics. Hence, the only differences are that there are no closed fermion loops in the present case and that the factor $\bar{v}$ for each interaction vertex is more complicated. The frequency summations are unchanged, except that we get Bose–Einstein (BE) thermal factors instead of the Fermi–Dirac (FD) factors. For diagram (a), in this case we find

$$-\beta \sum_{\mathbf{k}'} \bar{v}(\mathbf{k}, \mathbf{k}', 0) n_{\mathbf{k}'} = -\beta \sum_{\mathbf{k}'} \left[\frac{1}{2} \{ J(\mathbf{k}) + J(\mathbf{k}') \} - J(0) \right] n_{\mathbf{k}'},$$

while for diagram (b) the result is

$$-\beta \sum_{\mathbf{k}'} \bar{v}(\mathbf{k}, \mathbf{k}', \mathbf{k} - \mathbf{k}') n_{\mathbf{k}'} = -\beta \sum_{\mathbf{k}'} \left[\frac{1}{2} \{ J(\mathbf{k}) + J(\mathbf{k}') \} - J(\mathbf{k} - \mathbf{k}') \right] n_{\mathbf{k}'}.$$

Combining these two expressions, we see that the full result for the proper self-energy in first order is therefore

$$\Sigma(\mathbf{k}, i\omega_m) = -\beta \sum_{\mathbf{k}'} \left[J(\mathbf{k}) + J(\mathbf{k}') - J(\mathbf{k} - \mathbf{k}') - J(0) \right] n_{\mathbf{k}'}. \tag{9.29}$$

It follows from expressions in Section 8.5 that the renormalized SW energy is $E_{\mathbf{k}} + \Delta E_{\mathbf{k}}$, where the correction term due to the interactions is

$$\Delta E_{\mathbf{k}} = \beta^{-1} \mathrm{Re}\, \Sigma(\mathbf{k}, i\omega_m)$$

$$= \sum_{\mathbf{k}'} \left\{ J(0) + J(\mathbf{k} - \mathbf{k}') - J(\mathbf{k}) - J(\mathbf{k}') \right\} n_{\mathbf{k}'}. \tag{9.30}$$

We notice that the proper self-energy $\Sigma(\mathbf{k}, i\omega_m)$, as evaluated in first order, is *real* and has no dependence on the frequency label $i\omega_m$. Consequently, there is no damping in this order of perturbation. From the expression for the energy shift in Equation (9.30) it can be shown at small $k = |\mathbf{k}|$ that $\Delta E_{\mathbf{k}}$ is proportional to k^2 multiplied by a temperature-dependent factor (see Problem 9.6). A general theory of SW interactions was originally developed by Dyson [132] without the use of GF methods, and the combination of exchange terms in curly brackets in Equation (9.30) is often referred to as the Dyson vertex.

To obtain a contribution to the SW damping, it would be necessary to go to the next order of perturbation to consider self-energy diagrams that are of second order in the interaction vertices (e.g., by close analogy with the situation for the interacting fermion gas considered in Section 9.1).

9.5 Scattering by Static Impurities

As an example of a different kind of application of the diagrammatic perturbation theory, we next consider the scattering of particles, which may be either bosons or fermions, by a set of static impurities (see, e.g., [135, 136] for a discussion of the model). Some examples of applications would be in transport theory for the scattering of electrons in a metal by impurity sites *or* in the theory of randomly mixed alloys *or* certain collisions problems involving excitations and static (or quasistatic) centers.

9.5.1 General Formalism

In such cases as mentioned in the preceding text, we can start with a Hamiltonian in the second quantized form, which we may write rather generally as

$$\mathcal{H} \equiv \mathcal{H}_0 + \mathcal{H}_1 = \sum_{\mathbf{k}} E_{\mathbf{k}} a_{\mathbf{k}}^\dagger a_{\mathbf{k}} + \sum_{\mathbf{k}_1, \mathbf{k}_2} A(\mathbf{k}_1, \mathbf{k}_2) a_{\mathbf{k}_2}^\dagger a_{\mathbf{k}_1}. \tag{9.31}$$

Here the first term has the standard form for noninteracting particle with $a_{\mathbf{k}}^{\dagger}$ and $a_{\mathbf{k}}$ being the creation and annihilation operators, respectively, and $E_{\mathbf{k}} = (k^2/2m) - \mu$ as before. The extra term $\mathcal{H}_1$, which involves an interaction strength $A(\mathbf{k}_1, \mathbf{k}_2)$, describes a form of scattering for the particles. Specifically, it is taken to depend on the wave vectors $\mathbf{k}_1$ and $\mathbf{k}_2$ of the incident and scattered particles. Note that the total Hamiltonian for the particles no longer conserves the wave vector for the particles (because momentum can be transferred to the scattering centers), but it does conserve the number of particles. To study the dynamical properties of the system we define the interacting imaginary-time GF as

$$g_{M,\mathbf{k},\mathbf{k}'}(\tau_1 - \tau_2) = -\langle \hat{T}_W \check{a}_{\mathbf{k}}(\tau_1) \check{a}_{\mathbf{k}'}(\tau_2) \rangle, \tag{9.32}$$

where the thermal average is with respect to the full Hamiltonian $\mathcal{H}$ given in the preceding text, and we use the notation

$$\check{a}_{\mathbf{k}}(\tau) = e^{\mathcal{H}\tau} a_{\mathbf{k}} e^{-\mathcal{H}\tau}, \tag{9.33}$$

as previously in Chapters 3 and 8. Notice that the preceding definition of the GF is just a direct generalization of the previous case to include the possibility of $\mathbf{k} \neq \mathbf{k}'$.

It is important to realize that the previous proof of Wick's theorem still holds because we again have an even number of boson or fermion operators in the Hamiltonian. Also Fourier transforms can be made with respect to the τ-labels to a representation in terms of the imaginary frequency $i\omega_m$. Therefore, we will deal with a GF quantity denoted by $\mathcal{G}_{\mathbf{k},\mathbf{k}'}(i\omega_m)$. In the noninteracting case (where we take the Hamiltonian to be $\mathcal{H}_0$) we have the same result as previously, and necessarily we have $\mathbf{k} = \mathbf{k}'$ because $\mathcal{H}_0$ conserves wave vector. We may write

$$\mathcal{G}_{\mathbf{k},\mathbf{k}'}^0(i\omega_m) = \delta_{\mathbf{k},\mathbf{k}'} \frac{1}{i\omega_m - E_{\mathbf{k}}} \equiv \delta_{\mathbf{k},\mathbf{k}'} \mathcal{G}_{\mathbf{k}}^0(i\omega_m). \tag{9.34}$$

By contrast, in a diagrammatic representation that includes the effect of the perturbation $\mathcal{H}_1$ we now have interaction vertices that do not require wave-vector conservation and they can be represented as in Figure 9.10. We note that there are now just two GF lines associated with the vertex, one entering and one leaving. Although we have $\mathbf{k}_1 \neq \mathbf{k}_2$ in general, there is still frequency conservation for the GF lines. This can easily be seen by a simple modification of the arguments in Subsection 8.4.1.

The complete diagrams that contribute to the interacting GF now *appear* at first sight to be all very simple, consisting only of the infinite series on the right in Figure 9.11(a). The resulting renormalized GF will be drawn as a heavy line as on the left. The infinite series of diagrams is equivalent to the result shown

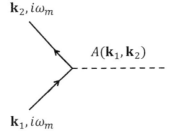

Figure 9.10 An interaction vertex for scattering from the static impurities. The factor $A(\mathbf{k}_1, \mathbf{k}_2)$ associated with the vertex (dashed line) describes the scattering of the particles, and there are two GF lines (with $\mathbf{k}_1 \neq \mathbf{k}_2$) entering and leaving the vertex.

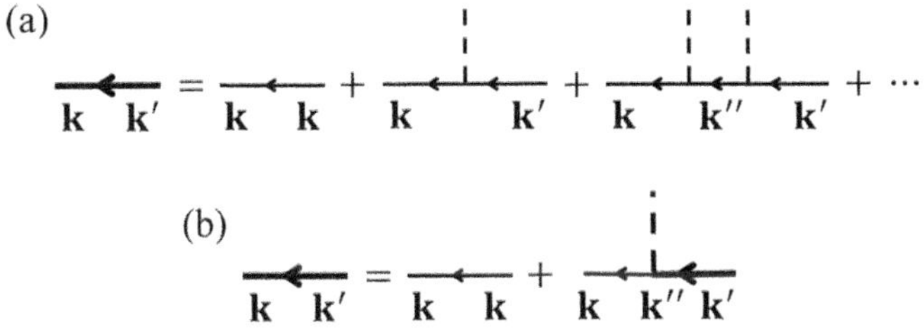

Figure 9.11 (a) The infinite series of diagrams contributing to the GF defined in Equation (9.32); and (b) the equivalent Dyson equation in diagrammatic form. In both cases, the renormalized GF is shown by the heavy line.

diagrammatically in Figure 9.11(b), as can be seen by an iteration process. The equivalent algebraic expressions to Figures 9.11(a) and (b) are, respectively,

$$\mathcal{G}_{\mathbf{k},\mathbf{k}'}(i\omega_m) = \mathcal{G}_{\mathbf{k}}^0(i\omega_m)\delta_{\mathbf{k},\mathbf{k}'} + \mathcal{G}_{\mathbf{k}}^0(i\omega_m)A(\mathbf{k},\mathbf{k}')\mathcal{G}_{\mathbf{k}'}^0(i\omega_m)$$
$$+ \sum_{\mathbf{k}''} \mathcal{G}_{\mathbf{k}}^0(i\omega_m)A(\mathbf{k},\mathbf{k}'')\mathcal{G}_{\mathbf{k}''}^0(i\omega_m)A(\mathbf{k}'',\mathbf{k}')\mathcal{G}_{\mathbf{k}'}^0(i\omega_m) + \cdots,$$

$$(9.35)$$

and

$$\mathcal{G}_{\mathbf{k},\mathbf{k}'}(i\omega_m) = \mathcal{G}_{\mathbf{k}}^0(i\omega_m)\delta_{\mathbf{k},\mathbf{k}'} + \sum_{\mathbf{k}''} \mathcal{G}_{\mathbf{k}}^0(i\omega_m)A(\mathbf{k},\mathbf{k}'')\mathcal{G}_{\mathbf{k}'',\mathbf{k}'}(i\omega_m). \qquad (9.36)$$

If the summation over $\mathbf{k}''$ is rewritten as an integration (in the usual way), the preceding result in Equation (9.36) becomes an integral equation for $\mathcal{G}_{\mathbf{k},\mathbf{k}'}(i\omega_m)$, and so it represents only an implicit result for the GF. In some special cases, the functional form of the scattering amplitude $A(\mathbf{k},\mathbf{k}'')$, which depends on the spatial distribution of the impurities (among other factors), might be sufficiently simple that this integral equation can be solved exactly. An example is when the scattering amplitude is a separable function of the wave vectors, i.e., when $A(\mathbf{k},\mathbf{k}'') = \alpha(\mathbf{k})\alpha(\mathbf{k}'')$ for a specified function α. However, in most cases, including the case of a *random* distribution of a large number of impurities, an exact solution will no longer be possible. An approximation scheme becomes necessary, and this can be developed using a modified diagrammatic perturbation approach as outlined in the following subsection.

9.5.2 Random Impurities

Here we consider the situation when the distribution of impurities can be taken as random in terms of their spatial coordinates. We will eventually work in terms of a *mean value* of the GF, taken over this random distribution of impurities. We shall indicate such configurational mean values by placing a curly bracket around them, e.g., by $\{\mathcal{G}\}$ for the GFs or by $\{A\}$ for the scattering amplitude.

The first step is to determine expressions for the configurational averaging over individual scattering amplitudes and products of these amplitudes. We keep in mind that the interaction (scattering) term at a general point $\mathbf{r}$, due to the impurities at different points $\mathbf{R}_i$, will have the form

$$\sum_i A(\mathbf{r} - \mathbf{R}_i)\, a_{\mathbf{r}}^{\dagger} a_{\mathbf{r}}$$

in the position representation. In the wave-vector representation, however, this becomes

$$A(\mathbf{k}_1, \mathbf{k}_2) = \frac{1}{V} \int \sum_i A(\mathbf{r} - \mathbf{R}_i) e^{i(\mathbf{k}_1 - \mathbf{k}_2)\cdot \mathbf{r}} d^3 r, \qquad (9.37)$$

where V is the volume of the system. With N denoting the total number of atoms this can be rewritten as

$$A(\mathbf{k}_1, \mathbf{k}_2) = \frac{1}{V} \int \sum_i A(\mathbf{r} - \mathbf{R}_i) e^{i(\mathbf{k}_1 - \mathbf{k}_2)\cdot (\mathbf{r} - \mathbf{R}_i)} e^{i(\mathbf{k}_1 - \mathbf{k}_2)\cdot \mathbf{R}_i} d^3 r$$

$$\equiv \frac{1}{N} A(\mathbf{k}_1 - \mathbf{k}_2) \sum_i e^{i(\mathbf{k}_1 - \mathbf{k}_2)\cdot \mathbf{R}_i}. \qquad (9.38)$$

Now we take an average of this quantity over a *random* distribution of impurities (meaning a random set of values for $\mathbf{R}_i$). This leads to

$$\{A(\mathbf{k}_1, \mathbf{k}_2)\} = \frac{1}{N} A(\mathbf{k}_1 - \mathbf{k}_2) \left\{ \sum_i e^{i(\mathbf{k}_1 - \mathbf{k}_2)\cdot \mathbf{R}_i} \right\}$$

$$= \frac{N_0}{N} A(\mathbf{k}_1 - \mathbf{k}_2)\, \delta_{\mathbf{k}_1, \mathbf{k}_2} = c_0 A(0)\, \delta_{\mathbf{k}_1, \mathbf{k}_2}, \qquad (9.39)$$

where N_0 is the (large) number of impurities. For some cases to be considered later we will assume the concentration $c_0 = N_0/N$ to be small.

Next, for an average over a product of two amplitude terms we find by following similar arguments that

$$\{A(\mathbf{k}_1, \mathbf{k}_2) A(\mathbf{k}'_1, \mathbf{k}'_2)\}$$

$$= \frac{1}{N^2} A(\mathbf{k}_1 - \mathbf{k}_2) A(\mathbf{k}'_1 - \mathbf{k}'_2) \left\{ \sum_{i,j} e^{i(\mathbf{k}_1 - \mathbf{k}_2) \cdot \mathbf{R}_i + i(\mathbf{k}'_1 - \mathbf{k}'_2) \cdot \mathbf{R}_j} \right\}$$

$$= \frac{1}{N^2} A(\mathbf{k}_1 - \mathbf{k}_2) A(\mathbf{k}'_1 - \mathbf{k}'_2) \left\{ \left(\sum_{i \neq j} + \sum_{i=j} \right) e^{i(\mathbf{k}_1 - \mathbf{k}_2) \cdot \mathbf{R}_i + i(\mathbf{k}'_1 - \mathbf{k}'_2) \cdot \mathbf{R}_j} \right\}$$

$$= \frac{N_0(N_0 - 1)}{N^2} [A(0)]^2 \, \delta_{\mathbf{k}_1, \mathbf{k}_2} \, \delta_{\mathbf{k}'_1, \mathbf{k}'_2} + \frac{N_0}{N} [A(\mathbf{k}_1 - \mathbf{k}_2)]^2 \, \delta_{\mathbf{k}_1 + \mathbf{k}'_1, \mathbf{k}_2 + \mathbf{k}'_2}.$$

In the third line of the preceding equation we have split the double sums over all i and j into two parts: one part corresponding to $i \neq j$, which has $N_0(N_0 - 1)$ terms, and the other part corresponding to $i = j$, which has N_0 terms. Then, because the number of impurities is large ($N_0 \gg 1$), we have to a good approximation

$$\{A(\mathbf{k}_1, \mathbf{k}_2) A(\mathbf{k}'_1, \mathbf{k}'_2)\} \simeq c_0^2 [A(0)]^2 \, \delta_{\mathbf{k}_1, \mathbf{k}_2} \, \delta_{\mathbf{k}'_1, \mathbf{k}'_2} + c_0 [A(\mathbf{k}_1 - \mathbf{k}_2)]^2 \, \delta_{\mathbf{k}_1 + \mathbf{k}'_1, \mathbf{k}_2 + \mathbf{k}'_2}. \tag{9.40}$$

This configurational averaging process could be continued for mean values involving three or more factors of $A(\mathbf{k}_1, \mathbf{k}_2)$. We would have a series of terms with various weighting factors, and with each depending on Kronecker deltas for the wave-vector labels. The algebra, however, becomes rapidly very complicated for products of more than two amplitude terms, and so we look for a modified diagram technique to represent the averaged terms.

To establish this technique we note that the nth order term (i.e., the one with n vertices) in the expansion of the GF $\mathcal{G}_{\mathbf{k}, \mathbf{k}'}$ is found from

$$\sum_{\mathbf{k}_2, \cdots, \mathbf{k}_n} \mathcal{G}^0_{\mathbf{k}} \mathcal{G}^0_{\mathbf{k}_2} \cdots \mathcal{G}^0_{\mathbf{k}_n} \mathcal{G}^0_{\mathbf{k}'} \, A(\mathbf{k}, \mathbf{k}_2) \cdots A(\mathbf{k}_n, \mathbf{k}'),$$

and so the corresponding term in the averaged $\{\mathcal{G}_{\mathbf{k}, \mathbf{k}'}\}$ will be

$$\sum_{\mathbf{k}_2, \cdots, \mathbf{k}_n} \mathcal{G}^0_{\mathbf{k}} \mathcal{G}^0_{\mathbf{k}_2} \cdots \mathcal{G}^0_{\mathbf{k}_n} \mathcal{G}^0_{\mathbf{k}'} \, \{A(\mathbf{k}, \mathbf{k}_2) \cdots A(\mathbf{k}_n, \mathbf{k}')\}. \tag{9.41}$$

This result follows because no averaging is needed over the GFs in Equation (9.41) because they are the $\mathcal{G}^0$ terms for the unperturbed system. Therefore, the expression involves the mean value over a product of n factors of the A terms.

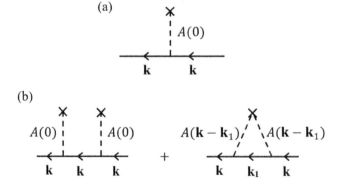

Figure 9.12 The contributions to the averaged interacting GF $\{\mathcal{G}_{\mathbf{k},\mathbf{k}'}\}$ for a system with random impurities in order (a) $n = 1$ and (b) $n = 2$.

We now represent the first-order ($n = 1$) term as in Figure 9.12 where the cross at one end of the vertex is used to indicate that a mean has been taken. Because $\{A(\mathbf{k},\mathbf{k}')\}$ is proportional to $\delta_{\mathbf{k},\mathbf{k}'}$ we have the convenient property of wave-vector conservation for the crossed (averaged) vertex. Also, from Equation (9.39) for $n = 1$, we see that the contribution of the first-order diagram is

$$\frac{1}{\beta}\mathcal{G}_{\mathbf{k}}^{0}(i\omega_m)\,\mathcal{G}_{\mathbf{k}'}^{0}(i\omega_m)\,c_0 A(0)\,\delta_{\mathbf{k},\mathbf{k}'} = \frac{1}{\beta}\left[\mathcal{G}_{\mathbf{k}}^{0}(i\omega_m)\right]^2 c_0 A(0).$$

Next we look at the modified diagrams for the averaged GF in the second order, where there are the two diagrammatic terms corresponding to the expression in Equation (9.40) for the $\{AA\}$ average. We see once again that the Kronecker deltas on the wave-vector labels are automatically taken care of by wave-vector conservation at the crossed vertices. This process of representation with crossed vertices can be continued to higher orders with $n \geq 3$, and some examples for diagrams of this type are shown in Figure 9.13. We note that the order of perturbation is given by the number of dashed interactions (the vertices) and not by the number of crosses.

We are now in a position to summarize the procedure by stating the *rules* for evaluating any GF diagram for the boson or fermion systems with random impurities:

1. Label the diagram so that wave vector is conserved at any crossed vertex and the imaginary frequency is continuous along each GF line.
2. Associate a factor of $(1/\beta)\mathcal{G}_{\mathbf{k}}^{0}(i\omega_m)$ with each line, where $\mathbf{k}$ is its wave-vector label and $i\omega_m$ is its frequency label.
3. Associate a factor of $\beta A(\mathbf{q})$ with each vertex line, where $\mathbf{q}$ is the wave vector along the line.
4. Associate a factor of the impurity concentration c_0 with each cross.

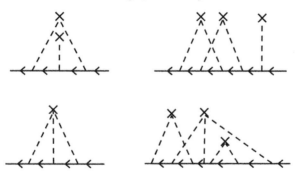

Figure 9.13 Some examples of higher-order ($n \geq 3$) diagrams contributing to the GF $\{\mathcal{G}_{\mathbf{k},\mathbf{k}'}\}$ for a system with random impurities.

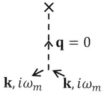

Figure 9.14 The first-order proper self-energy contribution to the GF for a system with random impurities. Choices for the wave-vector and frequency labels have been inserted.

5. Include a factor $(-1)^n/p$, where n is the number of vertices and p is the symmetry number.
6. Finally, sum over all internal wave-vector and frequency labels.

Example of Calculating the Interacting Green's Functions

As in most of the GF examples discussed previously, we follow a systematic approach to calculate the interacting GF by first choosing a proper self-energy, from which the renormalized energy (and the damping, if there is any contribution) can then be deduced.

The simplest proper self-energy diagram is that shown in Figure 9.14, which can readily be evaluated to give the result for the first-order proper self-energy that

$$\Sigma^{(1)}(\mathbf{k}, i\omega_m) = -\beta c_0 A(0). \tag{9.42}$$

We note that this is a real constant, which is independent of the wave vector $\mathbf{k}$ and the frequency $i\omega_m$. It follows that the renormalized energy in this first order is

$$E_{\mathbf{k}} + \frac{1}{\beta}\Sigma^{(1)} = E_{\mathbf{k}} - c_0 A(0), \tag{9.43}$$

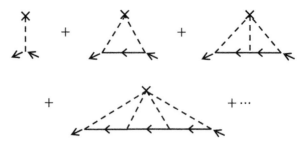

Figure 9.15 The series of "one-cross" diagrammatic proper self-energies to the GF for a system with random impurities.

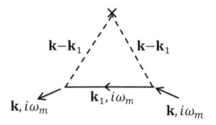

Figure 9.16 The second-order proper self-energy contribution to the GF for a system with random impurities. The wave-vector and frequency labels have been inserted.

which merely shifts the energies by a constant. This correction is equivalent to a mean-field approximation and the damping is zero for this level of approximation.

For a higher-order renormalization beyond that just considered, we could take account of the infinite set of "one-cross" diagrams as represented by the series shown in Figure 9.15. The diagrams can be thought of as a summing up all the scattering events (single scattering and multiple scattering) that occur at a single impurity. Because they are one-cross diagrams, they give an overall contribution that is proportional to c_0. If the concentration c_0 of impurities is small, as may sometimes be the case (e.g., for a dilute metallic alloy), these diagrams would be expected to provide the dominant contribution to the self-energy Σ.

We have already obtained the result for the first member of the series, and we have shown that it gives rise to the expression in Equation (9.42). As a further example we consider the evaluation of the next (second-order) diagram in the series, which is shown again in Figure 9.16 with the relevant labels inserted to help us in following the diagram rules. The result obtained for the self-energy contribution is

$$\Sigma^{(2)}(\mathbf{k}, i\omega_m) = \sum_{\mathbf{k}_1}(-1)^2 \left(\frac{1}{\beta}\right) \mathcal{G}_{\mathbf{k}}^0(i\omega_m)\beta^2 A^2(\mathbf{k} - \mathbf{k}_1) c_0$$

$$= c_0 \beta \sum_{\mathbf{k}_1} \frac{A^2(\mathbf{k} - \mathbf{k}_1)}{i\omega_m - E_{\mathbf{k}_1}}. \tag{9.44}$$

This expression depends explicitly on frequency, as well as on the external wave vector. Therefore, to obtain the renormalized energy and damping we need to evaluate the self-energy when an analytic continuation is made for the frequency. Specifically, we take as an approximation $i\omega_m \to \tilde{\omega} \to E_{\mathbf{k}} - i\eta$, as discussed in Chapter 8. This leads to

$$\Sigma^{(2)}(\mathbf{k}, E_{\mathbf{k}} - i\eta) = c_0\beta \sum_{\mathbf{k}_1} \frac{A^2(\mathbf{k} - \mathbf{k}_1)}{(E_{\mathbf{k}} - E_{\mathbf{k}_1} - i\eta)}$$

$$= c_0\beta \left[\sum_{\mathbf{k}_1} \frac{A^2(\mathbf{k} - \mathbf{k}_1)}{(E_{\mathbf{k}} - E_{\mathbf{k}_1})} + i\pi \sum_{\mathbf{k}_1} A^2(\mathbf{k} - \mathbf{k}_1)\delta(E_{\mathbf{k}} - E_{\mathbf{k}_1}) \right].$$

Therefore, combining the contributions from the pair of one-cross diagrams just considered, we conclude that the renormalized energy is

$$E_{\mathbf{k}} + \frac{1}{\beta}\left[\Sigma^{(1)} + \mathrm{Re}\,\Sigma^{(2)}(\mathbf{k}, E_{\mathbf{k}} - i\eta)\right] = E_{\mathbf{k}} - c_0 A(0) + c_0 \sum_{\mathbf{k}_1} \frac{A^2(\mathbf{k} - \mathbf{k}_1)}{(E_{\mathbf{k}} - E_{\mathbf{k}_1})},$$

(9.45)

and the reciprocal lifetime (damping) to the same order is

$$\frac{1}{T_{\mathbf{k}}} = \pi c_0 \sum_{\mathbf{k}_1} A^2(\mathbf{k} - \mathbf{k}_1)\delta(E_{\mathbf{k}} - E_{\mathbf{k}_1}).$$

(9.46)

More accurate results for the renormalized energy and damping could presumably be obtained by including all the higher-order diagrams in the one-cross series shown in Figure 9.15. This is more difficult because it usually requires the solution of a complicated integral equation. However, the preceding results as given will be good approximations provided the interaction (the scattering amplitude A) is weak enough.

A further type of generalization would be needed if the concentration of impurities is higher, requiring the evaluation of diagrams that are proportional to c_0^2, c_0^3, and so on. These are the diagrams where multiple crosses occur. Then, for example, the third-order diagram in Figure 9.17(a), which is proportional to c_0, would be comparable to that in Figure 9.17(b), which is proportional to c_0^2.

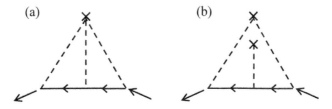

Figure 9.17 Comparison of two third-order proper self-energy diagrams: (a) a one-cross diagram and (b) a two-cross diagram.

9.6 Diagrammatic Techniques for Spin Operators

As a final topic we consider alternative diagrammatic formulations that are specific for spin systems. As mentioned previously, a major drawback of the previous diagrammatic formulation in Section 9.4 for the Heisenberg ferromagnet is that it employed the HP transformation to make a power-series expansion in terms of boson operators. Such an expansion will be useful only at relatively low temperatures below T_C, where the convergence of the operator expansions in Equation (9.22) is rapid. At higher temperatures (even below T_C) this is not the case, and so the applicability of this approach is restricted.

9.6.1 The Drone-Fermion Method

One way to overcome this problem is to employ a different kind of representation for the spin operators that does not require an expansion and for which a modified form of Wick's theorem can be established. An example of a technique for doing this makes use of the *drone-fermion* (DF) representation, which was originally proposed for spin $S = 1/2$ systems (see [27, 133]). The representation can be expressed as

$$S_j^+ = \phi_j^\dagger c_j \,, \qquad S_j^- = c_j^\dagger \phi_j \,, \qquad S_j^z = 1/2 - c_j^\dagger c_j. \qquad (9.47)$$

Here the c_j operators are fermion operators at site j, satisfying the usual anticommutation relationships, while the ϕ_j operators are subsidiary (or "drone") operators that anticommute with any of the c operators and obey among themselves

$$\phi_j^\dagger = \phi_j \quad \text{and} \quad \phi_j \phi_i + \phi_i \phi_j = 2\delta_{i,j}. \qquad (9.48)$$

It is a straightforward exercise to verify using Equations (9.47) and (9.48) that the spin commutation relationships, as quoted in Equation (1.84), are recovered (see Problem 9.7). It is also seen from Equation (9.47) that $c_j^\dagger c_j$ is the number operator for the longitudinal spin deviation at site j. The role of the ϕ_j operators is to enable the spin commutation relations to be reproduced.

The diagrammatic perturbation method in terms of the DF representation was introduced for Heisenberg ferromagnets by Spencer [133] and for dipole-exchange ferromagnets by Cottam [134]. Here we will take the simpler case of Heisenberg ferromagnets, as in Section 9.4, for which there are only exchange interactions. On rewriting the spin Hamiltonian in Equation (9.20) in terms of the DF operators using Equation (9.47) we obtain $\mathcal{H} = \mathcal{H}_0 + \mathcal{H}_1$, ignoring an unimportant constant term. Here $\mathcal{H}_0$ has a diagonalized form that involves c operators only (see Problem 9.8):

$$\mathcal{H}_0 = \left\{ b + \frac{1}{2} J(0) \right\} \sum_k c_k^\dagger c_k, \qquad (9.49)$$

where we have made a Fourier transform of the operators to a wave-vector representation as before. The interaction term $\mathcal{H}_1$ is

$$\mathcal{H}_1 = -\frac{1}{2N} \sum_{\mathbf{k}_1, \mathbf{k}_2, \mathbf{q}} \left[J(\mathbf{q}) c^\dagger_{\mathbf{k}_1 + \mathbf{q}} c^\dagger_{\mathbf{k}_2 - \mathbf{q}} c_{\mathbf{k}_2} c_{\mathbf{k}_1} + \frac{1}{2} J(\mathbf{q}) \left\{ \phi^\dagger_{\mathbf{k}_1 + \mathbf{q}} c^\dagger_{\mathbf{k}_2 - \mathbf{q}} \phi_{\mathbf{k}_2} c_{\mathbf{k}_1} + \text{H.c.} \right\} \right].$$

$$(9.50)$$

An important difference to note at this point compared with the boson expansion method in Section 9.4 is that there are no higher-order expansion terms being neglected, as was the case for Equation (9.26) where the products of more than four operators were neglected in a low-temperature approximation.

We introduce GFs associated with the c and ϕ operators by defining

$$g_{M,\mathbf{k}}(\tau) = -\langle \hat{T}_W \check{c}_{\mathbf{k}}(\tau) \check{c}^\dagger_{\mathbf{k}}(0) \rangle, \qquad d_{M,\mathbf{k}}(\tau) = -\langle \hat{T}_W \check{\phi}_{\mathbf{k}}(\tau) \check{\phi}^\dagger_{\mathbf{k}}(0) \rangle. \qquad (9.51)$$

The Fourier components of the corresponding quantities when transformed to the frequency representation as in Chapter 8 will be denoted by $\mathcal{G}_{\mathbf{k}}(i\omega_m)$ and $\mathcal{D}_{\mathbf{k}}(i\omega_m)$, respectively, where ω_m is a fermion frequency. The expressions for these GFs when they are evaluated with respect to the unperturbed Hamiltonian $\mathcal{H}_0$ are easily found to be

$$\mathcal{G}^0_{\mathbf{k}}(i\omega_m) = \frac{1}{i\omega_m - b - \frac{1}{2} J(0)}, \qquad \mathcal{D}^0_{\mathbf{k}}(i\omega_m) = \frac{2}{i\omega_m}. \qquad (9.52)$$

The first of these results follows straightforwardly by analogy with Equation (8.45). The form of the second result is a consequence of there being no $\phi^\dagger_{\mathbf{k}} \phi_{\mathbf{k}}$ term in $\mathcal{H}_0$, leading to a pole for $i\omega_m$ at zero, while the factor of 2 comes from Equation (9.48).

Because $\mathcal{H}_0$ and $\mathcal{H}_1$ both involve an even number of operators that satisfy simple anticommutation relationships, one would expect Wick's theorem to apply here, and this was indeed proved to be the case [133]. Hence a diagrammatic representation can be established in which there are two interaction vertices corresponding to the terms in $\mathcal{H}_1$. These vertices are depicted in Figure 9.18, where the black and gray lines correspond to the $\mathcal{G}^0_{\mathbf{k}}(i\omega_m)$ and $\mathcal{D}^0_{\mathbf{k}}(i\omega_m)$ GFs, respectively, and the dashed line represents an exchange interaction. The interaction vertices in (a) and (b) will be referred to as longitudinal and transverse, respectively, because they originate from either the $S^z S^z$ or the spin-flip $S^+ S^-$ parts of the Heisenberg Hamiltonian.

The rules for evaluating any diagrammatic contribution to $\mathcal{G}_{\mathbf{k}}(i\omega_m)$ or $\mathcal{D}_{\mathbf{k}}(i\omega_m)$, or any related self-energy term, are very similar to those given in Section 8.6 for $\mathcal{G}_{\mathbf{k}}(i\omega_m)$ in the single-field boson or fermion case. They may be stated as follows:

1. Label the diagrams so that wave vector and frequency are conserved at each vertex.
2. Associate factors $-\beta J(\mathbf{q})$ and $-\frac{1}{2} \beta J(\mathbf{q})$ with each longitudinal and transverse vertex, respectively, where $\mathbf{q}$ is the wave-vector transfer.

(a)   (b)

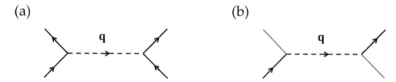

Figure 9.18 The (a) longitudinal and (b) transverse interaction vertices in the drone-fermion method. The black and gray lines denote the noninteracting GFs $\mathcal{G}_{\mathbf{k}}^0(i\omega_m)$ and $\mathcal{D}_{\mathbf{k}}^0(i\omega_m)$, respectively, while the dashed line denotes an exchange interaction vertex.

3. For each full line associate a factor $(1/\beta)\mathcal{G}_{\mathbf{k}}^0(i\omega_m)$ or $(1/\beta)\mathcal{D}_{\mathbf{k}}^0(i\omega_m)$, as appropriate, where $\mathbf{k}$ and $i\omega_m$ are the wave vector and frequency along the line. Also include an extra factor of $\exp(i\omega_m\eta)$ if the line is self-contracted.

4. Include a factor $(-1)^L/p$, where L is the number of closed GF loops and p is the symmetry number.

5. Include a factor $(-1)^n$, where n is the number of vertices in the diagram.

6. Finally, sum over all wave-vector and frequency labels within the restrictions imposed by rule 1.

9.6.2 The $1/z$ Expansion

We now come to the crucial matter of how to choose the important diagrams to evaluate. We stress that this cannot be done here in terms of the number of interaction vertices because the factors like $\beta J(\mathbf{q})$ are not generally small. In this problem a suitable parameter of smallness will be $1/z$, where z is the effective number of spins that interact with any given spin (e.g., $z = 8$ in a b.c.c. ferromagnet with nearest-neighbor exchange only). We shall see below that the lowest order $(1/z)^0$ results are equivalent to mean-field theory, and so the results in order $(1/z)^1$ and higher represent successive improvements to mean-field theory through the inclusion of spin-fluctuation effects. The use of $1/z$ expansions has a long history in the theory of ferromagnetism and antiferromagnetism, and it precedes the drone-fermion method (see, e.g., [138, 139]).

In the present context, the rule for determining the $1/z$ dependence is as follows [133, 134]: if there are m independent wave-vector labels that appear explicitly in the arguments of at least one of the vertex factors *and which are eventually summed over*, the dependence of that diagram is $(1/z)^m$.

Briefly, the justification comes by noting that the Curie temperature in mean-field theory corresponds to $k_B T_C = J(0)/4$ for a spin $S = 1/2$ Heisenberg ferromagnet (see Section 5.2.3). Now $J(0) = z\bar{J}$, where $\bar{J}$ denotes an average exchange interaction between neighboring spins, implying that $\bar{J}$ is of order $(1/z)T_C$. On

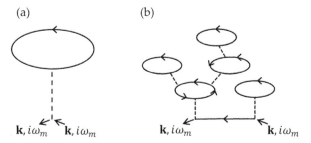

Figure 9.19 Examples of diagrammatic contribution to the GF $\mathcal{G}_{\mathbf{k}}^0(i\omega_m)$, showing (a) the simplest bubble diagram and (b) a treelike structure derived from the bubble diagram.

transforming to the site representation, each vertex factor proportional to $J(\mathbf{q})$ gives a $J_{i,j}$ term multiplied by an exponential factor and summed over the separation vector $\mathbf{r}_{ij}$. After integration with respect to τ, each vertex yields a factor $\beta \bar{J}$ which is of order $1/z$. There remain the summations over $\mathbf{r}_{ij}$ terms that occur in factors like $\exp(-i\mathbf{q} \cdot \mathbf{r}_{ij})$. If a particular wave-vector label appears in the argument of one (or more) of the $J(\mathbf{q})$, the summation over that wave vector will give a Kronecker delta relating the different $\mathbf{r}_{ij}$ and thereby removing one of the $\mathbf{r}_{ij}$ summations. We must keep in mind that the unperturbed GFs in Equation (9.52) have no explicit wave-vector dependence. Each independent $\mathbf{r}_{ij}$ summation that remains after the previously mentioned process contributes a factor of order z. Therefore, if there are n vertices in a diagram and m independent wave vectors to be summed over, the overall dependence is $(1/z)^n z^{n-m}$, or $(1/z)^m$ as stated earlier.

The lowest-order renormalization to consider for the GFs comes from the diagrammatic contributions with $(1/z)^0$ dependence, i.e., those diagrams that involve no wave-vector summations. It is easy to convince oneself that there are no diagrams of this type for $\mathcal{D}_{\mathbf{k}}^0(i\omega_m)$, whereas for $\mathcal{G}_{\mathbf{k}}^0(i\omega_m)$ the "bubble" diagram shown in Figure 9.19(a) provides the simplest contribution. Furthermore, the analogous "treelike" structures such as those shown in Figure 9.19(b) are also of order $(1/z)^0$ and must be included. A systematic way to incorporate all such contributions into $\mathcal{G}_{\mathbf{k}}^0(i\omega_m)$ is represented by Figure 9.20 which is a type of Dyson equation like that in Subsection 9.5.1.

In proceeding now to the self-consistent evaluation of the GF in order $(1/z)^0$, we note that the bubble diagram shown in Figure 9.19(a) corresponds formally to one of the HF diagrams already calculated in Section 9.1. By comparing with Equation (9.1) we see that the self-energy term here is

$$-\beta J(0) \sum_{\mathbf{k}'} \langle c_{\mathbf{k}'}^{\dagger} c_{\mathbf{k}'} \rangle. \qquad (9.53)$$

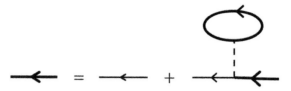

Figure 9.20 Representation of a Dyson equation for the lowest-order $(1/z)^0$ renormalization of the GF $\mathcal{G}_{\mathbf{k}}^0(i\omega_m)$. The renormalized GF is shown by the heavy line.

Then, from the Dyson equation represented in Figure 9.20, we have (see also Problem 9.9)

$$\left[\mathcal{G}_{\mathbf{k}}^0(i\omega_m)\right]^{-1} = \left[\mathcal{G}_{\mathbf{k}}(i\omega_m)\right]^{-1} - J(0)\sum_{\mathbf{k}'}\langle c_{\mathbf{k}'}^\dagger c_{\mathbf{k}'}\rangle, \tag{9.54}$$

which leads to the result that the GF $\mathcal{G}_{\mathbf{k}}(i\omega_m)$ in order $(1/z)^0$ is given by

$$\mathcal{G}_{\mathbf{k}}(i\omega_m) = \frac{1}{i\omega_m - \gamma}. \tag{9.55}$$

Here γ, which is real and independent of the external label $\mathbf{k}$, is defined by

$$\gamma = b + \left\{\frac{1}{2} - \sum_{\mathbf{k}'}\langle c_{\mathbf{k}'}^\dagger c_{\mathbf{k}'}\rangle\right\} J(0) = b + \langle S^z\rangle_0 J(0). \tag{9.56}$$

The simplification in the last step in the preceding equation occurs because the term in curly brackets represents the average $\langle S^z\rangle_0$ at any spin site in order $(1/z)^0$, as can be seen by averaging the expression for S^z in Equation (9.47). Also, using the FD distribution function, we arrive at a consistency condition that

$$\langle S^z\rangle_0 = \frac{1}{2} - \frac{1}{\exp(\beta\gamma) + 1} = \frac{1}{2}\tanh\left(\frac{1}{2}\beta\gamma\right). \tag{9.57}$$

Equation (9.57) is now recognizable as being the well-known result for the longitudinal spin average (proportional to the magnetization) according to mean-field theory (see, e.g., [19, 20]).

To summarize, at this stage we have carried out the renormalization of the drone-fermion GFs to lowest order $(1/z)^0$ and shown that the results are equivalent to mean-field theory. The interesting stage comes next when we proceed to higher order(s) and incorporate a description of the magnons (or SWs). We present only a brief overview of this method (and related techniques) in the following paragraphs; further details may be found, for example, in [134, 137, 140].

We have seen previously (e.g., in Section 5.2) that the SW excitations can be investigated by studying GFs of the form $\mathcal{G}(S^+; S^- | i\omega_m)$. Therefore, in view of

(a)

$\mathbf{k}, i\omega_m$ $\mathbf{k}, i\omega_m$

(b)

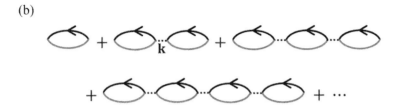

Figure 9.21 Diagrams contributing to the GF $\mathcal{G}(S_{\mathbf{k}}^+; S_{\mathbf{k}}^- \mid i\omega_m)$ defined in Equation (9.58) and leading to spin waves: (a) the general form of any contribution and (b) the series of lowest-order chain diagrams.

the definitions in Equation (9.47), we introduce the imaginary-time GF that has the wave-vector and frequency Fourier components as follows:

$$\mathcal{G}(S_{\mathbf{k}}^+; S_{\mathbf{k}}^- \mid i\omega_m) = \sum_{\mathbf{q},\mathbf{q}'} \mathcal{G}(\phi_{\mathbf{q}'}^\dagger c_{\mathbf{q}'+\mathbf{k}}; c_{\mathbf{q}+\mathbf{k}}^\dagger \phi_{\mathbf{q}} \mid i\omega_m). \tag{9.58}$$

It is important to realize that the overall $i\omega_m$ label attached to this GF is a *boson* frequency because it comes from the sum or difference between two fermion frequencies. The general diagrammatic form of a contribution to the preceding GF is as depicted in Figure 9.21(a), where the shaded area represents any allowed set of single-particle GF lines (black or gray lines) and exchange interaction vertices. Its form is chosen in accordance with the $1/z$ classification. In lowest order the contributions are simply those with no internal wave-vector label appearing in a vertex, keeping in mind that $\mathbf{k}$ in Equation (9.58) is a fixed external label. Hence it follows that the sequence of single transverse loop and chain diagrams in Figure 9.21(b) provides the required contribution.

We start by considering the first loop diagram, for which it follows from application of the diagrammatic rules that the contribution is

$$\frac{2}{\beta^2} \sum_{m'} \frac{1}{(i\omega_{m'} - \gamma)(i\omega_{m'} - i\omega_m)}. \tag{9.59}$$

Two noteworthy points here are that the preceding $i\omega_{m'}$ is a fermion frequency and that we have included the $(1/z)^0$ renormalization of the internal lines (so it is γ that appears in the denominator). The summation over $i\omega_{m'}$ is straightforward to carry out on noting that the preceding summation is formally analogous to that already analyzed for Equation (9.6), but with different values for the poles. Hence the result in the present case is found to be (see Problem 9.10)

$$\frac{2\langle S^z\rangle_0}{\beta(i\omega_m - \gamma)}, \tag{9.60}$$

where we have made use of the mean-field Equation (9.57) to simplify the numerator. Finally, we need to sum of the sequence of chain diagrams represented by Figure 9.21(b), which yields a geometric series leading to

$$\mathcal{G}(S_\mathbf{k}^+; S_\mathbf{k}^- \,|\, i\omega_m) = \frac{2\langle S^z\rangle_0}{\beta(i\omega_m - \gamma)} + \frac{2\langle S^z\rangle_0}{\beta(i\omega_m - \gamma)}\left(-\frac{1}{2}\beta J(\mathbf{k})\right)\frac{2\langle S^z\rangle_0}{\beta(i\omega_m - \gamma)} + \cdots$$

$$= \frac{2\langle S^z\rangle_0}{\beta\bigl(i\omega_m - \gamma + \langle S^z\rangle_0 \, J(\mathbf{k})\bigr)}. \tag{9.61}$$

We see now that the GF in Equation (9.61) has a simple pole at a quantity $E_\mathbf{k}$ given by

$$E_\mathbf{k} = \gamma - \langle S^z\rangle_0 \, J(\mathbf{k}) = b + \langle S^z\rangle_0 \bigl[J(0) - J(\mathbf{k})\bigr]. \tag{9.62}$$

This result can be recognized as being very similar to that for the SW energy found in Equation (5.20) using the GF equation-of-motion method with a RPA decoupling approximation. A point of difference between the two calculations is that in the present case we have employed the $1/z$ expansion, and so we can proceed to the next order of the expansion to renormalize the SW excitations (see, e.g., [137] for calculations of the energy shift and damping that are valid over a wide range of temperatures below T_C). Examples of two diagrams that have to be considered in the next order in $1/z$ are shown in Figure 9.22; they both have one internal wave vector $\mathbf{q}$ that needs to be summed over.

Some final considerations relating to this topic are as follows. The DF representation quoted in Equation (9.47) and the ensuing diagrammatic method are for spin systems with $S = 1/2$. Any direct extensions to higher spin values turned out to be difficult initially because of the complicated form of Wick's theorem, but an alternative generalization of the DF method to other spin values that satisfy $2S + 1 = 2^n$, where n is any positive integer, was made by Psaltakis and Cottam

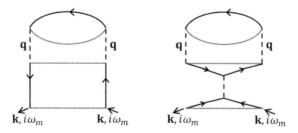

Figure 9.22 Two diagrams for the renormalization of the spin waves.

[140]. Another closely related diagrammatic method, applicable to spin values other than just $S = 1/2$, was introduced by Vaks et al. [141, 142].

Problems

9.1. Draw the set of distinct diagrams contributing to the proper self-energy (see Section 9.1) in the second order.

9.2. Consider the first-order diagram in Figure 9.2 for the proper self-energy, corresponding to HF theory for an interacting fermion gas. Show that the final result for the contribution to the proper self-energy is given by Equation (9.8).

9.3. Draw six diagrams as examples for contributions to the GF loop $F(\mathbf{q}, i\omega_m)$ in the first and the second order of perturbation.

9.4. Verify that the unperturbed GF for the phonon system has the form quoted in Equation (9.16). You may assume the symmetry property that $\omega_{\mathbf{q}} = \omega_{-\mathbf{q}}$.

9.5. A diagram for a proper self-energy contribution to the electronic GF due to the electron–phonon interactions is shown in Figure 9.9(b). A formal expression for that self-energy (after application of the diagram rules) is quoted in Equation (9.17). Now carry out the summation over the internal frequency $i\omega_{m'}$ and show that the expression for $\Sigma_{el}(\mathbf{k}, i\omega_m)$ has energy denominators of the form $(i\omega_m - E_{\mathbf{k}-\mathbf{q}} \pm \omega_{\mathbf{q}})$.

9.6. The energy shift for renormalized SWs in Heisenberg ferromagnets at $T \ll T_C$ is given by Equation (9.30). Taking the case of a b.c.c. structure, verify that in the limit of small wave vector $\mathbf{k}$ (such that $k^2 a^2 \ll 1$ where $k = |\mathbf{k}|$) the energy shift is proportional to k^2 multiplied by a temperature-dependent factor.

9.7. Verify using Equations (9.47) and (9.48) for the DF representation that the spin commutation relationships quoted in Equation (1.84) are satisfied.

9.8. Go through the steps of rewriting the Heisenberg Hamiltonian in Equation (9.20) in terms of the DF operators using Equation (9.47) to obtain $\mathcal{H} = \text{constant} + \mathcal{H}_0 + \mathcal{H}_1$, showing that $\mathcal{H}_0$ and $\mathcal{H}_1$ have the form quoted in Equations (9.49) and (9.50).

9.9. A representation of a Dyson equation for the lowest-order $(1/z)^0$ renormalization of the GF $\mathcal{G}_{\mathbf{k}}^0(i\omega_m)$ is presented in Figure 9.20. From the Dyson equation in the figure and Equation (9.53) for the self-energy term, verify that the GF $\mathcal{G}_{\mathbf{k}}$ has the form quoted in Equation (9.55), where γ is defined in Equation (9.56).

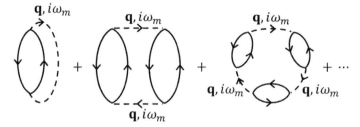

Figure 9.23 Series of electron-hole ring diagrams (see Problem 9.11).

9.10. Consider the infinite series of diagrams in Figure 9.21(b) that contribute to the GF $\mathcal{G}(S^+; S^- | i\omega_m)$ using the DF method. Show that the contribution arising from the first loop diagram (i.e., a single loop) has the form quoted in Equation (9.60). Then sum the Dyson series to verify Equation (9.61).

9.11. Consider the series of electron-hole ring diagrams depicted in Figure 9.23. Taking account of the symmetry factor p to be associated with a general diagram and using the result in Equation (9.6) for a single electron-hole loop F^0, obtain an expression for the sum of all ring diagrams. Show that it involves a logarithmic term that depends on the boson frequency $i\omega_m$ (but do not attempt to carry out this frequency summation).

References

[1] D. J. Griffiths, *Introduction to Quantum Mechanics*, 2nd edn. (Cambridge: Cambridge University Press, 2017).

[2] B. H. Bransden and C. J. Joachain, *Quantum Mechanics*, 2nd edn. (New York: Pearson Education, 2000).

[3] L. E. Reichl, *A Modern Course in Statistical Physics*, 3rd edn. (New York: Wiley, 2009).

[4] R. K. Pathria and P. D. Beale, *Statistical Physics*, 3rd edn. (Amsterdam: Academic Press, 2011).

[5] P. A. M. Dirac, *The Principles of Quantum Mechanics*, 4th edn. (Oxford: Clarendon Press, 1958).

[6] P. A. M. Dirac, The quantum theory of the emission and absorption of radiation. *Proc. Royal Soc. London A*, **114** (1927), 243–65.

[7] J. D. Jackson, *Classical Electrodynamics*, 3rd edn. (New York: Wiley, 2005).

[8] D. J. Griffiths, *Introduction to Electrodynamics*, 4th edn. (Cambridge: Cambridge University Press, 2017).

[9] C. Kittel, *Quantum Theory of Solids*, 2nd edn. (New York: Wiley, 1987).

[10] R. J. Glauber, Coherent and incoherent states of radiation field. *Phys. Rev.*, **131** (1963), 2766–88.

[11] R. J. Glauber, The quantum theory of optical coherence. *Phys. Rev.*, **130** (1963), 2529–39.

[12] R. J. Glauber, Photon correlations. *Phys. Rev. Lett.*, **10** (1963), 84–6.

[13] R. Loudon, *The Quantum Theory of Light* (Oxford: Oxford University Press, 2000).

[14] M. O. Scully and M. S. Zubairy, *Quantum Optics* (Cambridge: Cambridge University Press, 1997).

[15] C. Gerry and P. Knight, *Introductory Quantum Optics* (Cambridge: Cambridge University Press, 2005).

[16] L. Mandel and E. Wolf, *Optical Coherence and Quantum Optics* (Cambridge: Cambridge University Press, 1995).

[17] P. Meystre and M. Sargent, *Elements of Quantum Optics*, 3rd edn. (Berlin: Springer, 1999).

[18] W. Magnus, On the exponential solution of differential equations for a linear operator. *Commun. Pure Appl. Math.*, **7** (1954), 649–73.

[19] C. Kittel, *Introduction to Solid State Physics*, 8th edn. (New York: Wiley, 2004).

[20] N. W. Ashcroft, D. Wei, and N. D. Mermin, *Solid State Physics*, 2nd edn. (Singapore: CENGAGE Learning Asia, 2016).

[21] J. Hubbard, Electron correlations in narrow energy bands. *J. Proc. Roy. Soc. London A*, **276** (1963), 238–57.

[22] K. B. Davis, M. O. Mewes, M. R. Andrews, N. J. van Druten, D. S. Durfee, D. M. Kurn, and W. Ketterle, Bose–Einstein condensation in a gas of sodium atoms. *Phys. Rev. Lett.*, **75** (1995), 3969–73.

[23] D. R. Tilley and J. Tilley, *Superfluidity and Superconductivity*, 3rd edn. (Bristol: IoP Publishing, 1990).

[24] L. D. Landau, On the theory of superfluidity of helium II. *J. Phys. USSR*, **11** (1947), 91.

[25] D. G. Henshaw and A. D. B. Woods, Modes of atomic motions in liquid helium by inelastic scattering of neutrons. *Phys. Rev.*, **121** (1961), 1266–74.

[26] R. M. White, *Quantum Theory of Magnetism: Magnetic Properties of Materials*, 3rd edn. (Berlin: Springer, 2007).

[27] D. C. Mattis, *The Theory of Magnetism I: Statics and Dynamics* (Berlin: Springer, 1981).

[28] T. Holstein and H. Primakoff, Field dependence of the intrinsic domain magnetization of a ferromagnet. *Phys. Rev.*, **58** (1940), 1098–113.

[29] E. Merzbacher, *Quantum Mechanics*, 3rd edn. (New York: Wiley, 1998).

[30] R. A. Ferrell, Forced harmonic oscillator in the interaction picture. *Am. J. Phys.*, **45** (1977), 468–9.

[31] R. P. Feynman, Space-time approach to non-relativistic quantum mechanics. *Rev. Mod. Phys.*, **20** (1948), 367–87.

[32] K. Husimi, Miscellanea in elementary quantum mechanics, II. *Progr. Theor. Phys.*, **9** (1953), 381–402.

[33] E. H. Kerner, Note on the forced and damped oscillator in quantum mechanics. *Can. J. Phys.*, **36** (1957), 371–7.

[34] P. Carruthers and M. M. Nieto, Coherent states and the forced quantum oscillator. *Am. J. Phys.*, **33** (1965), 537–44.

[35] S. Blanes, F. Casasb, J. A. Oteoc, and J. Roscd, The Magnus expansion and some of its applications. *Phys. Rep.*, **470** (2009), 151–238.

[36] P. Bongaarts, *Quantum Theory: A Mathematical Approach* (Heidelberg: Springer, 2015).

[37] S. M. Rezende and N. Zagury, Coherent magnon states. *Phys. Lett. A*, **29** (1969), 47–8.

[38] N. Zagury and S. M. Rezende, Theory of macroscopic excitations of magnons. *Phys. Rev. B*, **4** (1971), 201–9.

[39] M. G. Cottam and D. R. Tilley, *Introduction to Surface and Superlattice Excitations*, 2nd edn. (Bristol: IoP Publishing, 2005).

[40] G. P. Srivastava, *The Physics of Phonons* (New York: Taylor and Francis, 1990).

[41] M. Born and K. Huang, *Dynamical Theory of Crystal Lattices* (Oxford: Clarendon Press, 1988).

[42] M. H. Cohen and F. Keffer, Dipolar sums in the primitive cubic lattices. *Phys. Rev.*, **99** (1955), 1128–34.

[43] K. S. Novoselov, A. K. Geim, S. V. Morozov, D. Jiang, Y. Zhang, S. V. Dubonos, I. V. Grigorieva, and A. A. Firsov, Electric field effect in atomically thin carbon films. *Science*, **306** (2004), 666–9.

[44] A. H. Castro Neto, F. Guinea, F. M. R. Peres, K. S. Novosolov, and A. K. Geim, The electronic properties of graphene. *Rev. Mod. Phys.*, **81** (2009), 109–62.

[45] S. R. Power and M. S. Ferreira, Indirect exchange and Ruderman–Kittel–Kasuya–Yosida (RKKY) interactions in magnetically-doped graphene. *Crystals*, **3** (2013), 49–78.

[46] G. F. Roach, *Green's Functions*, 2nd edn. (Cambridge: Cambridge University Press, 1982).

[47] G. D. Mahan, *Many-Particle Physics*, 3rd edn. (New York: Kluwer Academic/Plenum, 2000).

[48] E. N. Economou, *Green's Functions in Quantum Physics*, 3rd edn. (Berlin: Springer, 2006).

[49] G. Rickayzen, *Green's Functions and Condensed Matter* (London: Academic Press, 1980).

[50] P. Coleman, *Introduction to Many-Body Physics* (Cambridge: Cambridge University Press, 2015).

[51] A. A. Abrikosov, L. P. Gorkov, and I. E. Dzyaloshinski, *Methods of Quantum Field Theory in Statistical Physics* (New York: Dover Publications, 1963).

[52] R. D. Mattuck, *A Guide to Feynman Diagrams in the Many-Body Problem*, 2nd edn. (New York: Dover Publications, 1976).

[53] D. N. Zubarev, Double-time Green functions in statistical physics. *Sov. Phys. Uspekhi*, **3** (1960), 320–45.

[54] J. Mathews and R. L. Walker, *Mathematical Methods of Physics*, 2nd edn. (San Francisco: Benjamin, 1973).

[55] G. B. Arfken, H. J. Weber, and F. E. Harris, *Mathematical Methods for Physicists*, 7th edn. (Amsterdam: Elsevier, 2012).

[56] L. D. Landau and E. M. Lifschitz, *Statistical Physics*, 3rd edn. (Amsterdam: Elsevier, 1980).

[57] G. F. Mazenko, *Nonequilibrium Statistical Mechanics* (Weinheim: Wiley-VCH, 2006).

[58] T. Matsubara, A new approach to quantum-statistical mechanics. *Prog. Theor. Phys.*, **14** (1955), 351–78.

[59] H. Lehmann, On the properties of propagation functions and renormalization constants of quantized fields. *Nuovo Cimento*, **11** (1954), 342–57.

[60] E. T. Jaynes and F. W. Cummings, Comparison of quantum and semiclassical radiation theories with application to the beam maser. *Proc. IEEE.*, **51** (1963), 89–109.

[61] F. W. Cummings, Reminiscing about thesis work with E. T. Jaynes at Stanford in the 1950s. *J. Phys. B: At. Mol. Opt. Phys.*, **46** (2013), 220202 (3 pages).

[62] M. D. Crisp and E. T. Jaynes, Radiative effects in semiclassical theory. *Phys. Rev.*, **179** (1969), 1253–61.

[63] C. R. Stroud Jr. and E. T. Jaynes, Long-term solutions in semiclassical radiation theory. *Phys. Rev. A*, **1** (1970), 106–21.

[64] E. T. Jaynes, Survey of the present status of neoclassical radiation theory in coherence and quantum optics. In *Coherent and Quantum Optics*, ed. L. Mandel and E. Wolf (New York: Plenum, 1973), pp. 35–81.

[65] B. W. Shore and P. L. Knight, The Jaynes–Cummings model. *J. Mod. Opt.*, **40** (1993), 1195–238.

[66] M. Orszag, *Quantum Optics*, 2nd edn. (Berlin: Springer, 2007).

[67] L. Mandel and E. Wolf, *Optical Coherence and Quantum Optics* (Cambridge: Cambridge University Press, 1995).

[68] R. H. Dicke, Coherence in spontaneous radiation processes. *Phys. Rev.*, **93** (1954), 99–110.

[69] M. Tavis and F. W. Cummings, Exact solution for an N-molecule-radiation-field Hamiltonian. *Phys. Rev.*, **170** (1968), 379–84.

[70] B. M. Garraway, The Dicke model in quantum optics: Dicke model revisited. *Phil. Trans. Roy. Soc. A*, **369** (2011), 1137–55.

[71] O. Jedrkiewicz and R. Loudon, Atomic dynamics in microcavities: Absorption spectra by Green function method. *J. Opt. B: Quantum Semiclass. Opt.*, **2** (2000), R47–60.

[72] M. G. Cottam and A. N. Slavin, Fundamentals of linear and nonlinear spin-wave processes in bulk and finite magnetic samples. In *Linear and Nonlinear Spin Waves in Magnetic Films and Superlattices*, ed. M. G. Cottam (Singapore: World Scientific, 1994), pp. 1–88.

[73] J. M. D. Coey, *Magnetism and Magnetic Materials* (Cambridge: Cambridge University Press, 2010).

[74] H. E. Stanley, *Mean Field Theory of Magnetic Phase Transitions* (Oxford: Oxford University Press, 1971).

[75] R. A. Tahir-Kheli and D. ter Haar, Use of Green functions in the theory of ferromagnetism: I. General discussion of the spin-S case. *Phys. Rev.*, **127** (1962), 88–94.

[76] M. G. Cottam and D. J. Lockwood, *Light Scattering in Magnetic Solids* (New York: Wiley, 1986).

[77] O. Madelung, *Introduction to Solid-State Theory* (Berlin: Springer, 1978).

[78] N. F. Mott, *Metal-Insulator Transitions* (Oxford: Taylor and Francis, 1990).

[79] K. Yosida, *Theory of Magnetism* (Berlin: Springer, 1996).

[80] J. Hubbard, Electron correlations in narrow energy bands II. The degenerate band case. *J. Proc. Roy. Soc. London A*, **277** (1964), 237–59.

[81] J. Hubbard, Electron correlations in narrow energy bands III. An improved solution. *J. Proc. Roy. Soc. London A*, **281** (1964), 401–19.

[82] P. W. Anderson, Localized magnetic states in metals. *Phys. Rev.*, **115** (1961), 41–53.

[83] C. Herring, Exchange interactions among itinerant electrons. In *Magnetism, Volume IV*, ed. G. T. Rado and H. Suhl (New York: Academic Press, 1966), pp. 1–407.

[84] P. W. Anderson, New approach to the theory of superexchange interactions. *Phys. Rev.*, **124** (1959), 2–13.

[85] A. J. Heeger, Localized moments and nonmoments in metals. In *Solid State Physics, Vol. 23*, ed. F. Seitz and D. Turnbull (Amsterdam: Academic Press, 1969), pp. 283–411.

[86] T. Moriya, *Spin Fluctuations in Itinerant Electron Magnetism* (Berlin: Springer, 1985).

[87] S. Doniach and E. H. Sondheimer, *Green's Function for Solid State Physics*, 2nd edn. (London: Imperial College Press, 1998).

[88] J. Bardeen, L. N. Cooper, and J. R. Schrieffer, Theory of superconductivity. *Phys. Rev.*, **108** (1957), 1175–204.

[89] J. G. Valatin, Comments on the theory of superconductivity. *Il Nuovo Cimento*, **7** (1958), 843–57.

[90] P. W. Anderson, Random-phase approximation in the theory of superconductivity. *Phys. Rev.*, **112** (1958), 1900–16.

[91] N. N. Bogoljubov, On a new method in the theory of superconductivity. *Il Nuovo Cimento*, **7** (1958), 794–805.

[92] L. N. Cooper, Bound electron pairs in a degenerate Fermi gas. *Phys. Rev.*, **104** (1956), 1189–90.

[93] J. G. Bednorz and K. A. Müller, Possible high T_c superconductivity in the Ba-La-Cu-O system. *Zeitschrift für Physik B: Condensed Matter*, **64** (1986), 189–93.

[94] J. von Neumann, Probabilistic foundation theory of quantum mechanics. *Nachrichten von der Gesellschaft der Wissenschaften zu Göttingen, Mathematisch-Physikalische Klasse*, **1** (1927), 245–72.

[95] D. ter Haar, Theory and applications of the density matrix. *Rep. Prog. Phys.*, **24** (1961), 304–62.

[96] F. Schwabl, *Statistical Mechanics*, 2nd edn. (Berlin: Springer, 2006).

[97] F. Mandl, *Statistical Physics*, 2nd edn. (New York: Wiley, 1988).

[98] L. D. Landau and E. M. Lifschitz, *Theory of Elasticity*, 3rd edn. (Amsterdam: Elsevier, 1986).

[99] R. Kubo, The fluctuation-dissipation theorem. *Rep. Prog. Phys.*, **29** (1966), 255–84.

[100] W. Nolting, *Fundamentals of Many-Body Physics: Principles and Methods* (Berlin: Springer, 2009).

[101] W. Hayes and R. Loudon, *Scattering of Light by Crystals* (New York: Wiley, 1978).

[102] S. W. Lovesey, *Condensed Matter Physics: Dynamic Correlations*, 2nd edn. (San Francisco: Benjamin, 1986).

[103] R. A. Cowley and W. J. L. Buyers, The properties of defects in magnetic insulators. *Rev. Mod. Phys.*, **44** (1972), 406–50.

[104] A. D. Boardman (Ed.), *Electromagnetic Surface Modes* (New York: Wiley, 1982).

[105] V. M. Agranovich and D. L. Mills (Eds.), *Surface Polaritons: Electromagnetic Waves at Surfaces and Interfaces* (Amsterdam: North-Holland, 1982).

[106] V. M. Agranovich and R. Loudon (Eds.), *Surface Excitations* (Amsterdam: North-Holland, 1984).

[107] R. F. Wallis and G. I. Stegeman (Eds.), *Electromagnetic Surface Excitations* (Heidelberg: Springer, 1986).

[108] J. D. Joannopoulos, S. G. Johnson, J. N. Winn, and R. D. Meade, *Photonic Crystals: Molding the Flow of Light* (Princeton: Princeton University Press, 2008).

[109] R. Loudon, Theory of line shapes for normal-incidence Brillouin scattering by acoustic phonons. *J. Phys. C: Solid State Phys.*, **11** (1978), 403–17.

[110] R. F. Wallis, Effect of free ends on the vibration frequencies of one-dimensional lattices. *Phys. Rev.*, **105** (1957), 540–5.

[111] R. E. de Wames and T. Wolfram, Theory of surface spin waves in the Heisenberg ferromagnet. *Phys. Rev.*, **185** (1969), 720–7.

[112] N. Wax, *Selected Papers on Noise and Stochastic Processes* (New York: Dover Publications, 1954).

[113] A. Akbari-Sharbaf and M. G. Cottam, Finite-width effects for the localized edge modes in zigzag graphene nanoribbons. *Phys. Rev. B*, **93** (2016), 235136, (12 pages).

[114] M. Fujita, M. Yoshida, and K. Nakada, Polymorphism of extended fullerene networks: Geometrical parameters and electronic structures. *Fullerene Sci. Technol.*, **4** (1996), 565–82.

[115] G. Gubbiotti (Ed.), *Three-Dimensional Magnonics* (Singapore: Jenny Stanford Pub., 2019).

[116] S. M. Rytov, Acoustical properties of a thinly laminated medium. *Akust. Zh.*, **2** (1956), 71 [*Sov. Phys. Acoustics*, **2** (1956), 68–80].

[117] T. Wolfram and J. Callaway, Spin-wave impurity states in ferromagnets. *Phys. Rev.*, **130** (1963), 2207–17.

[118] J. W. Negele and H. Orland, *Quantum Many-Particle Systems* (Redwood City: Addison Wesley, 1988).

[119] W. E. Parry, *The Many-Body Problem* (Oxford: Clarendon Press, 1973).

[120] J. D. Bjorken and S. D. Drell, *Relativistic Quantum Fields* (New York: McGraw-Hill, 1965).

[121] L. H. Ryder, *Quantum Field Theory*, 2nd edn. (Cambridge: Cambridge University Press, 1996).

[122] T. Padmanabhan, *Quantum Field Theory: The Why, What and How* (Berlin: Springer, 2016).

[123] G. C. Wick, The evaluation of the collision matrix. *Phys. Rev.*, **80** (1950), 268–72.

[124] L. A. Fetter and J. D. Walecka, *Quantum Theory of Many-Particle Systems*, (New York: Dover Publications, 2003).

[125] R. P. Feynman, The theory of positrons. *Phys. Rev.*, **76** (1949), 749–59.

[126] R. P. Feynman, Space-time approach to quantum electrodynamics. *Phys. Rev.*, **76** (1949), 769–89.

[127] R. P. Feynman, *Photon-Hadron Interactions* (Boca Raton: CRC Press, 1973).

[128] D. W. Snoke, *Solid State Physics: Essential Concepts* (San Francisco: Addison-Wesley, 2009).

[129] H. Fröhlich, Interaction of electrons with lattice vibrations. *Proc. Roy. Soc. A*, **215** (1952), 291–8.

[130] J. Bardeen and D. Pines, Electron-phonon interaction in metals. *Phys. Rev.*, **99** (1955), 1140–50.

[131] J. T. Devreese and F. Peeters (Eds.), *Polarons and Excitons in Polar Semiconductors and Ionic Crystals* (New York: Plenum, 1984).

[132] F. J. Dyson, General theory of spin-wave interactions. *Phys. Rev.*, **102** (1956), 1217–30.

[133] H. J. Spencer, Quantum-field-theory approach to the Heisenberg ferromagnet. *Phys. Rev.*, **167** (1968), 434–44.

[134] M. G. Cottam, Theory of dipole-dipole interactions in ferromagnets. I. The thermodynamic properties. *J. Phys. C: Solid St. Phys.*, **4** (1971), 2658–72.

[135] S. F. Edwards, A new method for the evaluation of electric conductivity in metals. *Phil. Mag.*, **3** (1958), 1020–31.

[136] P. Soven, Contribution to the theory of disordered alloys. *Phys. Rev.*, **178** (1969), 1136–44.

[137] M. G. Cottam, Theory of dipole-dipole interactions in ferromagnets. II. The correlation functions. *J. Phys. C: Solid St. Phys.*, **4** (1971), 2673–83.

[138] R. Brout, Statistical mechanical theory of ferromagnetism: High density behavior. *Phys. Rev.*, **118** (1960), 1009–19.

[139] A. Brooks Harris, $1/z$ expansion for the Heisenberg antiferromagnet at low temperatures. *Phys. Rev. Lett.*, **21** (1968), 602–4.

[140] G. C. Psaltakis and M. G. Cottam, A generalisation of the drone-fermion representation and its application to Heisenberg ferromagnets. *J. Phys. C: Solid St. Phys.*, **13** (1980), 6009–23.

[141] V. G. Vaks, A. I. Larkin, and S. A. Pikin, Thermodynamics of an ideal ferromagnetic substance. *Sov. Phys. JETP*, **26** (1968), 188–99.

[142] V. G. Vaks, A. I. Larkin, and S. A. Pikin, Spin waves and correlation functions in a ferromagnetic. *Sov. Phys. JETP*, **26** (1968), 647–55.

Index